Proceedings of the International Neural Networks Society

Volume 1

Series Editors

Plamen Angelov, School of Computing and Communications,
University of Lancaster, Lancaster, UK
Robert Kozma, Optimization and Networks Department, University of Memphis,
Memphis, TN, USA

The "Proceedings of the International Neural Networks Society INNS" publishes research contributions on fundamental principles and applications related to neural networks and modeling behavioral and brain processes. Topics of interest include new developments, state-of-art theories, methods and practical applications, covering all aspects of neural networks and neuromorphic technologies for (artificially; replace with anthropomorphic) intelligent (designs; replace with systems). This series covers high quality books that contribute to the full range of neural networks research, from computational neuroscience, cognitive science, behavioral and brain modeling, (add machine) learning algorithms, mathematical theories, to technological applications of systems that significantly use neural network concepts and techniques.

The series publishes monographs, contributed volumes, lecture notes, edited volumes, and conference proceedings in neural networks spanning theoretical, experimental, computational, and engineering aspects. Submissions of highly innovative cutting-edge contributions are encouraged, which extend our understanding at the forefront of science going beyond mainstream approaches. Of particular value to both the contributors and the readership are the short publication timeframe and the world-wide distribution, which enable both wide and rapid dissemination of research output in this rapidly developing research field.

More information about this series at http://www.springer.com/series/16268

Luca Oneto · Nicolò Navarin ·
Alessandro Sperduti · Davide Anguita
Editors

Recent Advances in Big Data and Deep Learning

Proceedings of the INNS Big Data
and Deep Learning Conference, INNSBDDL2019,
held at Sestri Levante,
Genova, Italy, 16–18 April, 2019

 Springer

Editors
Luca Oneto
Department of Informatics, Bioengineering,
Robotics, and Systems Engineering
University of Genova
Genoa, Italy

Alessandro Sperduti
Department of Mathematics
University of Padova
Padua, Italy

Nicolò Navarin
Department of Mathematics
University of Padova
Padua, Italy

Davide Anguita
Department of Informatics, Bioengineering,
Robotics, and Systems Engineering
University of Genova
Genoa, Italy

ISSN 2661-8141   ISSN 2661-815X (electronic)
Proceedings of the International Neural Networks Society
ISBN 978-3-030-16840-7   ISBN 978-3-030-16841-4 (eBook)
https://doi.org/10.1007/978-3-030-16841-4

Library of Congress Control Number: 2019935979

This Springer imprint is published by the registered company Springer Nature Switzerland AG
The registered company address is: Gewerbestrasse 11, 6330 Cham, Switzerland

Preface

This book presents 39 original articles that have been accepted to the 2019 INNS Big Data and Deep Learning (INNS BDDL), a major event for researchers in the field of artificial neural networks, big data, and related topics, organized by the International Neural Network Society (INNS) and hosted by the University of Genoa.

In 2019, INNS BDDL will be held in Sestri Levante (Italy) from April 16 to April 18. Sestri Levante is a town in Liguria, Italy. Lying on the Mediterranean Sea, it is approximately 56 kilometers (35 mi) south of Genoa and is set on a promontory. While nearby Portofino and the Cinque Terre are probably the best-known touristic destinations on the Italian Riviera, Sestri Levante has several beautiful natural bays for visitors.

In addition to regular sessions, INNS BDDL regularly welcomes tutorials organized by renowned scientists in their respective fields and international renowned invited speakers.

The contributions in this book show that INNS BDDL covers a broad range of topics in big data and deep learning, from theoretical aspects to state-of-the-art applications. More than 80 researchers from 20 countries participated in the INNS BDDL in April 2019. Around 40 oral communications and 6 tutorials have been presented this year together with 4 invited plenary speakers.

The editors would like to thank all the authors for their interesting contributions and all reviewers for their excellent work. Authors and reviewers were asked to respect a very tight schedule, which allowed this book to be published close to the end of the conference. We would also like to thank Springer for giving us the opportunity to publish this book and for the very efficient and seamless management of the publication procedure. Finally, we would like to thank the sponsors, the program committee, and the INNS for their precious and fundamental support.

The Editors

Contents

**On the Trade-Off Between Number of Examples and Precision
of Supervision in Regression** 1
Giorgio Gnecco and Federico Nutarelli

Distributed SmSVM Ensemble Learning 7
Jeff Hajewski and Suely Oliveira

**Size/Accuracy Trade-Off in Convolutional Neural Networks:
An Evolutionary Approach** 17
Tomaso Cetto, Jonathan Byrne, Xiaofan Xu, and David Moloney

Fast Transfer Learning for Image Polarity Detection 27
Edoardo Ragusa, Paolo Gastaldo, and Rodolfo Zunino

Dropout for Recurrent Neural Networks 38
Nathan Watt and Mathys C. du Plessis

**Psychiatric Disorders Classification with 3D Convolutional
Neural Networks** 48
Stefano Campese, Ivano Lauriola, Cristina Scarpazza, Giuseppe Sartori,
and Fabio Aiolli

**Perturbed Proximal Descent to Escape Saddle Points for Non-convex
and Non-smooth Objective Functions** 58
Zhishen Huang and Stephen Becker

**Deep-Learning Domain Adaptation Techniques for Credit Cards
Fraud Detection** 78
Bertrand Lebichot, Yann-Aël Le Borgne, Liyun He-Guelton,
Frédéric Oblé, and Gianluca Bontempi

**Selective Information Extraction Strategies for Cancer Pathology
Reports with Convolutional Neural Networks** 89
Hong-Jun Yoon, John X. Qiu, J. Blair Christian, Jacob Hinkle,
Folami Alamudun, and Georgia Tourassi

An Information Theoretic Approach to the Autoencoder   99
Vincenzo Crescimanna and Bruce Graham

Deep Regression Counting: Customized Datasets
and Inter-Architecture Transfer Learning   109
Iam Palatnik de Sousa, Marley Maria Bernardes Rebuzzi Vellasco,
and Eduardo Costa da Silva

Improving Railway Maintenance Actions with Big Data and
Distributed Ledger Technologies   120
Roberto Spigolon, Luca Oneto, Dimitar Anastasovski, Nadia Fabrizio,
Marie Swiatek, Renzo Canepa, and Davide Anguita

Presumable Applications of Deep Learning for Cellular
Automata Identification   126
Anton Popov and Alexander Makarenko

Restoration Time Prediction in Large Scale Railway Networks:
Big Data and Interpretability   136
Luca Oneto, Irene Buselli, Paolo Sanetti, Renzo Canepa, Simone Petralli,
and Davide Anguita

Train Overtaking Prediction in Railway Networks:
A Big Data Perspective   142
Luca Oneto, Irene Buselli, Alessandro Lulli, Renzo Canepa,
Simone Petralli, and Davide Anguita

Cavitation Noise Spectra Prediction with Hybrid Models   152
Francesca Cipollini, Fabiana Miglianti, Luca Oneto, Giorgio Tani,
Michele Viviani, and Davide Anguita

Pseudoinverse Learners: New Trend and Applications to Big Data   158
Ping Guo, Dongbin Zhao, Min Han, and Shoubo Feng

Innovation Capability of Firms: A Big Data Approach with Patents . . .   169
Linda Ponta, Gloria Puliga, Luca Oneto, and Raffaella Manzini

Predicting Future Market Trends: Which Is the Optimal Window? . . .   180
Simone Merello, Andrea Picasso Ratto, Luca Oneto, and Erik Cambria

F_0 Modeling Using DNN for Arabic Parametric Speech Synthesis   186
Imene Zangar, Zied Mnasri, Vincent Colotte, and Denis Jouvet

Regularizing Neural Networks with Gradient Monitoring   196
Gavneet Singh Chadha, Elnaz Meydani, and Andreas Schwung

**Visual Analytics for Supporting Conflict Resolution in Large
Railway Networks** 206
Udo Schlegel, Wolfgang Jentner, Juri Buchmueller, Eren Cakmak,
Giuliano Castiglia, Renzo Canepa, Simone Petralli, Luca Oneto,
Daniel A. Keim, and Davide Anguita

**Modeling Urban Traffic Data Through Graph-Based
Neural Networks** 216
Viviana Pinto, Alan Perotti, and Tania Cerquitelli

Traffic Sign Detection Using R-CNN 226
Philipp Rehlaender, Maik Schroeer, Gavneet Chadha,
and Andreas Schwung

Deep Tree Transductions - A Short Survey 236
Davide Bacciu and Antonio Bruno

**Approximating the Solution of Surface Wave Propagation Using
Deep Neural Networks** 246
Wilhelm E. Sorteberg, Stef Garasto, Chris C. Cantwell,
and Anil A. Bharath

**A Semi-supervised Deep Rule-Based Approach for Remote Sensing
Scene Classification** 257
Xiaowei Gu and Plamen P. Angelov

**Comparing the Estimations of Value-at-Risk Using Artificial Network
and Other Methods for Business Sectors** 267
Siu Cheung, Ziqi Chen, and Yanli Li

**Using Convolutional Neural Networks to Distinguish Different Sign
Language Alphanumerics** 276
Stephen Green, Ivan Tyukin, and Alexander Gorban

**Mise en abyme with Artificial Intelligence: How to Predict
the Accuracy of NN, Applied to Hyper-parameter Tuning** 286
Giorgia Franchini, Mathilde Galinier, and Micaela Verucchi

Asynchronous Stochastic Variational Inference 296
Saad Mohamad, Abdelhamid Bouchachia,
and Moamar Sayed-Mouchaweh

Probabilistic Bounds for Binary Classification of Large Data Sets 309
Věra Kůrková and Marcello Sanguineti

Multikernel Activation Functions: Formulation and a Case Study 320
Simone Scardapane, Elena Nieddu, Donatella Firmani, and Paolo Merialdo

Understanding Ancient Coin Images 330
Jessica Cooper and Ognjen Arandjelović

Effects of Skip-Connection in ResNet and Batch-Normalization on Fisher Information Matrix 341
Yasutaka Furusho and Kazushi Ikeda

Skipping Two Layers in ResNet Makes the Generalization Gap Smaller than Skipping One or No Layer 349
Yasutaka Furusho, Tongliang Liu, and Kazushi Ikeda

A Preference-Learning Framework for Modeling Relational Data 359
Ivano Lauriola, Mirko Polato, Guglielmo Faggioli, and Fabio Aiolli

Convolutional Neural Networks for Twitter Text Toxicity Analysis 370
Spiros V. Georgakopoulos, Sotiris K. Tasoulis, Aristidis G. Vrahatis, and Vassilis P. Plagianakos

Fast Spectral Radius Initialization for Recurrent Neural Networks 380
Claudio Gallicchio, Alessio Micheli, and Luca Pedrelli

Author Index 391

On the Trade-Off Between Number of Examples and Precision of Supervision in Regression

Giorgio Gnecco$^{(\boxtimes)}$ and Federico Nutarelli

IMT School for Advanced Studies, Piazza S. Francesco 19, 55100 Lucca, Italy
{giorgio.gnecco,federico.nutarelli}@imtlucca.it

Abstract. We investigate regression problems for which one is given the additional possibility of controlling the conditional variance of the output given the input, by varying the computational time dedicated to supervise each example. For a given upper bound on the total computational time, we optimize the trade-off between the number of examples and their precision, by formulating and solving a suitable optimization problem, based on a large-sample approximation of the output of the ordinary least squares algorithm. Considering a specific functional form for that precision, we prove that there are cases in which "many but bad" examples provide a smaller generalization error than "few but good" ones, but also that the converse can occur, depending on the "returns to scale" of the precision with respect to the computational time assigned to supervise each example. Hence, the results of this study highlight that increasing the size of the dataset is not always beneficial, if one has the possibility to collect a smaller number of more reliable examples.

Keywords: Ordinary least squares · Large-sample approximation · Variance control

1 Introduction

In several engineering applications, one has to approximate a function from a finite set of noisy examples, where the noise level can be controlled to some extent by the researcher, by varying the computational time dedicated to the supervision of each example. For instance, the output could be the solution of an eigenvalue problem parametrized by the input (this is the case, e.g., in the analysis of dispersion curves/dispersion surfaces in wave propagation models [1,2]). This is equivalent to finding the roots of a polynomial, which is a well-known ill-conditioned problem [6]. In this case, to reduce the level of the computational noise (due, e.g., to repeated round-offs in the calculations), one can resort to extended-precision computations. These, however, can increase significantly the computational time needed for each function evaluation, possibly reducing also the number of training examples available. Hence, the investigation of an optimal trade-off between the number of supervised examples and their precision

© Springer Nature Switzerland AG 2020
L. Oneto et al. (Eds.): INNSBDDL 2019, INNS 1, pp. 1–6, 2020.
https://doi.org/10.1007/978-3-030-16841-4_1

is needed. A similar issue arises when the output is the optimal solution of an optimization problem, which is parametrized by the input, and such a solution is approximated by a controllable number of steps of an iterative optimization algorithm, possibly including one or several random initializations obtained, e.g., by Monte-Carlo techniques [2]. Both round-off errors and random initializations motivate modeling the computational noise as a random variable [3], with possibly controllable variance.

In this framework, in the work we consider a modification of the classical linear regression model, in which one is additionally given the possibility of controlling the conditional variance of the output given the input, by varying the computational time dedicated to the supervision of each training example. Here, an upper bound on the available total computational time is given. Hence, increasing the computational time per example typically decreases the number of such examples. Using this model, we formulate and solve an optimization problem, minimizing the generalization error as a function of the computational time per example. Two main cases are considered in the analysis: for "decreasing returns of scale", the precision of the supervision (defined here as the reciprocal of the conditional variance of the output) increases less than proportionally when increasing the computational time per example; for "increasing returns of scale", instead, it increases more than proportionally. In summary, the results of the analysis highlight, from a theoretical point of view, that increasing the number of data is not always beneficial, if it is feasible to collect less but more reliable data. Not only their number, but also their quality matters. This looks particularly relevant in the current era of Big Data, and should not be overlooked when designing data collection processes. An additional feature of the paper is its combination of methods from machine learning and econometrics, which is in accordance with a very recent trend of joint research in these two fields [5].

The paper is structured as follows. Section 2 introduces the model under investigation. Section 3 presents the optimization problem used to represent the trade-off between number of examples and precision, and its optimal solution. Finally, Sect. 4 discusses possible extensions.

2   Model

We consider the following linear model for an input-output relationship:

$$y = f_\beta(\underline{x}) = \underline{\beta}'\underline{x}, \tag{1}$$

where $\underline{x} \in \mathbb{R}^{p\times 1}$ is a feature vector, $\underline{\beta} \in \mathbb{R}^{p\times 1}$ is a parameter vector, and $y \in \mathbb{R}$ is the output of the model. Suppose that, for each $\underline{x} \in \mathbb{R}^{p\times 1}$, evaluating $f_\beta(\underline{x})$ exactly is computationally demanding (e.g., because this may involve solving a complex physical or optimization subproblem) and that, in a finite computational time, one can get only an approximation of $f_\beta(\underline{x})$, corrupted by additive noise. Hence, to approximate the parameter vector $\underline{\beta}$ with an estimate $\hat{\underline{\beta}}$ and predict the output y^{test} for a new test example $\underline{x}^{\text{test}}$, only a noisy training set can be used.

It is assumed here that this is made of a finite number N of independent and identically distributed supervised examples $(\underline{x}_n, \tilde{y}_n)$ $(n = 1, \ldots, N)$, where $\tilde{y}_n$ is an approximation of $y_n = f_\beta(\underline{x}_n)$, which is modeled as follows: $\tilde{y}_n = y_n + \varepsilon_{n, \Delta T_n}$. Here, $\varepsilon_{n, \Delta T_n}$ represents a computational noise, modeled as a random variable, independent from $\underline{x}_n$, with mean 0 and variance $\sigma^2_{\varepsilon_{n, \Delta T_n}}$. The latter depends on the amount of computational time ΔT_n used to evaluate $f_\beta(\underline{x}_n)$ approximately. By increasing ΔT_n, the variance $\sigma^2_{\varepsilon_{n, \Delta T_n}}$ decreases. Hence, a reasonable model for this variance, proposed in the following, is

$$\sigma^2_{\varepsilon_{n, \Delta T_n}} = k_1 \left(\frac{k_2}{\Delta T_n} \right)^\alpha = C(\Delta T_n)^{-\alpha}, \tag{2}$$

where $\alpha, k_1 > 0$, k_2 is a positive dimensional constant, and $C = k_1 k_2^\alpha$. For $0 < \alpha < 1$, there are "decreasing returns of scale" of the precision of each supervision with respect to its computational time because, if one doubles ΔT, then the precision $1/\sigma^2_{\varepsilon_{n, \Delta T_n}}$ becomes less than two times its initial value. Conversely, for $\alpha > 1$, one has "increasing returns of scale" because, if one doubles ΔT, then the precision $1/\sigma^2_{\varepsilon_{n, \Delta T_n}}$ becomes more than two times its initial value. In any case, it is possible to increase the precision of each supervision by increasing ΔT_n.

However, in practice, computational resources are limited. Hence, the following "budget constraint" is supposed to hold: the total available computational time is smaller than or equal to a given $T > 0$. Assuming for simplicity the same computational time $\Delta T_n = \Delta T$ for each supervised example, it follows that the number $N(\Delta T)$ of such examples is bounded from above by $N^\star(\Delta T) = \lfloor \frac{T}{\Delta T} \rfloor$, where $\lfloor \cdot \rfloor$ denotes the largest integer smaller than or equal to its argument. In other words, the supervised examples are acquired with a "sampling frequency" smaller than or equal to $\frac{1}{\Delta T}$. Here, it is also assumed that ΔT can vary on a closed and bounded interval, i.e., $\Delta T \in [\Delta T_{\min}, \Delta T_{\max}]$, with $\Delta T_{\min} > 0$.

In the next section, our interest is in choosing ΔT and $N(\Delta T)$ in such a way to minimize the generalization error $\mathbb{E}\left\{ \left(\hat{\beta}' \underline{x}^{\text{test}} - y^{\text{test}} \right)^2 \middle| X \right\}$ conditioned on the training input data matrix $X \in \mathbb{R}^{N \times p}$ (i.e., the matrix whose n-th row is $\underline{x}'_n$), where $\underline{x}^{\text{test}}$ is assumed to be independent from the training examples and to have the same probability distribution as $\underline{x}$, and $\hat{\beta}$ is generated by a specific algorithm. As discussed later, this conditional generalization error depends typically on both the number of examples and the precision of their supervision. Since in the proposed framework they are related by the constraint $N(\Delta T) \leq N^\star(\Delta T) = \lfloor \frac{T}{\Delta T} \rfloor$ and Eq. (2), one expects the presence of an optimal trade-off between such precision and the number of training examples.

Finally, in order to simplify the notation, in the following we assume T to be a multiple of both $\Delta T_{\min}$ and $\Delta T_{\max}$. In this case, it follows immediately from the analysis of the generalization error made in the next section that optimal choices ΔT° and $N(\Delta T^\circ)$ for ΔT and $N(\Delta T)$ necessarily satisfy $N(\Delta T^\circ) = N^\star(\Delta T^\circ) = \frac{T}{\Delta T^\circ} \in \mathbb{N}$. Indeed, if $N(\Delta T) < \frac{T}{\Delta T}$, then one can either increase $N(\Delta T)$ by keeping ΔT constant, or increase ΔT by keeping $N(\Delta T)$ constant. In both cases, either the number or the precision of the examples increases. Hence, in the following, we replace the constraint $N(\varDelta T) \leq N^{\star}(\varDelta T)$ with $N(\varDelta T) = \frac{T}{\varDelta T}$.

3 Optimal Trade-Off Between Number of Examples and Precision of Supervision Under Ordinary Least Squares

In this section, it is assumed that the classical Ordinary Least Squares (OLS) regression algorithm is used to produce the estimate $\hat{\beta}$ of β, here denoted by $\hat{\underline{\beta}}_{OLS}$. This is a reasonable choice, since under the assumptions stated in Sect. 2, for each $\varDelta T$, OLS is the best linear unbiased estimator of β, due to Gauss-Markov theorem [4, Sect. 9.4]. When X has full column rank p, then the OLS estimate of the parameter vector β (provided, e.g., in [4, Chap. 2]) is[1]

$$\hat{\underline{\beta}}_{OLS} = (X'X)^{-1}X'\underline{\tilde{y}}. \tag{3}$$

It follows from [4, Sect. 8.2] that, being all the variances $\sigma^2_{\varepsilon_n, \varDelta T_n}$ equal to a positive constant denoted by $\sigma^2_\varepsilon(\varDelta T)$, the covariance matrix of the estimate $\hat{\underline{\beta}}_{OLS}$, conditioned on the training input data matrix, has the following expression:

$$\mathrm{Var}\left(\hat{\underline{\beta}}_{OLS}\big|X_{N(\varDelta T)}\right) = \sigma^2_\varepsilon(\varDelta T)(X'_{N(\varDelta T)}X_{N(\varDelta T)})^{-1}, \tag{4}$$

where the notation $X_{N(\varDelta T)}$ is used here to recall the dependence from $\varDelta T$ of the number of rows in the training input data matrix.

For a new test example $\underline{x}_{test}$, the prediction error on the associated y_{test} is $\hat{e}^{OLS}_{test} = \hat{y}^{OLS}_{test} - y_{test} = (\hat{\underline{\beta}}_{OLS} - \underline{\beta})'\underline{x}_{test}$, which, conditioned on $X_{N(\varDelta T)}$, is unbiased and has conditional variance

$$\mathrm{Var}\left(\hat{e}^{OLS}_{test}\big|X_{N(\varDelta T)}\right) = \mathbb{E}\left\{\underline{x}'_{test}(\hat{\underline{\beta}}_{OLS} - \underline{\beta}_{OLS})(\hat{\underline{\beta}}_{OLS} - \underline{\beta}_{OLS})'\underline{x}_{test}\big|X_{N(\varDelta T)}\right\}, \tag{5}$$

Equation (5) represents the generalization error conditioned on $X_{N(\varDelta T)}$. Since, for every scalar a, one has $a = \mathrm{Tr}(a)$, Eq. (5) can be re-written as

$$\mathrm{Var}\left(\hat{e}^{OLS}_{test}\big|X_{N(\varDelta T)}\right) = \mathrm{Tr}\left(\mathbb{E}\left\{\underline{x}'_{test}(\hat{\underline{\beta}}_{OLS} - \beta)(\hat{\underline{\beta}}_{OLS} - \beta)'\underline{x}_{test}\big|X_{N(\varDelta T)}\right\}\right)$$

$$= \mathrm{Tr}\left(\mathbb{E}\left\{\underline{x}\underline{x}'\right\}\mathrm{Var}\left(\hat{\underline{\beta}}_{OLS}\big|X_{N(\varDelta T)}\right)\right). \tag{6}$$

When both the expectation $\mathbb{E}\left\{\underline{x}\underline{x}'\right\}$ and the fourth-order moments of $\underline{x}$ are finite, the sample mean $\frac{X'_{N(\varDelta T)}X_{N(\varDelta T)}}{N(\varDelta T)} = \frac{1}{N(\varDelta T)}\sum_{n=1}^{N(\varDelta T)}\underline{x}_n\underline{x}'_n$ converges in probability[2] to $\mathbb{E}\left\{\underline{x}\underline{x}'\right\}$ by Chebychev's law of large numbers [4, Sect. 13.4.2]. If $\mathbb{E}\left\{\underline{x}\underline{x}'\right\}$ is an invertible matrix, too, then $\left(\frac{X'_{N(\varDelta T)}X_{N(\varDelta T)}}{N(\varDelta T)}\right)^{-1}$ converges in

---

[1] One can notice that the invertibility of $X'X$ in Eq. (3) requires $N \geq p$.

[2] I.e., for every $\varepsilon > 0$, $\mathrm{Prob}\left(\left\|\frac{X'_{N(\varDelta T)}X_{N(\varDelta T)}}{N(\varDelta T)} - \mathbb{E}\left\{\underline{x}\underline{x}'\right\}\right\| > \varepsilon\right)$ (where $\|\cdot\|$ is an arbitrary matrix norm) tends to 0 as $N(\varDelta T)$ tends to $+\infty$.

probability to $(\mathbb{E}\{\underline{x}\underline{x}'\})^{-1}$ by the generalization of Slutsky's lemma reported in [4, Lemma 13.4]. This also justifies the invertibility assumption in Eq. (3). Since $\Delta T_{\max} < +\infty$, the convergence arguments above can be applied for every $\Delta T \in [\Delta T_{\min}, \Delta T_{\max}]$. Hence, one can assume that T is sufficiently large so that the large-sample approximation $\left(\frac{X'_{N(\Delta T)} X_{N(\Delta T)}}{N(\Delta T)}\right)^{-1} \simeq (\mathbb{E}\{\underline{x}\underline{x}'\})^{-1}$ can be made for each $\Delta T \in [\Delta T_{\min}, \Delta T_{\max}]$. Finally, combining Eqs. (4) and (6) with the large-sample approximation above, including ΔT in the notation, and using Eq. (2) and the expression $N(\Delta T) = \frac{T}{\Delta T}$, one gets

$$
\begin{aligned}
\mathrm{Var}\left(\hat{e}_{test}^{OLS}\big| X_{N(\Delta T)}\right)(\Delta T) &\simeq \mathrm{Tr}\left(\mathbb{E}\{\underline{x}\underline{x}'\} \sigma_\varepsilon^2(\Delta T) \frac{1}{N(\Delta T)} (\mathbb{E}\{\underline{x}\underline{x}'\})^{-1}\right) \\
&= \frac{p\sigma_\varepsilon^2(\Delta T)}{N(\Delta T)} = \frac{pC(\Delta T)^{-\alpha}}{\frac{T}{\Delta T}} = \frac{pC}{T}(\Delta T)^{1-\alpha}, \quad (7)
\end{aligned}
$$

which has to be minimized with respect to $\Delta T \in [\Delta T_{\min}, \Delta T_{\max}]$, with $\frac{T}{\Delta T} \in \mathbb{N}$. Clearly, the optimal solution ΔT° to the optimization problem above is:

1. $\Delta T_{\min}$, for $0 < \alpha < 1$ ("decreasing returns of scale");
2. any $\Delta T \in [\Delta T_{\min}, \Delta T_{\max}]$ with $\frac{T}{\Delta T} \in \mathbb{N}$, for $\alpha = 1$ ("constant returns");
3. $\Delta T_{\max}$, for $\alpha > 1$ ("increasing returns").

Concluding, the results of the analysis show that, for "decreasing returns of scale", "many but bad" examples provide a smaller generalization error than "few but good" ones. The converse occurs for "increasing returns of scale".

4 Extensions

The analysis can likely be extended to the case in which different examples are associated with possibly different computational times, giving rise to multiplicative heteroskedastic computational noise. In this case, Gauss-Markov theorem does not apply anymore to OLS, which potentially loses its efficiency. Hence, it would be worth investigating also its performance compared with alternative methods, such as Weighted Least Squares (WLS) [4, Chap. 18].

References

1. Bacigalupo, A., Gnecco, G., Lepidi, M., Gambarotta, L.: Optimal design of low-frequency band gaps in anti-tetrachiral lattice meta-materials. Compos. Part B Eng. **115**, 341–359 (2017)
2. Bacigalupo, A., Lepidi, M., Gnecco, G., Gambarotta, L.: Optimal design of auxetic hexachiral metamaterials with local resonators. Smart Mater. Struct. **25**(5), 19 (2016). Article ID. 054009
3. Hamming, R.: Numerical Methods for Scientists and Engineers, 2nd edn. McGraw-Hill, New York (1973)

4. Ruud, P.A.: An Introduction to Classical Econometric Theory, 1st edn. Oxford University Press, Oxford (2000)
5. Varian, H.R.: Big data: new tricks for econometrics. J. Econ. Perspect. **28**, 3–28 (2014)
6. Wilkinson, J.H.: The evaluation of the zeros of ill-conditioned polynomials. Part I. Numer. Math. **1**, 150–166 (1959)

Distributed SmSVM Ensemble Learning

Jeff Hajewski$^{(\boxtimes)}$ and Suely Oliveira

Department of Computer Science, University of Iowa, Iowa City, IA, USA
{jeffrey-hajewski,suely-oliveira}@uiowa.edu

Abstract. Traditional ensemble methods are typically performed with models that are fast to construct and evaluate, such as random trees and Naive Baye's. More complex models frequently suffer from increased computational load in both training and inference. In this work, we present a distributed ensemble method using SmoothSVM, a fast support vector machine (SVM) algorithm. We build and evaluate a large ensemble of SVMs in parallel, with little overhead when compared to a single SVM. The ensemble of SVMs trains in less time than a single SVM while maintaining the same test accuracy and, in some cases, even exhibits improved test accuracy. Our approach also has the added benefit of trivially scaling to much larger systems.

Keywords: Support vector machine · Parallel ensemble learning · Distributed SVM · SmoothSVM

1 Introduction

Support Vector Machines (SVMs) are commonly known for their CPU intensive workloads and large memory requirements. In the big data setting, where it is common for data size to exceed available memory, these issues quickly become major bottlenecks. Because of their memory requirements, it is desirable to parallelize the SVM algorithm for large datasets, allowing them to take advantage of greater compute resources and a larger memory pool. In this work, we implement a distributed SmoothSVM (SmSVM) ensemble model. SmSVM [11] is an SVM algorithm that uses a smooth approximation to the hinge-loss function and an active-set approximation to the ℓ_1 norm. It has two major advantages over the standard SVM formulation: (1) the smoothness of the loss function permits use of Newton's method for optimization and (2) the active-set approximation of the ℓ_1 norm smoothes the loss function but not the ℓ_1 penalty, yielding a sparse solution. This reduces the computational requirements and memory footprint.

Efficient communication patterns in large-scale distributed systems is a complex and challenging problem in both design and implementation. In the distributed machine learning setting, unnecessary communication can consume valuable network bandwidth and memory resources with little impact on training time and test accuracy. Conversely, too little communication can result in poorly performing models due to reduced information transfer from the dataset to the

© Springer Nature Switzerland AG 2020
L. Oneto et al. (Eds.): INNSBDDL 2019, INNS 1, pp. 7–16, 2020.
https://doi.org/10.1007/978-3-030-16841-4_2

models. We circumvent these issues by scaling the amount of data distributed to each node inversely with the number of worker nodes and use a bootstrap sample of the dataset rather than disjoint subsets. This approach improves the training speed due to improved convergence during optimization while training on fewer data points per model. Our method is able to maintain test accuracy via the improved generalizability of the ensemble models (an artifact of the models being trained on different subsets of the original dataset) while decreasing training time. This work investigates the parallelization problem via a distributed ensemble of SmSVM models, each trained on different, but possibly overlapping, subsets of the original dataset (known as bagging [5]). The advantages of this technique in the distributed machine learning setting are:

- Improved generalization of the aggregated models
- Reduced network utilization via sub-sampling the data
- Reduced training time due to the concurrent construction of SVMs on sub-sampled datasets

We parallelize SmSVM using message passing via MPI, rather than a MapReduce-based framework such as Hadoop or Spark. This approach allows our system to easily scale from running locally on multiple cores to running on a large cluster without requiring reconfiguration. Additionally, it avoids the startup overhead commonly associated with MapReduce-based frameworks (see, for example, [17]). The drawback of this approach is having to handle communication between nodes at a much lower level; however, in our experience this is typically a worthwhile trade-off when performance is the primary goal.

2   Background

In this section we describe the SmSVM algorithm in detail. The standard SVM loss function with both ℓ_1 and ℓ_2 regularizers is shown in (1).

$$\frac{1}{n}\sum_{i=1}^{n}\max(0, 1 - y_i\boldsymbol{\omega}\cdot\boldsymbol{x}_i) + \mu||\boldsymbol{\omega}||_1 + \frac{\lambda}{2}||\boldsymbol{\omega}||_2^2 \tag{1}$$

We use a different formulation, called SmoothSVM (SmSVM) [11], which uses a smooth approximation to the hinge-loss and maintains an active-set to approximate the ℓ_1 regularizer. The hinge-loss function, $\psi(\boldsymbol{x}_i, y_i, \boldsymbol{\omega}) = \max(0, 1 - y_i\boldsymbol{\omega}\cdot\boldsymbol{x}_i)$, is non-smooth and thus not globally differentiable. Defining $u = 1 - y\boldsymbol{\omega}\cdot\boldsymbol{x}$, the smoothed hinge-loss is given by Eq. (2).

$$\psi_\epsilon(u) = \frac{1}{2}(u + \sqrt{\epsilon^2 + u^2}) \tag{2}$$

The choice of ϵ is determined via a relaxation method, where we choose a larger value for ϵ initially, find an optimal solution $\boldsymbol{\omega}$, and then scale ϵ by a relaxation factor, β. In practice we found $\beta = 100$ to work sufficiently well. Using this definition, and defining $\phi(\boldsymbol{\omega})$ as in (3)

$$\phi(\boldsymbol{\omega}) = \frac{1}{n} \sum_{i=1}^{n} \psi_\epsilon(1 - y_i(\boldsymbol{\omega} \cdot \boldsymbol{x}_i)) + \mu||\boldsymbol{\omega}||_1 + \frac{\lambda}{2}||\boldsymbol{\omega}||_2^2 \tag{3}$$

the problem we aim to solve is given by (4).

$$\min_{\boldsymbol{\omega}} \phi(\boldsymbol{\omega}) \tag{4}$$

The advantage of this formulation is that the optimization problem is smooth, which allows us to use Newton's method, rather than solving the dual problem. Solving (4) is equivalent to solving (5) under the requirement that the Hessian, $H(\boldsymbol{\omega})$, is positive semi-definite.

$$\nabla_{\boldsymbol{\omega}} \phi(\boldsymbol{\omega}) = 0 \tag{5}$$

However, SmSVM does not require the entire Hessian matrix because the active indices of $\boldsymbol{\omega}$ are maintained via the active-set. Thus, we reduce the dimensionality of $H(\boldsymbol{\omega})$ by restricting it to the non-zero components of $\boldsymbol{\omega}$. With this formulation we can use Newton's method [16], which is an iterative solver based on the recurrence relation

$$\boldsymbol{\omega}_{n+1} = \boldsymbol{\omega}_n - f'(\boldsymbol{\omega}_n)^{-1} f(\boldsymbol{\omega}_n)$$

where $f(\boldsymbol{\omega}) = \nabla\phi(\boldsymbol{\omega})$, $f'(\boldsymbol{\omega}) = H(\boldsymbol{\omega})$, and the weight update is given via

$$\boldsymbol{\omega} \leftarrow \boldsymbol{\omega} + s\boldsymbol{d}$$

In the general Newton's method implementation we typically have $s = 1$. Computing $H(\boldsymbol{\omega})$ is expensive – we limit the cost of computing $H(\boldsymbol{\omega})$ by using an Armijo linesearch [3] algorithm to reduce the number of required steps. This done by solving (6).

$$\operatorname{argmin}_{s \geq 0} \psi(\boldsymbol{\omega} + s\boldsymbol{d}) + \mu||\boldsymbol{\omega} + s\boldsymbol{d}||_1 \tag{6}$$

Replacing $\psi(\boldsymbol{\omega} + s\boldsymbol{d})$ via a second-order approximation yields

$$\psi(\boldsymbol{\omega} + s\boldsymbol{d}) \approx as^2 + bs + c + \mu||\boldsymbol{\omega} + s\boldsymbol{d}||_1$$

with $s \geq 0$. For $a \geq 0$, $\mu \geq 0$ this is a coercive convex function, which means it has a global minimum. We find this global minimum via binary search. The binary search is computationally efficient because it is with respect to s and thus does not require any additional function or gradient evaluations.

Bootstrap aggregation (commonly referred to as *bagging*) is a statistical technique that improves a model's ability to generalize to unseen data. Bagging consists of a bootstrap phase and an aggregation phase. During the bootstrap phase, data is sub-sampled uniformly with replacement from the original dataset. In this work, we sample a subset that is inversely proportional to the number of nodes (i.e., for n nodes, and a dataset with cardinality equal to $|D|$, each subset has cardinality equal to $\lfloor |D|/|n| \rfloor$) and train a different model in parallel for each subset. During the inference/aggregation phase, each model generates a prediction for the same data point, the mode of these predictions is computed and used as the final prediction of the bagged model for the given data point. If there are an equal number of votes for each class, the master can choose a class at random or train a model of its own to use as a tie-breaker.

3   Related Work

Yu et al. [22] propose a multiagent SVM ensemble in a sequential compute environment. Their agents are designed to favor diversity by using disjoint subsets from the training set. This approach improves the generalization of their model but inhibits its scalability. As the number of agents grows, the size of the training set per agent decreases, eventually resulting in an agent-based model that has little resemblance to the actual data distribution. Chen et al. [7] propose a sequential, bagged SVM ensemble in the small data setting. Claesen et al. [8] introduce an ensemble SVM library called `EnsembleSVM` with a focus on sequential efficiency via avoiding data duplication as well as duplicate support vector evaluations. While this library achieves very impressive results, it is designed for computational efficiency in the multicore setting rather than the distributed, big data setting.

3.1   Hadoop and Spark

Distributed SVM architectures are widely studied. Perhaps the most popular approach is a MapReduce-based architecture, using either Hadoop [18] or Spark [23]. A number of implementations use a MapReduce framework via Hadoop [1,2,9,12]. While MapReduce works well for many big data applications, it has two bottlenecks: (i) data is stored on disk during intermediate steps and (ii) a shuffle stage where data is shuffled between servers in a distributed sort These two issues lead to decreased performance when running the iterative style algorithms common to many machine learning algorithms. Specifically, highly iterative algorithms will go through several phases of reading data from disk, processing, writing data to disk, and shuffling *for each iteration*. Other work utilizes the MapReduce framework via Spark [13,15,20,21]. Spark is built on top of Hadoop and attempts to keep its data in memory, which alleviates the disk IO bottleneck. However, Spark can still incur communication overhead during the shuffle stage of MapReduce and, if the data is large enough or there are too few nodes, Spark may spill excessive data to disk (when there is more data than available RAM). While this prior work shows promising results, the use of a MapReduce framework can be sub-optimal. Reyes-Ortiz et al. [17] show that MPI generally outperforms Spark in the distributed SVM setting, seeing as much as a 50× speedup.

3.2   Message Passing-Based Approaches

Chang et al. [6] show that two core bottlenecks in solving the SVM optimization problem (via interior point methods) are the required computational and
memory resources. Graf et al. [10] develop a parallel Incomplete Cholesky Factorization to reduce memory usage, and then solve the dual SVM problem via
a parallel interior point method [14]. Although they achieve promising results,
their communication overhead scales with the number of nodes. In our approach,
communication overhead is constant with respect to the number of nodes.

The advantage of message passing-based approaches over MapReduce-based
approaches is the avoidance of the unnecessary disk IO, the shuffle stage, and
start-up overhead. Message passing frameworks are typically lower-level than
a MapReduce type framework such as Hadoop. The benefit of this lower-level
approach is the ability to explicitly control how and when communication occurs,
avoiding unnecessary communication. However, the lower-level nature leads to
longer development times, more bugs, and generally more complex software.

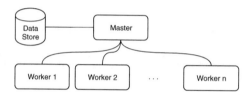

Fig. 1. Overview of system architecture.

4   Distributed Ensemble SmSVM

Distributed ensemble SmSVM is a message-passing based distributed algorithm,
detailed in Algorithms 1, 2, 3, and 4. Figure 1 shows an overview of the system
architecture.

Algorithms 1 and 2 detail the training stage of the distributed ensemble model
for the master and worker nodes, respectively. Algorithms 3 and 4 detail the
inference phase for the master and worker nodes, respectively. The core intuition
behind this algorithm is to train a number of different SVM models using the
SmSVM algorithm in parallel with bootstrapped subsets of the data and use the
resulting models as voters during the inference phase. Consider the case where
$n_{workers} = 5$, after the training phase the master node will have five different
weight vectors $\boldsymbol{\omega}$. During the inference phase the master node will (possibly)
send a weight vector to each of the worker nodes along with the prediction data,
each worker will send its prediction(s) back to the master node, and finally the
master node will compute the mode prediction for each data point, using this as
the final output. If there are an even number of workers and the results during
voting are split, the master may randomly select a class as the final prediction

---

Algorithm 1. Distributed ensemble training algorithm for master node.

Master Node

Require: $X \in \mathbb{R}^{m \times n}$, $y \in \mathbb{R}^m$ - Training data
Require: n_{nodes} - Number of nodes
1: **for** $i = 1$ **to** n_{nodes} **do**
2:     $\mathcal{I}_i = \{j : j \sim \mathcal{U}(0, m)\}$, with $|\mathcal{I}_i| = m/n_{\text{node}}$
3:     send_to$(X[\mathcal{I}_i, :], y[\mathcal{I}_i], i)$
4: **end for**
5: **for** $i = 1$ **to** n_{nodes} **do**
6:     $\boldsymbol{\omega}^{(i)} \leftarrow$ receive_from(i)
7: **end for**
8: **return** $W \in \mathbb{R}^{n_{\text{nodes}} \times n}$

Algorithm 2. Distributed ensemble training algorithm for worker node.

Worker Node

1: $X, y \leftarrow$ receive_from(i_{master})
2: $\boldsymbol{\omega} \leftarrow$ SmSVM(X, y)
3: send_to$(\boldsymbol{\omega}, i_{\text{master}})$

Algorithm 3. Distributed ensemble inference algorithm for master node.

Master Node

Require: $W \in \mathbb{R}^{n_{\text{nodes}} \times n}$
Require: $X \in \mathbb{R}^{m \times n}$ - Input data (to be classified)
1: **for** $i = 1$ **to** n_{nodes} **do**
2:     $\boldsymbol{\omega} \leftarrow W[i, :]$
3:     send_to(X, i) {Send X to process i}
4:     send_to$(\boldsymbol{\omega}, i)$ {Send $\boldsymbol{\omega}$ to process i}
5: **end for**
6: **for** $i = 1$ **to** n_{nodes} **do**
7:     $\mathbf{y}_i \leftarrow$ receive_from(i)
8: **end for**
9: **results** $\leftarrow$ mode$_i(\mathbf{y}_i)$
10: **return results**

Algorithm 4. Distributed ensemble inference algorithm for worker node.

Worker Node

Require: $X \in \mathbb{R}^{m \times n}$
1: $X \leftarrow$ receive_from(i_{master})
2: $\boldsymbol{\omega} \leftarrow$ receive_from(i_{master})
3: **for** $i = 1$ **to** m **do**
4:     $z \leftarrow \max\{0, 1 - X_i \boldsymbol{\omega}\}$
5:     $y_i \leftarrow 1$ **if** $z > 0$ **else** -1
6: **end for**
7: send_to$(\mathbf{y}, i_{\text{master}})$

or train its own model (during the training phase) and to use as a tie-breaker. It is not strictly required that the master node uses workers during the inference stage of the algorithm. Specifically, for small enough inference datasets it is more efficient for the master to evaluate the models and voting itself. For large inference datasets, however, it is more efficient for the master to distribute the data and weight vectors to the workers.

The SmSVM algorithm was chosen over other SVM algorithms due to its efficient usage of system resources. A core benefit of the active-set approach to ℓ_1 regularization used by SmSVM is that it allows us to create an optimized implementation of the algorithm. Because the active indices are tracked throughout the computation, we are able to avoid unnecessary multiplications by reducing the data matrix and weight vector to only the active dimensions. This results in reduced memory and CPU usage. The low compute and memory requirements of the SmSVM algorithm allows our distributed ensemble algorithm to run just as effectively on a single many-core machine as it does in a multi-node setting while improving training times.

Table 1. Datasets used in experiments

Dataset	Dimension	Size (GB)	Source
Synthetic	$2,500,000 \times 1,000$	20	N/A
Epsilon	$400,000 \times 2,000$	12	[19]
CoverType	$581,000 \times 54$	0.25	[4]

5   Discussion

We evaluate our model by measuring the speed-up scale factor (t_0/t_n) and the test accuracy, as a function of node count, using the three datasets described in Table 1. The test accuracy is evaluated on a hold-out set and the model hyperparameters are tuned using a validation set. Table 1 lists the datasets used in our experiments. For all experiments, each node is sent a subset of the original dataset D. The cardinality of this subset is equal to $\lfloor |D|/n \rfloor$ where n is the number of worker nodes. For each node count, we run an experiment where the bootstrap step is performed with replacement (referred to as "bagged") and another experiment where the bootstrap step is performed without replacement (referred to as "disjoint").

5.1   Results

Figures 2(a), (c), and (e) show the change in test accuracy compared to a single node (core) run as a function of the number of nodes used to train the distributed SmSVM model. Figure 2(a) shows a drop-off in test accuracy for the Epsilon dataset as the number of nodes increases. This is a result of the training subsets containing too little information (due to their small size) to accurately capture the prior probability distribution – at 50 nodes, each node training the Epsilon dataset gets 8,000 data points while each node under the Synthetic dataset gets 50,000 data points. Figures 2(c) and (e), CoverType and Synthetic, respectively, do not exhibit this behavior at the given node counts because there were not enough compute resources (cores or nodes) to create small enough training sets.

Figures 2(b), (d), and (f) show the training speed-up results for the Epsilon, CoverType, and Synthetic datasets. Figure 2(b) shows the optimal node count for

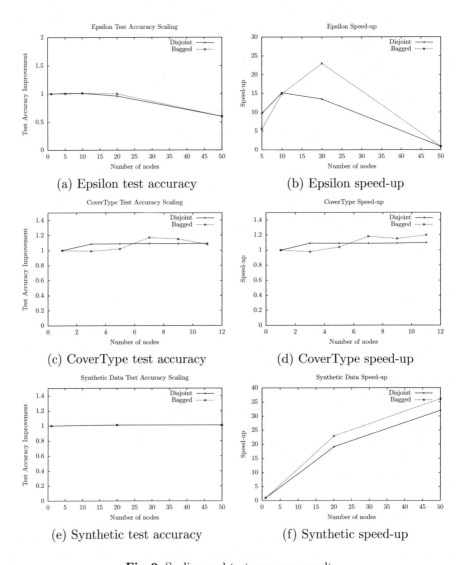

(a) Epsilon test accuracy   (b) Epsilon speed-up

(c) CoverType test accuracy   (d) CoverType speed-up

(e) Synthetic test accuracy   (f) Synthetic speed-up

Fig. 2. Scaling and test accuracy results.

the Epsilon dataset is 20 nodes using the bagged model, achieving about a 24× speed-up compared to a single core. The disjoint model exhibits slower training times for larger node counts due to the reduced training set size. Because the sub-sampled datasets are disjoint, at a certain point the datasets fail to accurately capture the prior distribution of the aggregate dataset, which results in an increased training time. The CoverType results, seen in Fig. 2(d) show a similar speed-up behavior to the Epsilon dataset, with the bagged model achieving the best speed-up results. Figure 2(f) shows the most impressive training speed-up results of the three datasets. Due to compute resource constraints, we were only able to run the synthetic dataset on a maximum of 50 nodes. Despite the limitation on resources, the synthetic dataset shows very strong scalability, which is due to information-rich subsets. The synthetic data requires fewer data points than the Epsilon or CoverType datasets to accurately reconstruct its prior distribution. Figure 2(f) clearly shows a decrease in slope from one node to 20 nodes and 20 nodes to 50 nodes, indicating at a larger node count the training speed-up will eventually plateau.

6 Conclusion

In this paper we introduce distributed ensemble SmSVM – a fast, robust distributed SVM model. We take advantage of SmSVM's performance and resource efficiency to achieve significant speed-ups in training time while maintaining test accuracy. These characteristics make our framework suitable for many-core single machines as well as large clusters of machines.

References

1. Alham, N.K., Li, M., Liu, Y., Hammoud, S.: A MapReduce-based distributed SVM algorithm for automatic image annotation. Comput. Math. Appl. **62**(7), 2801–2811 (2011). Computers & Mathematics in Natural Computation and Knowledge Discovery
2. Alham, N.K., Li, M., Liu, Y., Qi, M.: A MapReduce-based distributed SVM ensemble for scalable image classification and annotation. Comput. Math. Appl. **66**(10), 1920–1934 (2013)
3. Armijo, L.: Minimization of functions having Lipschitz continuous first partial derivatives. Pac. J. Math. **16**(1), 1–3 (1966)
4. Blackard, J.A., Dean, D.J.: Comparative accuracies of neural networks and discriminant analysis in predicting forest cover types from cartographic variables. In: Second Southern Forestry GIS Conference (1998). UCI Machine Learning Repository: https://archive.ics.uci.edu/ml/datasets/covertype
5. Breiman, L.: Bagging predictors. Mach. Learn. **24**(2), 123–140 (1996)
6. Chang, E.Y., Zhu, K., Wang, H., Bai, H., Li, J., Qiu, Z., Cui, H.: PSVM: parallelizing support vector machines on distributed computers. In: NIPS (2007)
7. Chen, S., Wang, W., van Zuylen, H.: Construct support vector machine ensemble to detect traffic incident. Expert Syst. Appl. **36**(8), 10976–10986 (2009)

8. Claesen, M., De Smet, F., Suykens, J.A.K., De Moor, B.: EnsembleSVM: a library for ensemble learning using support vector machines. J. Mach. Learn. Res. **15**(1), 141–145 (2014)
9. Dean, J., Ghemawat, S.: MapReduce: simplified data processing on large clusters. Commun. ACM **51**(1), 107–113 (2008)
10. Graf, H.P., Cosatto, E., Bottou, L., Dourdanovic, I., Vapnik, V.: Parallel support vector machines: the cascade SVM. In: Saul, L.K., Weiss, Y., Bottou, L. (eds.) Advances in Neural Information Processing Systems, vol. 17, pp. 521–528. MIT Press, Amsterdam (2005)
11. Hajewski, J., Oliveira, S., Stewart, D.E.: Smoothed hinge loss and ℓ1 support vector machines. In: 2018 International Conference on 2018 Workshop on Optimization Based Techniques for Emerging Data Mining Problems (OEDM) (2018)
12. Ke, X., Jin, H., Xie, X., Cao, J.: A distributed SVM method based on the iterative MapReduce. In: Proceedings of the 2015 IEEE 9th International Conference on Semantic Computing (IEEE ICSC 2015), pp. 116–119, February 2015
13. Liu, C., Wu, B., Yang, Y., Guo, Z.: Multiple submodels parallel support vector machine on spark. In: 2016 IEEE International Conference on Big Data (Big Data), pp. 945–950, December 2016
14. Mehrotra, S.: On the implementation of a primal-dual interior point method. SIAM J. Optim. **2**(4), 575–601 (1991)
15. Nguyen, T.D., Nguyen, V., Le, T., Phung, D.: Distributed data augmented support vector machine on spark. In: 2016 23rd International Conference on Pattern Recognition (ICPR), pp. 498–503, December 2016
16. Nocedal, J., Wright, S.J.: Numerical Optimization, 2nd edn. Springer, New York (2006)
17. Reyes-Ortiz, J.L., Oneto, L., Anguita, D.: Big data analytics in the cloud: spark on Hadoop vs MPI/OpenMP on Beowulf. Procedia Comput. Sci. **53**, 121–130 (2015). iNNS Conference on Big Data 2015 Program San Francisco, CA, USA 8-10
18. Shvachko, K., Kuang, H., Radia, S., Chansler, R.: The Hadoop distributed file system. In: Proceedings of the 2010 IEEE 26th Symposium on Mass Storage Systems and Technologies (MSST), MSST 2010, pp. 1–10. IEEE Computer Society (2010)
19. Sonnenburg, S., Franc, V., Yom-Tov, E., Sebag, M.: Pascal large scale learning challenge, vol. 10, pp. 1937–1953, January 2008
20. Wang, H., Xiao, Y., Long, Y.: Research of intrusion detection algorithm based on parallel SVM on spark. In: 2017 7th IEEE International Conference on Electronics Information and Emergency Communication (ICEIEC), pp. 153–156, July 2017
21. Yan, B., Yang, Z., Ren, Y., Tan, X., Liu, E.: Microblog sentiment classification using parallel SVM in apache spark. In: 2017 IEEE International Congress on Big Data (BigData Congress), pp. 282–288, June 2017
22. Yu, L., Yue, W., Wang, S., Lai, K.K.: Support vector machine based multiagent ensemble learning for credit risk evaluation. Expert Syst. Appl. **37**(2), 1351–1360 (2010)
23. Zaharia, M., Chowdhury, M., Franklin, M.J., Shenker, S., Stoica, I.: Spark: cluster computing with working sets. In: Proceedings of the 2Nd USENIX Conference on Hot Topics in Cloud Computing, HotCloud 2010, pp. 10–10. USENIX Association, Berkeley (2010)

Size/Accuracy Trade-Off in Convolutional Neural Networks: An Evolutionary Approach

Tomaso Cetto[✉], Jonathan Byrne, Xiaofan Xu, and David Moloney

Advanced Architecture Group, Intel Corporation, Leixlip, Ireland
tomaso.cetto@icloud.com,
{jonathan.byrne,xiaofan.xu,david.moloney}@intel.com

Abstract. In recent years, the shift from hand-crafted design of Convolutional Neural Networks (CNN's) to an automatic approach (AutoML)
has garnered much attention. However, most of this work has been concentrated on generating state of the art (SOTA) architectures that set
new standards of accuracy. In this paper, we use the NSGA-II algorithm
for multi-objective optimization to optimize the size/accuracy trade-off
in CNN's. This approach is inspired by the need for simple, effective,
and mobile-sized architectures which can easily be re-trained on any
datasets. This optimization is carried out using a Grammatical Evolution
approach, which, implemented alongside NSGA-II, automatically generates valid network topologies which can best optimize the size/accuracy
trade-off. Furthermore, we investigate how the algorithm responds to an
increase in the size of the search space, moving from strictly topology
optimization (number of layers, size of filter, number of kernels,etc.) and
then expanding the search space to include possible variations in other
hyper-parameters such as the type of optimizer, dropout rate, batch size,
or learning rate, amongst others.

Keywords: CNN · Grammatical evolution

1 Introduction

Since their introduction in 1998 [16], CNNs have steadily increased in popularity
for machine vision applications, and in the last few years have set benchmarks
in image classification and detection tasks [14,25,28,30]. Throughout the years,
the increase in depth/complexity of CNNs has been accompanied by a growing difficulty in identifying the interactions between architecture choices and
their effect on the accuracy of the models. In recent years, there has been growing interest in methods which seek to automatically search for optimal network
architectures (AutoML). Approaches have been varied and include reinforcement
learning [31,32], NeuroEvolution (NE) [18,20,21,24,27], sequentially structured
search [17,22], and Bayesian optimization [10,26]. In this paper, we deal with the

© Springer Nature Switzerland AG 2020
L. Oneto et al. (Eds.): INNSBDDL 2019, INNS 1, pp. 17–26, 2020.
https://doi.org/10.1007/978-3-030-16841-4_3

NE approach, which applies evolutionary algorithms (EA) to the optimization of artificial neural networks (ANNs).

In the past, the vast majority of AutoML methods which implement EA optimization have aimed (and succeeded) at finding state-of-the-art architectures [4,18,24], and as such have used validation accuracy on the test set as the fitness function for the algorithm. This can be very computationally expensive, for example in [24], the optimal network structure was found after 3150 GPU days of evolution. In this paper, we carry out multi-objective optimization using the NSGA-II algorithm [7] to optimize the performance of our networks with respect to both accuracy *and* size. The goal of this approach is to search for smaller, mobile-sized networks which can be easily re-trained, if need be, without the need for very substantial GPU power.

This search is carried out using Grammatical Evolution (GE) [23], which is a grammar-based form of Genetic Programming (GP) [12], where a formal grammar is used to map from phenotype to genotype. More precisely, we use PonyGE2 [8], which is a GE implementation in Python that uses a Backaus-Naur Form (BNF) grammar. The appeal of grammars are their flexibility, where one can easily change the grammar to accommodate changes in network structure, as well as the fact that grammatical evolution facilitates the encoding of domain specific knowledge.

In the first experiment, we seek to optimize only the network topology (number of layers, number of kernels, and filter size). In the second experiment, we expand to search space to include learning parameters such as learning rate, batch size, dropout rate, etc. The networks evolved in both these cases are trained and tested on two datasets: CIFAR10 [13] and STL10 [6]. Because of the fact all our experiments were run on a single GPU[1], it would have been unfeasible to train on the ImageNet dataset. It is also important to note that no data augmentation was applied to these datasets to generate more images.

The rest of the paper is organized as follows: Sect. 2 introduces related work in the field of NeuroEvolution as well as investigations in the size/accuracy trade-off of ANNs. Next, Sect. 3 details our experimental setup, and Sects. 4 and 5 compile and discuss the results. Finally, Sect. 6 draws conclusions and discusses ideas for future work.

2   Literature Review

2.1   NeuroEvolution/Grammar-Based Approaches

In recent years, the evolutionary approach has rapidly become one of the most popular ways to automatically optimize ANN performance, rivaling state-of-the-art manually crafted networks [4,18,24]. All NE methods are built around the same backbone: the evolutionary algorithm. However, we distinguish between the evolutionary approaches which evolve the 'building blocks', or modules of

---

[1] NVIDIA-SMI GeForce GTX TITAN.

the networks (hierarchical representations [18], NEAT/Co-NEAT methods [21, 27]), and those which specify these modules beforehand (for example, the CMA-ES method described in [20]), and only evolve structures which optimize the sequence they create within the network, as well as the hyper-parameters linked to them. Our approach is the latter; we use two different hand-crafted modules, which will be detailed in Sect. 3.

A variant of NeuroEvolution, and the one used in this paper, is one which makes use of a pre-defined grammar to carry out the phenotype-to-genotype mapping process, initially described in [1,29]. However, both of these works only evolve ANNs with one hidden layer. The work done in this paper is closely related to that set forward in a series of papers by Assuncao et al. [2–4]. Whilst our work shares some common aspects with DENSER (solutions are encoded through derivations of a Context Free Grammar (CFG)), it exhibits two major differences. First, we use a publicly available implementation of classical GE called PonyGE2, as opposed to DSGE, which is the genotypic representation used in the DENSER method. Secondly, the ANN architectures evolved using the DENSER method are optimized only in terms of validation accuracy on a test set. Because this paper searches for architectures which optimize a trade-off between size and accuracy, the evolved networks will exhibit different qualities to those evolved using the DENSER method.

2.2   Size/Accuracy Trade-Off in ANNs

The relationship between size and accuracy of ANNs has been of interest since their creation due to the lack of available compute power [15]. In recent years, different methods have been implemented to optimize this trade-off. In [9], a parameter server based distributed computing framework called 'Rudra' was developed, which focuses on the learning aspect of the networks. In [19], the trade-off is optimized by having ANNs allowing for selective execution, which was shown to vastly improve efficiency.

One of the first works attempting to explicitly optimize the size-accuracy trade-off (using multi objective optimization techniques) with respect to topological hyper-parameter selection is [11]; however, only the sequence of layers was varied, resulting in a naively small search space. Furthermore, the networks required very significant compute power during training (60 Tesla M40 GPU's). More recently, researchers at Tsing-Hua University in Taiwan proposed a Device-aware Progressive Search for Pareto-optimal Neural Architectures, or DPP-Net [5]. Whilst the approach to finding optimal architectures is very similar to the one employed here, there are key differences: First, the search space is restricted to parameters regarding the structures of the cells; both the overall architecture and the learning parameters are kept constant. Next, the search algorithm used is the Sequential Model Based Optimization (SMBO) algorithm; the grammatical evolution approach is not used. Finally, the emphasis of the research is on hardware-specific optimization with regard to performance on different computing devices; our paper focuses on a more general approach.

3   Experimental Setup

The algorithm used to carry out multi-objective optimization is the NSGA-II algorithm [7]. Full details of the algorithm and parameters can be found in Appendix-A. All networks evolved are constructed using a sequence of one of two building blocks, or 'modules'. We will refer to these as convolution blocks (CB) or inception blocks (IB) throughout the rest of this paper; figures showing their sequence of layers can be found in Appendix-B. The convolution block (using the ReLU activation function) was first proposed in [14], while the inception block is from the GoogLeNet network architecture [28].

There is the possibility of fully-connected (FC) layers between the output of the last module and the softmax layer. To ensure dimensionality reduction, a max-pooling layer is inserted after the first block, and then after every 2nd block for the CB networks, and after every 3rd block for the IB networks. Specifications regarding each of these variables are cataloged in Table 1, and an example grammar defining how they are chosen can be found in Appendix-C. We also have the following specifications regarding the size of the filters applied at each layer: (i) all MaxPool filters are of size 2×2 with stride 2; (ii) all convolutional filters in the ConvolutionBlockNets are either 3×3 or 5×5 (choosen at random by the grammar with a 50% chance for each) with stride 1; (iii) all convolutional filters in the InceptionBlockNets have size/properties described in [28].

The grammar also defines the number of channels going in and out of each convolutional layer. We only fix these values for the first block, and then allow them to vary as part of the phenotype constructed by the grammar. Information regarding output channel values for each convolution in the IB networks is found in Appendix-D. We also note that the 3-channeled RGB images are not fed directly into the first inception block; a preliminary 3×3 convolutional block is inserted in between these two, with the number of output channels equal to 128, 164, or 196. In experiment 2, the search space is expanded to include variations in learning parameters. The topology of the networks evolved remains identical, as do the evolutionary parameters (15 generations of 15 individuals).

Table 1. Network architecture specifications

Network type	Dataset	CB range	IB range	FC range	Kernels - FC1	Kernels - FC2	MaxPool layer location
ConvolutionBlockNet	CIFAR10	1–3	-	0–2	100–400	40–70	Every 2nd block
ConvolutionBlockNet	STL10	3–6	-	0–2	100–400	40–70	Every 2nd block
InceptionBlockNet	CIFAR10	-	4–8	0–1	100–500	-	Every 3rd block
InceptionBlockNet	STL10	-	5–9	0–1	100–500	-	Every 3rd block

The goal is to determine whether the introduction of variation with regard to learning parameters causes the nature of the solutions on the first front after 15 generations to change, and if so, which of these parameters caused the change. The parameters and their allowed values are shown in the Appendix-E. All the CB networks generated across both experiments are trained for 15 epochs, whilst the IB networks are trained for 20 epochs.

4   Results

The graphs show the evolution of the first non dominated front (or 'Pareto front') at each generation. The x-axis is the size of the network, in millions of parameters, and the y-axis is the validation accuracy of the network on the test set. It is important to note that due to the functioning of the NSGA-II algorithm the values on the x-axis decrease as we move from left to right along it. The evolution of the first pareto front at each generation is shown in Fig. 1 for the IB architectures in the first experiment. The rest of the graphs for the different architectures and experiments can be found in Appendix-G. The figures show that the quality of the solutions represented by the first front is increasing over time, with regard to both size and accuracy. In Fig. 1, the smallest network on the first front tested on CIFAR-10 (∼2M parameters) shows a validation accuracy of 60%, whilst the one on the last front, of approximately the same size, has a validation accuracy of 83%. Similarly, the network with the highest accuracy from the first front tested on CIFAR-10 (84.5%) has ∼5.7M parameters, and this goes down to ∼2.7M parameters on the last front, whilst retaining the accuracy.

For each of the 8 runs, we have compiled details of the two best networks (or solutions) with regard to validation accuracy. The description of these network architectures is shown in Tables 2 and 3. A table giving the learning parameters associated to each network in the second experiment, as encoded by the grammar, can be found in Appendix-F. In these tables, **C_X** represents a convolution block with filter of size **X**, and **FC_X** represents a fully connected layer with **X** kernels. For the inception blocks, only the total number of blocks is indicated (equal to **X** in the **XIB** term); the number of channels inside each of the

Table 2. Best architectures - Experiment 1

Network type	Dataset	Network	Architecture	Size (number of params)	Validation accuracy (%)
ConvolutionBlockNet	CIFAR10	1	C_5_C_3_C_5_FC_206_FC_65	662,925	73.71
-	-	2	C_3_C_5_C_5_FC_262_FC_62	577,082	72.76
ConvolutionBlockNet	STL10	1	C_3_C_3_C_5_C_3_C_5_FC_296	11,228,690	60.65
-	-	2	C_3_C_5_C_3_C_5_C_5_FC_107	6,198,393	60.16
InceptionBlockNet	CIFAR10	1	PRE_128_8IB_FC_133	4,739,221	85.11
-	-	2	PRE_128_5IB	2,617,182	85.02
InceptionBlockNet	STL10	1	PRE_160_8IB	3,271,638	69.51
-	-	2	PRE_160_8IB	2,792,826	68.63

Table 3. Best architectures - Experiment 2

Network type	Dataset	Network	Architecture	Size (number of params)	Validation accuracy (%)
ConvolutionBlockNet	CIFAR10	1	C_3_C_3_C_3_FC_186	691,324	73.60
-	-	2	C_3_C_3_C_3	125,706	73.42
ConvolutionBlockNet	STL10	1	C_5_C_3_C_5_C_5	1,701,642	58.43
-	-	2	C_3_C_5_C_3	638,474	55.21
InceptionBlockNet	CIFAR10	1	PRE_192_4IB	1,697,070	83.85
-	-	2	PRE_160_4IB	1,692,710	83.83
InceptionBlockNet	STL10	1	PRE_160_6IB	2,707,510	64.21
-	-	2	PRE_128_7IB	2,698,382	60.66

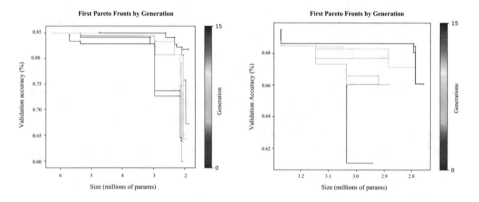

Fig. 1. Experiment 1 - IB architectures - CIFAR10 (left) and STL10 (right)

blocks can be found in Appendix-H. Furthermore, the number of channels out of the preliminary layer is denoted by **X** in the **PRE_X** part of the inception architectures.

5 Discussion

5.1 Experiment 1

The analysis of the information in Table 2 yields two crucial insights about the nature of the solutions generated. The first is that in 3 out of 4 cases, the two networks which show the best validation accuracy on the test set have the exact same sequence of layers, which indicates our algorithm is converging towards the best architecture. For example, both the top networks in the first run (CB - CIFAR10) are built of the maximum allowed number of both convolution blocks (3) and fully-connected layers (2). While it may seem intuitive that a network which makes use of as many layers as possible will yield the highest accuracy, the second run (CB -STL10) gives networks which only make use of a single fully-connected layer, whilst the IB networks 'choose', more often than not, to omit fully-connected layers altogether. This reinforces the idea that the best network architecture is not always the most intuitive one, and that trial-and-error is the most effective, if not only, way to determine its characteristics. Whilst none of the evolved networks in Table 2 exhibit state of the art results in terms of accuracy, they provide rapidly-generated preliminary insight as to which design choices work well on the given dataset; insight which can then be employed by the human expert.

Our second observation deals with the size of the networks rather than their accuracy. In the second (CB - STL10) and third (IB - CIFAR10) runs, we see that whilst we only lose a small amount of accuracy between the two networks (0.49% and 0.09%, respectively), the size of the second network is almost half that of the first. In the case of the second run especially, we have two networks which look almost identical in their topology, yet have a very different number of parameters. We can see that the implementation of multi-objective optimization, rather than yield state of the art architectures, has revealed small changes which can be made to a network to drastically reduce its size, at the expense of very little accuracy. Therein lies the danger of optimizing strictly with regard to validation accuracy: the algorithm, having no interest in the size of the network, misses opportunities to very easily cut a great deal of parameters and improve the efficiency of the networks.

5.2   Experiment 2

The first thing which is immediately apparent when analyzing the results presented in Table 3 is that the observations made about the nature of the networks in Experiment 1 also hold true in Experiment 2. Namely, we see a convergence towards an optimal architecture, with the top two solutions, especially in the third (IB - CIFAR10) and fourth (IB - STL10) runs, sharing almost identical topologies and sizes. We also observe that this similarity extends to the learning parameters: the top two networks in each of the four runs share identical optimizers, learning rates and momentum. We are also presented with a prime example of the second observation of Experiment 1: in the first run (CB - CIFAR10), the second network is only 0.18% less accurate than the first, but it is over 5x smaller. The reason behind this decrease in size is the removal of the fully connected layer from the networks architecture; this, along with the absence of fully-connected layers in 7 out of the 8 inception networks shown in Tables 2 and 3, reinforces the idea that fully-connected layers can be removed from ANN's to reduce their size, at the expense of very little accuracy.

The other main observation about the learning parameter selections of the networks has to do with accepted conventions with regard to these parameters. For example, in most conventional CNN's, SGD is used with a momentum of 0.9. However, half the networks evolved in this experiment 'choose' a momentum value of 0.85. We also see no constant trend with regard to weight decay. As a matter of fact, every allowed value of weight decay is present in the top architectures; yet again, our results show that optimal parameter choices for ANN's are largely problem-dependent.

When comparing networks evolved in both experiments, two things are immediately apparent: (i) the networks from Experiment 2 are less accurate than those from Experiment 1 and (ii) the networks from Experiment 2 are smaller than those from Experiment 1. We thus have a situation where the networks from the second experiment are converging towards a different part of the trade-off search space than those from the first experiment. Taking as an example the second run (CB - STL10), the average accuracy/size of the top two networks from Experiment 1 is 60.41%/8.71M, and for experiment 2 these averages become 56.82%/1.17M. It is interesting to note the difference in the nature of the solutions between both experiments, when it could have been reasonably hypothesized that, sharing unbiased initial conditions and identical evolutionary parameters, the algorithm would converge to similar solutions. This difference only exacerbates the advantage of the evolutionary approach; when dealing with a particular problem/dataset, the user can examine the solutions generated by **both** experiments, and make use of the architecture/learning choices of the solution which best satisfies the network requirements with regard to size and accuracy.

6   Conclusions and Future Work

In this paper, we proposed a novel approach combining grammatical evolution (GE) and multi-objective optimization techniques to automatically search for optimal ANN topologies. We set up experiments using two different types of cells, or blocks (convolution blocks and inception blocks), two different search spaces (one involving strictly topological parameters, the other involving both topological and learning parameters), and two different datasets (CIFAR10 and STL10).

We show that for every evolutionary run, the quality of the solutions generated increases at every generation, and converges towards an optimal architecture for each given problem/dataset. Our results reinforce some conventions with regard to ANN architecture choices, such as the fact that fully-connected layers can be omitted at the expense of very little accuracy loss, and are at odds with others, such as the commonly accepted value of 0.9 for SGD momentum. Furthermore, the optimization of the networks with regard to size, as well as accuracy, uncovers instances where small changes in the network architecture can drastically reduce the size of the network, whilst conserving the major part of its accuracy.

Our evolution is limited to the generation of a 'mere' 240 networks, and has only explored a small proportion of the entire search space. Our work consists of a proof of concept, and as such, more runs should improve the quality of the solutions. These solutions could be improved by enlarging the grammar, allowing for more flexibility in the sequences formed by these cells, varying further leaning parameters (learning rate scheduler, activation functions, etc.), as well as training for more epochs. These ideas are left for future work.

Appendix

All mentioned appendices can be found in the full paper at www.tomasocetto.com.

References

1. Ahmadizar, F., Soltanian, K., AkhlaghianTab, F., Tsoulos, I.: Artificial neural network development by means of a novel combination of grammatical evolution and genetic algorithm. Eng. Appl. Artif. Intell. **39**, 1–13 (2015)
2. Assunçao, F., Lourenço, N., Machado, P., Ribeiro, B.: Automatic generation of neural networks with structured grammatical evolution. In: 2017 IEEE Congress on Evolutionary Computation (CEC), pp. 1557–1564. IEEE (2017)
3. Assunçao, F., Lourenço, N., Machado, P., Ribeiro, B.: Towards the evolution of multi-layered neural networks: a dynamic structured grammatical evolution approach. In: Proceedings of the Genetic and Evolutionary Computation Conference, pp. 393–400. ACM (2017)
4. Assunçao, F., Lourenço, N., Machado, P., Ribeiro, B.: DENSER: deep evolutionary network structured representation. arXiv preprint arXiv:1801.01563 (2018)
5. Cheng, A.C., Dong, J.D., Hsu, C.H., Chang, S.H., Sun, M., Chang, S.C., Pan, J.Y., Chen, Y.T., Wei, W., Juan, D.C.: Searching toward pareto-optimal device-aware neural architectures. arXiv preprint arXiv:1808.09830 (2018)
6. Coates, A., Ng, A., Lee, H.: An analysis of single-layer networks in unsupervised feature learning. In: Proceedings of the Fourteenth International Conference on Artificial Intelligence and Statistics, pp. 215–223 (2011)
7. Deb, K., Pratap, A., Agarwal, S., Meyarivan, T.: A fast and elitist multiobjective genetic algorithm: NSGA-II. IEEE Trans. Evol. Comput. **6**(2), 182–197 (2002)
8. Fenton, M., McDermott, J., Fagan, D., Forstenlechner, S., Hemberg, E., O'Neill, M.: PonyGE2: grammatical evolution in Python. In: Proceedings of the Genetic and Evolutionary Computation Conference Companion, pp. 1194–1201. ACM (2017)
9. Gupta, S., Zhang, W., Wang, F.: Model accuracy and runtime tradeoff in distributed deep learning: a systematic study. In: 2016 IEEE 16th International Conference on Data Mining (ICDM), pp. 171–180. IEEE (2016)
10. Kandasamy, K., Neiswanger, W., Schneider, J., Poczos, B., Xing, E.: Neural architecture search with bayesian optimisation and optimal transport. arXiv preprint arXiv:1802.07191 (2018)
11. Kim, Y.H., Reddy, B., Yun, S., Seo, C.: NEMO: neuro-evolution with multiobjective optimization of deep neural network for speed and accuracy. In: ICML 2017, AutoML Workshop (2017)
12. Koza, J.R.: Genetic programming as a means for programming computers by natural selection. Stat. Comput. **4**(2), 87–112 (1994)
13. Krizhevsky, A., Hinton, G.: Learning multiple layers of features from tiny images. Technical report, Citeseer (2009)
14. Krizhevsky, A., Sutskever, I., Hinton, G.E.: ImageNet classification with deep convolutional neural networks. In: Advances in Neural Information Processing Systems, pp. 1097–1105 (2012)
15. Lawrence, S., Giles, C.L., Tsoi, A.C.: What size neural network gives optimal generalization? Convergence properties of backpropagation. Technical report (1998)

16. LeCun, Y., Bottou, L., Bengio, Y., Haffner, P.: Gradient-based learning applied to document recognition. Proc. IEEE **86**(11), 2278–2324 (1998)

17. Liu, C., Zoph, B., Shlens, J., Hua, W., Li, L.J., Fei-Fei, L., Yuille, A., Huang, J., Murphy, K.: Progressive neural architecture search. arXiv preprint arXiv:1712.00559 (2017)

18. Liu, H., Simonyan, K., Vinyals, O., Fernando, C., Kavukcuoglu, K.: Hierarchical representations for efficient architecture search. arXiv preprint arXiv:1711.00436 (2017)

19. Liu, L., Deng, J.: Dynamic deep neural networks: optimizing accuracy-efficiency trade-offs by selective execution. arXiv preprint arXiv:1701.00299 (2017)

20. Loshchilov, I., Hutter, F.: CMA-ES for hyperparameter optimization of deep neural networks. arXiv preprint arXiv:1604.07269 (2016)

21. Miikkulainen, R., Liang, J., Meyerson, E., Rawal, A., Fink, D., Francon, O., Raju, B., Shahrzad, H., Navruzyan, A., Duffy, N., et al.: Evolving deep neural networks. arxiv 2017. arXiv preprint arXiv:1703.00548

22. Negrinho, R., Gordon, G.: DeepArchitect: automatically designing and training deep architectures. arXiv preprint arXiv:1704.08792 (2017)

23. O'Neill, M., Ryan, C.: Grammatical evolution: Evolutionary automatic programming in a arbitrary language, volume 4 of genetic programming (2003)

24. Real, E., Aggarwal, A., Huang, Y., Le, Q.V.: Regularized evolution for image classifier architecture search. arXiv preprint arXiv:1802.01548 (2018)

25. Simonyan, K., Zisserman, A.: Very deep convolutional networks for large-scale image recognition. arXiv preprint arXiv:1409.1556 (2014)

26. Snoek, J., Rippel, O., Swersky, K., Kiros, R., Satish, N., Sundaram, N., Patwary, M., Prabhat, M., Adams, R.: Scalable Bayesian optimization using deep neural networks. In: International Conference on Machine Learning, pp. 2171–2180 (2015)

27. Stanley, K.O., Miikkulainen, R.: Evolving neural networks through augmenting topologies. Evol. Comput. **10**(2), 99–127 (2002)

28. Szegedy, C., Liu, W., Jia, Y., Sermanet, P., Reed, S., Anguelov, D., Erhan, D., Vanhoucke, V., Rabinovich, A.: Going deeper with convolutions. In: Proceedings of the IEEE Conference on Computer Vision and Pattern Recognition, pp. 1–9 (2015)

29. Tsoulos, I., Gavrilis, D., Glavas, E.: Neural network construction and training using grammatical evolution. Neurocomputing **72**(1–3), 269–277 (2008)

30. Zeiler, M.D., Fergus, R.: Visualizing and understanding convolutional networks. In: European Conference on Computer Vision, pp. 818–833. Springer (2014)

31. Zoph, B., Le, Q.V.: Neural architecture search with reinforcement learning. arXiv preprint arXiv:1611.01578 (2016)

32. Zoph, B., Vasudevan, V., Shlens, J., Le, Q.V.: Learning transferable architectures for scalable image recognition. arXiv preprint arXiv:1707.07012 **2**(6) (2017)

Fast Transfer Learning for Image Polarity Detection

Edoardo Ragusa$^{(\boxtimes)}$, Paolo Gastaldo, and Rodolfo Zunino

Department of Electrical, Electronic and Telecommunications Engineering, and Naval
Architecture, DITEN, University of Genoa, Genoa, Italy
edoardo.ragusa@edu.unige.it, {paolo.gastaldo,rodolfo.zunino}@unige.it

Abstract. Convolutional neural networks (CNNs) provide an effective
tool to extract complex information from images. In the area of image
polarity detection, CNNs are utilized in combination with transfer learning to tackle a major problem: the unavailability of large sets of labeled
data. Accordingly, polarity predictors in general exploit a pre-trained
CNN that in turn feeds a classification layer. While the latter layer is
trained from scratch, the pre-trained CNN is subject to fine tuning. In
the actual implementation of such configuration, the specific CNN architecture indeed sets the performances of the predictor both in terms of
generalization abilities and in terms of computational complexity. The
latter attribute becomes critical when considering that polarity predictors -in the era of social network and custom profiles- might need to
be updated within a short time interval (i.e., hours or even minutes).
Thus, the paper proposes a design of experiment that supports a fair
comparison between predictors that rely on different architectures.

Keywords: Convolutional neural networks · Transfer learning ·
Image polarity

1 Introduction

In the era of social networks, images are expected to convey affective information. Thus, in the recent years applications of sentiment analysis started covering
multimedia resources [15,19] in addition to text-only resources. The literature
shows that deep learning played a major role in the development of sentiment
analysis applied to images, i.e., image polarity detection, which can be accomplished only by understanding the interaction between different components of
an image. In this regard, Convolutional neural networks (CNNs) can effectively
support the task of object recognition, which is a prerequisite to characterizing interaction between objects. Polarity, though, also depends on the context
in which the image is examined; similar objects and even similar images can
induce different polarities in different scenarios. In general, in the era of social
media, content marketing, and custom profiles, this means that one might need
to re-train the polarity detection strategy within a short time interval (minutes

© Springer Nature Switzerland AG 2020
L. Oneto et al. (Eds.): INNSBDDL 2019, INNS 1, pp. 27–37, 2020.
https://doi.org/10.1007/978-3-030-16841-4_4

rather than days) to address the change of circumstances (e.g., cultural background). In fact, re-training a deep network in the context of polarity detection poses major problems. The availability of huge amount of labeled data cannot be taken for granted and the complexity of the learning procedure might hinder fast training, unless massive computational resources are available.

Transfer learning techniques [23] provide an effective solution to both the problems. The literature indeed shows that image polarity detection can achieve excellent performance by exploiting transfer learning [13, 15, 19]. In the standard design, a CNN pre-trained on object recognition provides the low-level features for the eventual polarity detector. Then, by adopting the fine tuning process, one updates the parameters of the pre-trained network by means of a learning procedure involving the new domain, thus avoiding training from scratch. To the best of authors' knowledge, though, the literature also leaves a few crucial issues open. The first issue is the lack of a fair comparison between the various implementations that have been derived from such design. Second, little attention has given to the trade-off between the computational cost associated to the transfer learning process and the accuracy of the eventual predictor.

This paper aims at addressing these open issues by defining a unambiguous design of experiment for the comparison of different implementations of image polarity detectors. In this regard, the evaluation process focused on the role played the pre-trained CNNs. Accordingly, the experimental protocol compared polarity detectors that differ in the CNN architecture adopted for feature extraction; the experiments involved all the CNNs that proved outstanding in the area of object recognition. The eventual predictors were mainly compared according to two attributes: classification accuracy and computational cost of the training process. Three well-known datasets provided the benchmarks for the experimental evaluation.

2   Image Polarity Detection with Deep Learning

Deep learning has changed the approach to image processing and consequently to image polarity detection [13, 15, 19]. In general, polarity detection inherits the core structure from object recognition frameworks. Accordingly, such approach leads to two main different designs, which exploit the low-level features extracted by a CNN trained on object recognition. In the first design (Fig. 1(a)), the last fully connected layer of the pre-trained CNN is removed to get the low-level features. Such features feed either a new fully connected layer that includes as many neurons as the classes involved in the polarity detection problem or a given classifier. By taking advantage of fine tuning, one can also re-train the CNN to tailor its parameters (and the eventual features) on the specific polarity detection problem. In the second design (Fig. 1(b)) a two-step procedure leads to the final predictor. First, the pre-trained CNN is augmented by replacing the last fully connected layer with a new architecture, which should model an ontology. The eventual layout is re-trained as a whole by using a Visual Sentiment Ontology (VSO), e.g., adjective noun pairs (ANPs) [22, 24]. In the second step, such new layout becomes the building block of the polarity detector. Hence, one may replace the last fully connected layer, analogously to the first design. A new training process then is applied to the final layout. It is worth noting that usually the first step relies on weak labels in the learning process, as VSOs are generated by automated algorithms that add noise to the labeling action.

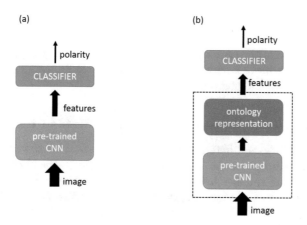

Fig. 1. Image polarity detector. In the design (a) the classifier receives as input the low-level features directly from the pre-trained CNN; in the design (b) the classifier receives as input the low-level features from a block that should model an ontology

2.1   CNNs for Object Recognition

CNNs have been obtaining outstanding results in the task of object recognition. Table 1 reviews the most significant CNN architectures, which indeed may represent the building block in both the designs presented above: *AlexNet* [12], *Vgg19* [18], *GoogLeNet* [20], *Inc_V3* [21], *ResNet101* [7], *ResNet152* [7], *DenseNet* [8], and *SqueezeNet* [9]. For each architecture the table provides: the number of weight layers; the top-1 accuracy achieved on ImageNet [17] competition (in percentage); the number of floating point operations performed for a single image classification [4]; the total amount of parameters. The number of floating point operations provides a suitable approximation of the overall computational complexity that characterizes an architecture. The amount of parameters basically affects two aspects: memory occupation and expected size of the training set. When starting with a pre-trained network, transfer learning can reduce the amount of patterns required to successfully complete fine tuning. On the other hand, the size of such training set is still strictly related to the number of parameters. *SqueezeNet* was included in this list because -in its standard implementation- it can achieve a 50× compression in model size with respect to *AlexNet* while scoring comparable performances in terms of accuracy.

The table shows that *SqueezeNet* -as expected- is the most convenient architecture in terms of computational complexity and number of parameters. However, the other architectures (except for *AlexNet*) proved able to outperform

Table 1. Attributes of the CNN architectures exploited in the area of object recognition

–	AlexNet	Vgg19	GoogLeN	Inc3	Res101	Res152	Dense201	SqueezeN
Layers	8	19	58	46	101	152	201	18
Top-1 acc (%)	57	71	58	78	77	79	77	57
GFlops	0.7	19.6	1.6	6.0	7.6	11.3	4.0	0.8
Par ($\cdot 10^6$)	61	144	7		24	45	60	20	1

SqueezeNet in terms of accuracy. In fact, *DenseNet* possibly represents the best choice, as it can achieve a satisfactory accuracy while limiting both the computational complexity and the number of parameters.

2.2   Image Polarity Detection with CNNs: State of the Art

The literature provides different actual implementations of the two designs presented above. Principally, the implementations differ in the choice of the CNN to be adopted, the transfer learning technique, and the training data domain. In the same year two distinct works exploited *AlexNet* as a low-level feature extractor in the design of Fig. 1(a) [16,22]. Campos et al. [3] indeed exploited the *AlexNet* architecture as feature extractor to complete an interesting analysis on the effects of layer ablation and layer addition in the eventual accuracy of the polarity detection framework. A design involving ontology representation (Fig. 1(b)) was implemented in [5]; the authors proposed a layout for sentiment ontologies recognition based on the *AlexNet* architecture, in which the last fully connected layer had one neuron for each ANP. Jou et al. [11] improved the predictor presented in [5] by exploiting a multilingual visual sentiment ontology (MVSO). In [24] the authors proposed a custom architecture for image polarity prediction: the model stacked two convolutional layers and four fully connected layers. An empirical comparison between such architecture, the architectures analyzed in [3], and the architecture proposed in [11] was presented in [2]. In this work, the authors indeed evaluated the role played by the specific ontology in the transfer learning process; the best performance was obtained by models relying on the *AlexNet* architecture and the English version of MVSO. The *Vgg19* architecture and the *GoogLeNet* architecture were exploited, respectively, in [6,10]. Both papers showed that such architectures proved able to support predictors that improved over state-of-the-art predictors based on the *AlexNet* architecture.

3   A Compared Analysis

CNNs are widely utilized as feature extractor -in the transfer learning setup- in all the frameworks that targets image sentiment analysis. Indeed, in general, these frameworks differ both in terms of specific design of the feature extraction process and of overall design of the predictor. As a major consequence, it is never easy to evaluate the actual contribution provided by a given architecture in boosting the performance of image polarity detection. The present paper wants to address this issue by proposing a formal experimental protocol that should support a fair comparison between different configurations of the feature extraction process (i.e., different pre-trained networks). To this purpose, the design of Fig. 1(a), *Layer Replacement*, is used as reference.

3.1   Image Polarity Detector: The Design

In the Layer Replacement design, the last fully connected layer of a CNN trained on object recognition is replaced by a new fully connected layer, which serves as classification layer for the eventual predictor and includes as many neurons as the number of classes represented in the polarity problem. Accordingly, the eventual predictor is a universal approximator combining a feature extraction block that inherits its parameterization from object recognition and a classification layer that should be trained from scratch. In the learning procedure, the predictor is trained on the new domain by utilizing a small number of epochs. Training is indeed extended to the feature extraction block, thus adopting fine tuning; by suitably setting the learning rate, the involved parameters are only subject to small perturbations [23]. Actually, similar, yet alternative configurations might be obtained by exploiting different strategies of layer ablation. Campos et al. [2], though, has showed that replacing the last fully connected layer represents the most convenient choice when addressing image sentiment analysis.

The design of Fig. 1(a) allows one to assess the role played by the single pre-trained architecture in terms of two parameters: the accuracy of the overall polarity predictor and the computational load involved in the corresponding learning procedure. Actually, the design of Fig. 1(b) would not be as useful in supporting a fair comparison between different architectures. The intermediate layer set to model ontologies could actually introduce biases brought about both by the configuration of this layer and by the adopted ANP, which might also add a cultural bias [11]. In the proposed design, an unbiased background has been set by relying only on CNNs pre-trained using the ImageNet dataset [17], a well-established benchmark for the object recognition area.

3.2   Computational Complexity

The Layer Replacement configuration can be firstly analyzed according to the computational load involved in the learning process of the eventual predictor. The computational cost of the training procedure O_{lr} can be approximated by defining the following parameters: n_{tg} is the number of training samples; n_{ep} is the number of epochs; n_{lay} is number of layers; $n_{par,i}$ is the number of parameters in the ith layer; O_f is computational cost of a feedforward step; O_b is the computational cost of a backward step, which involves gradient computation and parameters' update.

The architecture directly impacts on the last four quantities. Let P be the quantity defined as follows

$$P = \sum_{i=1}^{n_{lay}} n_{par,i}^3; \tag{1}$$

then, O_{lr} can be expressed as

$$
\begin{aligned}
O_{lr} &= n_{ep} \cdot n_{tg} \cdot [(O_f(P) + O_b(P)] \\
&= n_{ep} \cdot n_{tg} \cdot [(O_f(P) + \alpha \cdot O_f(P)] \approx 3 \cdot n_{ep} \cdot n_{tg} \cdot O_f(P).
\end{aligned} \tag{2}
$$

Empirical evidence suggests that the value of the coefficient α in (2) can be roughly approximated with 2 [4]. Equation (2) shows that both O_f and O_b scale cubically with the number of parameters. Indeed, (1) indicates that an architecture is not only characterized by the total number of parameters. In the case of uneven distribution of the parameters, the computational cost is mostly determined by the largest layers (in terms of weights).

A discussion on the computational cost should also take into account a few elements that do not emerge in Eq. (2). First, the issue of sub-optimal solutions affects heuristic optimization techniques such as stochastic gradient descent. As transfer learning relies on small datasets, such issue may plausibly perturb the training of polarity predictor. Besides, the risk of being trapped in local minima is increased by the adoption of early stopping strategies, which on the other side prevent overfitting. Actually, multi-start optimization represents a suitable solution. This solution also inflates the computational load of the learning procedure, which might require multiple runs of the training algorithm.

Second, the execution time of a CNN might not depend only on P. On the one hand, GPUs are designed to fully exploit data and model parallelism. CNNs, though, are realized as a stack of layers that should be completed sequentially. Hence, the execution time of *Vgg19* might be lower than the execution time of *Res152*. While the latter architecture is lighter in terms of parameters, *Vgg19* involve a sensibly smaller number of layers. Accordingly, GPUs supported by large memories can suitably take advantage of the configuration {small number of layers, several parameters per layer} to boost computational time.

4   Experimental Results

The experimental campaign aimed at assessing the generalization ability of the design presented above. Indeed, the goal was also to evaluate how the specific architecture of feature-extraction module influenced the performance of the over-all predictor. For the sake of consistency, all the experiments were implemented in MATLAB® with the Neural Network Toolbox, which provided the pre-trained versions of the architectures listed in Table 1. Actually, *Res_152* was the only architecture not included in the experiments, as the Neural Network Toolbox did not provide its implementation.

4.1   Experimental Setup

The experiments involved four datasets: Twitter1 (Tw_1) [24], Twitter2 (Tw_2a and Tw_2b) [14], and ANP40 [1]. Twitter1 is the most common benchmark for image polarity recognition. The dataset collected 1269 images obtained from image tweets, i.e., Twitter messages that also contain an image. All images have been labeled via an Amazon Mechanical Turk (AMT). This paper actually utilized the "5 agree" version of the dataset. Such version includes only the images for which all the five human assessors agreed on the same label. Thus, the eventual dataset included a total of 882 images (581 labeled as "positive" and 301 labeled as "negative").

Twitter2 also provides a collection of image tweets [14]. In this case, tweets were filtered according to a predetermined vocabulary of 406 emotional words. Thus, the dataset included only the tweets that contained in the text message at least one of those words. Three annotators labeled the pairs {text, image} by using a three-value scale: positive, negative, and neutral; text and image were annotated separately. This paper utilized a pruned version of the dataset: only images for which all the annotators agreed on the same label were employed, thus obtaining a total of 4109 images. As only 351 images out of 4109 were labeled as negative, such pruned version of Twitter2 resulted very unbalanced. Thus, two different balanced datasets (Tw_2a and Tw_2b) were generated. Each dataset included a total of 702 images: the 351 "negative" images and 351 images randomly extracted from the total amount of "positive" images.

The ANP dataset implements an ontology and is composed of Flickr images [1]. The dataset includes 3,316 adjective noun pairs; for each pair, at most 1,000 images were provided, thus leading to a total of about one million images. The ANP tags assigned from Flickr users have been utilized as labels for the images; thus, noise affected severely this dataset. To tackle this issue, this paper used a pruned version of the dataset. Accordingly, the 20 ANPs with the highest polarity values and the 20 ANPs with the lowest polarity values were selected (the website of the dataset[1] was exploited as reference). Eventually, the dataset involved in the experiments included 11857 "positive" images and 5257 "negative" images.

The first three benchmarks were utilized to assess the performance of the proposed design under the most common scenario: a small training set is available. ANP40 covered the case in which a larger dataset is available. In all the experiment, the setup of the predictor was organized as follows. Stochastic Gradient Descent with Momentum was exploited as optimization strategy, with momentum $= 0.9$ and learning rate $= 10^{-4}$ for all the layers in the pre-trained CNN. In the classification layer the learning rate was set to 10^{-3} and the regularization parameter was set to 0.5. Early stopping was adopted, with validation patience set to 2. Training involved a maximum of 10 epochs.

---

[1] http://visual-sentiment-ontology.appspot.com/.

4.2 Results and Comments

Seven different pre-trained architectures were utilized to generate as many implementations of the predictor. The experimental session aimed at assessing and comparing the generalization performances of such implementations.

The performances of the seven predictors have been evaluated by exploiting a 5-fold strategy. For each dataset, the classification accuracy of a predictor has been measured via five different experiments, corresponding to as many different training/test pairs. Indeed, five separate runs of each single experiment were completed; each run involved a different composition of the mini-batch. As a result, given a predictor and a benchmark, 25 measurements of the classification accuracy on the test set were eventually available.

Table 2 reports on the performance obtained with the Layer Replacement design on Tw_1, Tw_2a, Tw_2b, and ANP40. In each column, the first row indicates the pre-trained architecture adopted in the implementation of the predictor; the second row gives -for the Tw_1 dataset- the average accuracy obtained by the predictor over the 25 experiments, along with its standard uncertainty (between brackets); the third, fourth, and fifth rows give the same quantities for the experiments involving Tw_2a, Tw_2b, and ANP40, respectively. Experimental outcomes showed that *Dense201* was able to slightly outperform the other architectures on Tw_1. *Dense201* again scored the best average accuracy on Tw_2a; the gap between such predictor and *Res101*, though, was almost inappreciable, especially if one consider the corresponding standard uncertainties. Likewise, *Vgg19*, *Inc3*, and *Dense201* shared the role of best predictor on Tw_2b. On the ANP40 dataset, seven predictors out of nine achieved an average accuracy ranging between 78.7% and 79.9%; in practice, only the predictors based on *AlexNet* and *SqueezeNet* proved to be slightly less effective on ANP40. Such balance possibly stems from the availability of a larger dataset.

Table 2. Experimental results: classification accuracy

	AlexNet	Vgg19	GoogLeN	Inc3	Res101	Dense201	SqueezeN
Tw_1	82.5 (0.6)	86.4 (0.5)	84.2 (0.5)	86.5 (0.7)	88.2 (0.4)	89.4 (0.5)	82.5 (0.6)
Tw_2a	66.2 (0.5)	69.2 (0.7)	66.0 (0.6)	68.1 (0.7)	70.8 (0.6)	71.3 (0.8)	65.2 (0.8)
Tw_2b	65.4 (0.7)	71.0 (0.8)	68.2 (0.5)	70.5 (0.9)	69.2 (0.8)	70.6 (0.6)	66.8 (0.7)
ANP40	76.3 (0.3)	78.7 (0.3)	79.2 (0.3)	79.9 (0.2)	79.7 (0.2)	79.3 (0.3)	76.7 (0.3)

The experiments can also reveal which architecture has been the most consistent over the different datasets and the different training/set pairs. Thus, for each benchmark and for each training/set pair, the seven architectures have been ranked according to the classification accuracy scored by the corresponding predictor. In this case, the classification accuracy associated to a predictor was the best accuracy over the five runs completed for a given training/set pair. In each rank, one point was assigned to the best predictor, two points were assigned to the second best predictor, and so on until the worst predictor, which took seven points. Accordingly, Fig. 2 provides, for each architecture, the total points marked over the 20 ranks (5 training/test pairs × 4 benchmarks); a predictor could not mark less than 20 points, which meant first position in the rank (i.e., best predictor) for every rank. The plot shows that the predictor based on *Vgg19* proved to be the most consistent, i.e., this predictor often occupied the highest positions in the rank over the different experiments. Such outcome is compatible with the results reported in Table 2, which showed that *Vgg19* was always able to score effective performance over the different benchmarks.

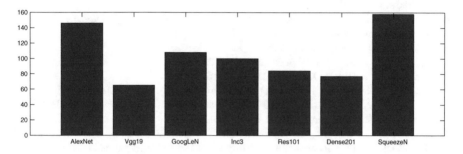

Fig. 2. Consistency evaluation of the nine architectures; the plot gives the architectures on the *x*-axis and the corresponding total amount of collected points on the *y*-axis.

Overall, *Vgg19*, *Res101*, and *Dense201* proved to be the most powerful architectures. In terms of generalization abilities they seemed almost comparable. Indeed, computational aspects may represent a discriminating factor. *Vgg19* requires the storing of a number of parameters that is 3 time bigger than *Res101* and 7 time bigger than *Dense201*; besides, the forward phase of *Vgg19* involves 3 or 4 times the number of floating point operation. However, most of the parameters and operations are introduced in the last fully connected layers of *Vgg19*. Thus, such layers may take advantage of parallel computing, which can boost the overall execution time. Hence, *Vgg19* may represent the best option when a GPU with a considerable amount of memory is available, while *Res101* and *Dense201* should be preferred when one has to deal with memory constraints.

References

1. Borth, D., Ji, R., Chen, T., Breuel, T., Chang, S.F.: Large-scale visual sentiment ontology and detectors using adjective noun pairs. In: Proceedings of the 21st ACM International Conference on Multimedia, pp. 223–232. ACM (2013)
2. Campos, V., Jou, B., Giro-i Nieto, X.: From pixels to sentiment: fine-tuning CNNs for visual sentiment prediction. Image Vis. Comput. **65**, 15–22 (2017)
3. Campos, V., Salvador, A., Giro-i Nieto, X., Jou, B.: Diving deep into sentiment: understanding fine-tuned CNNs for visual sentiment prediction. In: Proceedings of the 1st International Workshop on Affect & Sentiment in Multimedia, pp. 57–62. ACM (2015)

4. Canziani, A., Paszke, A., Culurciello, E.: An analysis of deep neural network models for practical applications. arXiv preprint arXiv:1605.07678 (2016)
5. Chen, T., Borth, D., Darrell, T., Chang, S.F.: DeepSentiBank: visual sentiment concept classification with deep convolutional neural networks. arXiv preprint arXiv:1410.8586 (2014)
6. Fan, S., Jiang, M., Shen, Z., Koenig, B.L., Kankanhalli, M.S., Zhao, Q.: The role of visual attention in sentiment prediction. In: Proceedings of the 2017 ACM on Multimedia Conference, pp. 217–225. ACM (2017)
7. He, K., Zhang, X., Ren, S., Sun, J.: Deep residual learning for image recognition. In: Proceedings of the IEEE Conference on Computer Vision and Pattern Recognition, pp. 770–778 (2016)
8. Huang, G., Liu, Z., Van Der Maaten, L., Weinberger, K.Q.: Densely connected convolutional networks. In: CVPR, vol. 1, p. 3 (2017)
9. Iandola, F.N., Han, S., Moskewicz, M.W., Ashraf, K., Dally, W.J., Keutzer, K.: SqueezeNet: AlexNet-level accuracy with 50x fewer parameters and <0.5 mb model size. arXiv preprint arXiv:1602.07360 (2016)
10. Islam, J., Zhang, Y.: Visual sentiment analysis for social images using transfer learning approach. In: 2016 IEEE International Conferences on BDCloud-SocialCom-SustainCom, pp. 124–130. IEEE (2016)
11. Jou, B., Chen, T., Pappas, N., Redi, M., Topkara, M., Chang, S.F.: Visual affect around the world: a large-scale multilingual visual sentiment ontology. In: Proceedings of the 23rd ACM International Conference on Multimedia, pp. 159–168. ACM (2015)
12. Krizhevsky, A., Sutskever, I., Hinton, G.E.: Imagenet classification with deep convolutional neural networks. In: Advances in Neural Information Processing Systems, pp. 1097–1105 (2012)
13. Luo, J., Borth, D., You, Q.: Social multimedia sentiment analysis. In: Proceedings of the 2017 ACM on Multimedia Conference, pp. 1953–1954. ACM (2017)
14. Niu, T., Zhu, S., Pang, L., El Saddik, A.: Sentiment analysis on multi-view social data. In: International Conference on Multimedia Modeling, pp. 15–27. Springer (2016)
15. Poria, S., Cambria, E., Bajpai, R., Hussain, A.: A review of affective computing: from unimodal analysis to multimodal fusion. Inf. Fusion 37, 98–125 (2017)
16. Razavian, A.S., Azizpour, H., Sullivan, J., Carlsson, S.: CNN features off-the-shelf: an astounding baseline for recognition. In: 2014 IEEE Conference on Computer Vision and Pattern Recognition Workshops (CVPRW), pp. 512–519. IEEE (2014)
17. Russakovsky, O., Deng, J., Su, H., Krause, J., Satheesh, S., Ma, S., Huang, Z., Karpathy, A., Khosla, A., Bernstein, M., et al.: Imagenet large scale visual recognition challenge. Int. J. Comput. Vis. 115(3), 211–252 (2015)
18. Simonyan, K., Zisserman, A.: Very deep convolutional networks for large-scale image recognition. arXiv preprint arXiv:1409.1556 (2014)
19. Soleymani, M., Garcia, D., Jou, B., Schuller, B., Chang, S.F., Pantic, M.: A survey of multimodal sentiment analysis. Image Vis. Comput. 65, 3–14 (2017)
20. Szegedy, C., Liu, W., Jia, Y., Sermanet, P., Reed, S., Anguelov, D., Erhan, D., Vanhoucke, V., Rabinovich, A., et al.: Going deeper with convolutions. In: CVPR (2015)
21. Szegedy, C., Vanhoucke, V., Ioffe, S., Shlens, J., Wojna, Z.: Rethinking the inception architecture for computer vision. In: Proceedings of the IEEE Conference on Computer Vision and Pattern Recognition, pp. 2818–2826 (2016)
22. Xu, C., Cetintas, S., Lee, K.C., Li, L.J.: Visual sentiment prediction with deep convolutional neural networks. arXiv preprint arXiv:1411.5731 (2014)

23. Yosinski, J., Clune, J., Bengio, Y., Lipson, H.: How transferable are features in deep neural networks? In: Advances in Neural Information Processing Systems, pp. 3320–3328 (2014)
24. You, Q., Luo, J., Jin, H., Yang, J.: Robust image sentiment analysis using progressively trained and domain transferred deep networks. In: AAAI, pp. 381–388 (2015)

Dropout for Recurrent Neural Networks

Nathan Watt$^{(\boxtimes)}$ and Mathys C. du Plessis

Nelson Mandela University, Port Elizabeth 6031, South Africa
natewatt0@gmail.com, mc.duplessis@mandela.ac.za

Abstract. Neural networks are computational structures which can be
trained to perform tasks based on training examples or patterns. Recurrent neural networks are a type of network designed to process time-series
data. Dropout is a neural network regularization technique. The literature advises that Dropout should not be directly applied to recurrent
neural networks as its effects are too dramatic when applied recurrently.
This direct approach is described as naive. Instead, there are two specialised recurrent neural network Dropout algorithms proposed by different authors. However, these specialised Dropout algorithms have not
been tested against one another and the naive algorithm under identical experimental conditions. This paper compares all of these algorithms
and finds that the naive approach performed as well as or better than
the specialised Dropout algorithms.

Keywords: Deep learning · Recurrent Neural Networks · Dropout

1 Introduction

A neural network (NN) is a computational structure consisting of layers of nodes
stacked one after another. These layers of nodes have connections between them
referred to as weights. A NN can be trained to reproduce a pattern or computation by adjusting these weights until the output node layer produces the desired
values.

Recurrent neural networks (RNNs) are networks where a node layer can have
connections back to the same layer or to previous layers. This forms a loop of
connections in the RNN. These nodes are referred to as recurrent nodes.

Dropout is a popular and simple regularization technique which provides two
separate benefits to NNs. Firstly, Dropout reduces the problem of overfitting,
which affects NNs trained on small datasets. Secondly, Dropout encorages the
NN to internally solve its task in multiple ways, resulting in the trained network
operating similar to an ensemble NN. This allows the NN to reap the benefits
of ensemble training without the need to retrain the network [7].

Dropout operates by disabling a different randomly selected subset of connections during each training step. This adds distortion to the network during the training process. RNNs contain recurrent connections. This means that
the distortion effect is recursively amplified when applying Dropout to a RNN.

© Springer Nature Switzerland AG 2020
L. Oneto et al. (Eds.): INNSBDDL 2019, INNS 1, pp. 38–47, 2020.
https://doi.org/10.1007/978-3-030-16841-4_5

This makes the RNN difficult to train. Applying Dropout directly to a RNN in this manner will be referred to as Naive Dropout [10]. To counteract the dramatic distortion, two Dropout algorithms specialised for RNNs have been proposed by Gal et al. [2] and Zaremba et al. [10]. These algorithms are designed to prevent Dropout's distortion from being recurrently amplified, while at the same time preserving the benefits of Dropout. These three Dropout algorithms have not been tested under the same experimental conditions. Furthermore, previous researchers incorporated algorithms unrelated to Dropout, such as weight decay in their experiments, where the weight decay parameters were tuned differently between algorithms.

A preliminary study [9] briefly investigated the performance of these specialised Dropout algorithms and found they did not perform as the literature suggested. The goal of this paper is to test the performance, specifically the accuracy of the RNNs under the same experimental conditions using additional benchmarks. The results are then used to rank the Dropout algorithms using the appropriate statistical tests.

The paper is structured as follows. Section 2 provides a litreature review of each RNN Dropout approach. Section 3 describes the benchmarks used to evaluate each algorithm. Section 4 describes the selected experimental conditions. Section 5 discusses the experimental results. Lastly, Sect. 6 draws the conclusion.

2   Literature Review

The simplest way to apply Dropout to an RNN is to randomly drop connections at each timestep. This approach or algorithm is referred to as Naive Dropout [2]. The effects of Dropout on a RNN are dramatic [1]. It is believed that when Dropout is applied directly to a RNN like this, the recurrent connections servery amplify the noise and distortion of Dropout. Making it too difficult to train [10]. With the exception of the preliminary study, this algorithm has not yet been applied successfully to RNNs [2,10].

In order to prevent the over amplification of the noise, Pachitariu et al. [5] and Zaremba et al. [10] recommend only applying Dropout to non-recurrent connections. This algorithm will be referred to as Half Dropout, as it only drops out half of the networks connections.

Lastly, Gal et al. [2] criticized Half Dropout for still showing signs of overfitting with a small dataset. Instead, they proposed a theoretically-based approach to applying Dropout to RNNs. This algorithm is named Variational Dropout. As with Naive Dropout, Variational Dropout does apply Dropout to every connection, including recurrent connections. Where Variational Dropout differs from both previous algorithms, is that it applies the same Dropout mask at all timesteps. Meaning that the same nodes are dropped out at each timestep. Figure 1 shows how each Dropout algorithm operates.

While Gal et al. [2] empirically showed that Variational Dropout outperforms Half Dropout, the experiments were not conducted under the same conditions. Weight decay was used in conjunction with Variational Dropout, but it was not included with Half Dropout. It is not clear whether Variational Dropout's superiority is dependent on being run in conjunction with specific weight decay hyperparameters parameters. In addition to this, for larger RNN models, Half Dropout and Variational Dropout were using different Dropout probabilities.

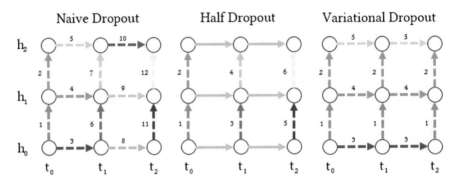

Fig. 1. All three Dropout algorithms applied to the same unfolded recurrent network. Each connection has a colour (alternatively, a number) which represents its Dropout mask. Connections with the same colour (or number) have the same Dropout mask applied, meaning the same nodes are dropped out. Solid lines do not have Dropout applied.

3   Benchmarks

A benchmark is a baseline problem used for comparison purposes. The RNN's performance is measured by the accuracy of its solution for a benchmark problem. For each benchmark, a RNN will be trained using each of the different Dropout algorithms and without any form Dropout. This process is repeated multiple times to allow for a statistical test to be used to compensate for the noise caused by the random weight initialisation, Dropout masks, and training pattern order. These scores then allow for the statistical test to rank the algorithms from best to worst. Two benchmarks are used for these experiments.

3.1   Bouncing Ball Benchmark

The first benchmark requires the RNN to predict frames of a video which shows a ball bouncing in a 2D box. It is a modified version of the problem proposed by Martens et al. [4]. The video consists of a 1-bit, 15 by 15 pixel resolution footage of the bouncing ball. Figure 2 shows an example. A video is considered one training pattern for the RNN. The movement of the ball is deterministic and can be predicted based on only the first video frame. The RNN is given the first frame and predicts all the remaining frames in the video. Since there are 225 pixels, there are 225 unique starting positions, and 225 unique videos, which result in a tiny training set with less than 200 patterns.

3.2   Language Modelling Benchmark

The second benchmark requires a RNN to predict the next word after been given a sequence of the previous words. This model can be constructed using a RNN that receives a sequence of n input words over n time-steps, and predicts next word as an output at final timestep. Language modelling was used as a benchmark by both Gal et al. [2] and Zaremba et al. [10] for their Dropout algorithms.

The text data consisted of the first 100 million charters of Wikipedia, which was provided by Pennington et al. [6] at nlp.stanford.edu/projects/glove. All punctuation was removed. The 2000 most frequently occurring words were used to construct a vocabulary. Each word in the vocabulary has an associated index number. These words will be presented to the network as one-hot encoded vectors of length 2000, where element $i = 1$ for the word with index i. The extracted words were divided up into 15-grams (sequences of length 15) where each 15-gram only contains words included in the vocabulary. This gives the network 14 words of context for each required output word. An 'unknown word' or 'UNK' symbol was not included in the vocabulary. A dataset of 91 393 15-grams was compiled.

4   Experimental Procedure

The following RNN configurations are tested for each benchmark.

1. **Classic LSTM:** No Dropout algorithm is applied.
2. **Naive Dropout:** Dropout is applied to every connection in the RNN randomly for each timestep, as would be done when applying Dropout to a fully connected NN, where the Dropout probability is 50%.
3. **Half Dropout:** Dropout only applied to non-recurrent connections, again applied randomly for each timestep, where the Dropout probability is 50%.
4. **Optimised Half Dropout:** Half Dropout where the Dropout probability is 65%.
5. **Variational Dropout:** Dropout is applied to every connection in the RNN, where the same connections are Dropped out at each timestep, where the Dropout probability is 50%.

For consistency, each Dropout algorithm is tested with a Dropout probability of 50%. This was the Dropout probability recommended by the authors of both Half and Variational Dropout [2, 10]. With the exception of a large RNN model, specifically 1500 units. Zaremba et al. [10] instead recommended a Dropout probability of 65% for Half Dropout. When the benchmark is run on the large RNN model, the additional Optimised Half Dropout algorithm is also tested. Allowing each Dropout algorithm to be tested using the same Dropout probability, as well as their recommended Dropout probability.

The RNN made use of the popular Long Short-Term Memory (LSTM) [3] structure. LSTM RNNs consisting of two hidden layers and a softmax output layer were constructed. Two LSTM sizes were investigated. The medium LSTM consisted of 650 cells per hidden layer. The large LSTM consisted of 1500 cells per layer. This matches the model sizes used by Gal et al. [2] and Zaremba et al. [10].

Each LSTM cell consists of multiple gates. Each gate has its own set of connections leading to it. Gal et al. [2] defines dropping out connections with tied weights as dropping the same connections for all gates in a single cell. This could be considered as using the same Dropout mask for each gate. Untied weights are defined as each gate having its connections Dropped out individually.

Gal et al. [2] determined that untied weights perform better than tied weights when using Variational Dropout. Zaremba et al. [10] did not specify whether Half Dropout uses tied or untied weights. For these experiments, untied weights are used for all Dropout algorithms.

When a NN is trained with Dropout, during the testing phase, connections which could be dropped out are scaled down by $1 - D$, where D is the Dropout probability. This is to compensate for more nodes being active during testing. In these experiments, this scale down was performed during testing. In addition, any connection that was never allowed to be dropped out, was instead scaled down during both testing and training. Connections scaled include the recurrent connections in Half Dropout, and all connections in this classic LSTM. Scaling connections ensures that the distribution of the net value in the nodes remains the same between all algorithms to allow for a fair comparison.

Ensemble averaging is used by both Gal et al. [2] and Zaremba et al. [10], however the amount of models trained is not consistent. Ensemble averaging is not used for these experiments. Instead, each RNN is trained with multiple runs, and the scores at the end of each run are supplied to the statistical test.

As done by Gal et al. [2], connections between the input layer and the first hidden layer can be dropped. This results in pixels missing from a video frame and words missing from the 14-gram input sequence.

All LSTMs were trained with RMSProp optimisation algorithm [8]. The dataset was divided into a testing (10%), validation (10%) and training set (80%). An initial learning rate of 0.01 was used for the medium RNN, and 0.001 for the large LSTM. Each training step trained the network with a minibatch of 128 patterns. If the validation score did not improve for three validation tests, the learning rate is multiplied by 0.1. Weights were initialised from a normal distribution. Weight decay was not included.

The bouncing ball benchmark was trained for 1000 epochs, where each epoch refers training with the entire training dataset once. Resulting in roughly 1400 training steps per run. The language modelling benchmark was trained for 15 epochs. Resulting in roughly 8500 training steps per run. After all epochs were completed, the model was restored to its state when it scored its best validation score, and the testing score was recorded.

Each configuration described at the beginning of Sect. 4 was run 30 times for each algorithm-benchmark-model configuration. Every test score set was paired with every other test score set, and two one-tailed Mann-Whitney U tests were performed, one in each direction. The result of the U tests would indicate if there was a statistically significant difference between the test score sets, and if so it would indicate its direction, which allowed the configurations to be ranked from best to worst for each benchmark. A p value of 0.05 was selected, since two tests are being performed per pair, after applying Bonferroni correction, each one-tailed test used a p value of 0.025.

5   Results and Discussion

Figure 2 shows an example run of the network predictions for the bouncing ball benchmark. At the beginning of the video, the prediction matches the target frames perfectly, but as the error compounds over time, the error grows larger and larger, making it more noticeable towards the end of the video.

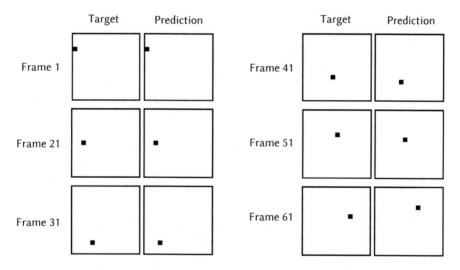

Fig. 2. Selected frames from a single bouncing ball video, which is a single training pattern. For each frame, Target shows the ball's desired position, and Prediction shows the network's ball prediction. The ball is represented by the black pixel. The closer the Prediction image to the Target image, the lower the network's error.

The language modelling RNN can have its output given back to it as an input and hence continuously predict words. The words will make less and less sense as the error builds up. The phrase "the most popular aircraft in use in the state much of this service can *be found in the early one nine th century the most important of the two zero th century*" was generated in this manner, where words in italics were generated from the RNN. Punctuation was not included and numerical digits are represented with words. Meaning "one nine th century" would be "19th century". Other phrases generated from the network include "in one nine five eight and re elected in one nine six four there *was a first of the first of the new city in the united states*" and "is the largest state by area in the united states it is larger in *the one nine th century the most of the one nine th century*".

Optimised Half Dropout refers to Half Dropout with a Dropout probability recommended by the authors specifically for large LSTM RNNs, meaning that it is only featured in benchmarks using a large LSTM RNN, specifically Table 3.

Table 1 shows the rankings for the bouncing ball benchmark experiment, which do not conform to the literature's prediction. Naive Dropout and Half Dropout are both at ranked 1 as there was no statistically significant difference between their scores. Naive Dropout was expected to make the RNN too difficult to train, and to perform very poorly. Instead, it either performed as well or better than both Half and Variational Dropout. On the other hand, Variational Dropout was expected to out perform all other Dropout algorithms, which was not the case for this benchmark.

Table 1. Dropout algorithms' ranks for the bouncing ball benchmark using a medium LSTM RNN.

Rank	Dropout algorithm	Average loss
1 (Joint)	Half	112
1 (Joint)	Naive	121
3	Variational	138
4	Classic LSTM	174

There are two advantages to using Dropout. Firstly, it adds distortion to reduce overfitting. Secondly, it trains the network in such a way that it gains partial benefits of ensemble training [7]. Applying any form of Dropout improved the LSTM's performance. Figure 3 (left) shows the loss curve of a single run of each algorithm for the bouncing ball benchmark. While this graph does not contain any significant information, it does suggest that the Classic LSTM was overfitting. Considering that the RNN contained two LSTM layers containing 650 cells each, and that the training set had under 200 training patterns, overfitting was likely to occur. Hence applying Dropout improved performance.

Table 2, the language modelling benchmark, shows similar results to the bouncing ball experiment. Again, Naive Dropout was expected to perform the worst, but performed the best. It appears that the distortion effects of Naive Dropout were over-exaggerated in the literature. This time, Variation Dropout did place first, but again it did not out-perform Naive Dropout. This network had just under 75 000 training patterns, while still using the medium LSTM of only 2 LSTM layers of each 650 units. Hence, overfitting was significantly less of a concern. This is confirmed by Fig. 3 (right) where there is no indication of overfitting. As before, any form of Dropout improved the LSTM's performance. This suggests that the networks were benefiting from the second advantage of Dropout, the pseudo ensemble training.

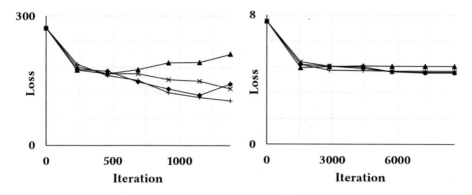

Fig. 3. The validation cross entropy loss of one run with each algorithm for the bouncing ball benchmark (left) and language modelling benchmark (right). As the loss is cross entropy, differently sized output layers produce different trained and untrained losses.

Table 2. Dropout algorithms' ranks for the language modelling benchmark using a medium LSTM RNN.

Rank	Dropout algorithm	Average loss
1 (Joint)	Variational	4.78
1 (Joint)	Naive	4.79
3	Half	4.97
4	Classic LSTM	5.16

Table 3 shows the language modelling benchmark when using a large LSTM network. This LSTM network did not benefit from any form of Dropout. Larger networks require that more weights need to be concurrently optimised during training. Dropout has the disadvantage in that by adding noise and distorting the network, it slows down training. It may have been that because this network is large and difficult to train, adding Dropout hindered the training too severely. The literature warned about this effect for Naive Dropout. However, the Half and Variational Dropout seem to suffer from the same problem, and they handled it no better than Naive Dropout.

All of the benchmark results were unexpected. They showed that there was little to no difference between these Dropout algorithms. If the RNN could benefit from Dropout, then it would benefit from all Dropout algorithms. Naive Dropout was expected to perform poorly, but instead it consistently performed well.

The contradiction in the results found in this study in comparison to the reported results in the literature, could be due to a number of factors. Firstly, it may be the case that the performance of Half Dropout and Variational Dropout is very sensitive to the networks hyperparameters, such as weight initialisation, learning rate, weight decay, and Dropout probability. If this is the case, then it

Table 3. Dropout algorithms' ranks for the language modelling benchmark using a large LSTM RNN.

Rank	Dropout algorithm	Average loss
1 (Joint)	Half	4.69
1 (Joint)	Naive	4.70
1 (Joint)	Classic LSTM	4.70
4	Variational	4.73
5	Optimised Half	4.88

still calls into question their suggested superiority, as selecting optimal hyper-parameters is a time consuming and imperfect process. Secondly, though the statistical tests did rank the Dropout algorithms, it did not provide an indication of the scale of the difference between the algorithms. Figure 3 suggests that the difference between the Dropout algorithms is minimal compared to the difference between any Dropout algorithm and the Classic LSTM. The curves for the Dropout algorithms intercept and overlap at many points during the training process. This is further backed up by the fact that the rankings of the Dropout algorithms themselves were slightly different between benchmarks, with none of the Dropout algorithms consistently out-performing the rest. While our experiments did show that it can be beneficial to use Dropout, they did not show any benefit of using a particular Dropout algorithm over the others. This suggests that in the cases where Dropout was beneficial, the Dropout algorithm selected was inconsequential.

6   Conclusion

The experiment results did not conform to the results reported in the literature. The distortion effects of Naive Dropout appear to have been exaggerated. It performed as well or better than the RNN specialized Dropout algorithms. There was no indication that Variational Dropout is superior to other Dropout algorithms.

The conclusion drawn from these experiments was: Applying any form of Dropout to a LSTM RNN can be beneficial. However, using a RNN specialised Dropout algorithm, such as Half and Variational Dropout, offers no statistically significant benefit over Naive Dropout.

References

1. Bayer, J., Osendorfer, C., Korhammer, D., Chen, N., Urban, S., van der Smagt, P.: On fast dropout and its applicability to recurrent networks. arXiv preprint arXiv:1311.0701 (2013)
2. Gal, Y., Ghahramani, Z.: A theoretically grounded application of dropout in recurrent neural networks. In: Advances in Neural Information Processing Systems, pp. 1019–1027 (2016)
3. Hochreiter, S., Schmidhuber, J.: Long short-term memory. Neural Comput. **9**(8), 1735–1780 (1997)
4. Martens, J., Sutskever, I.: Learning recurrent neural networks with hessian-free optimization. In: Proceedings of the 28th International Conference on Machine Learning (ICML-11), pp. 1033–1040. Citeseer (2011)
5. Pachitariu, M., Sahani, M.: Regularization and nonlinearities for neural language models: when are they needed? arXiv preprint arXiv:1301.5650 (2013)
6. Pennington, J., Socher, R., Manning, C.: Glove: Global vectors for word representation. In: Proceedings of the 2014 Conference on Empirical Methods in Natural Language Processing (EMNLP), pp. 1532–1543 (2014)
7. Srivastava, N., Hinton, G., Krizhevsky, A., Sutskever, I., Salakhutdinov, R.: Dropout: a simple way to prevent neural networks from overfitting. J. Mach. Learn. Res. **15**(1), 1929–1958 (2014)
8. Tieleman, T., Hinton, G.: Lecture 6.5—RmsProp: divide the gradient by a running average of its recent magnitude. In: COURSERA: Neural Networks for Machine Learning (2012)
9. Watt, N., du Plessis, M.C.: Dropout algorithms for recurrent neural networks. In: SAICSIT (2018)
10. Zaremba, W., Sutskever, I., Vinyals, O.: Recurrent neural network regularization. arXiv preprint arXiv:1409.2329 (2014)

Psychiatric Disorders Classification with 3D Convolutional Neural Networks

Stefano Campese[1], Ivano Lauriola[1,2(✉)], Cristina Scarpazza[3],
Giuseppe Sartori[3], and Fabio Aiolli[1]

[1] Department of Mathematics, University of Padova,
Via Trieste, 63, 35121 Padova, Italy
stefano.campese.90@gmail.com, ivano.lauriola@phd.unipd.it
[2] Bruno Kessler Foundation, Via Sommarive, 18, 38123 Trento, Italy
[3] Department of General Psychology, University of Padova,
Via Venezia, 8, 35121 Padova, Italy

Abstract. Recently, the literature showed that psychiatric disorders, such as Schizophrenia and Bipolar disorder, cause abnormalities in some brain regions. Therefore, several automatic mechanisms based on classical Machine Learning techniques have been used to recognize these diseases by means of the study of neuroimaging. A serious drawback of these approaches is that they consider only the intensity value of the points from neuroimages, without taking into account the spatiality information. Convolutional Neural Networks have subsequently applied to overcome the aforementioned issue, showing their empirical effectiveness on these tasks. However, generally Convolutional Neural Networks operate on 2D slices of the brain instead of the whole 3D structure.

This work aims to analyze the behavior of classical machine learning techniques against 2D and novel 3D Convolutional Neural Network models. An exhaustive empirical assessment has been performed to evaluate these methods on 4 real-world neuroimaging tasks, including Schizophrenia and Bipolar Disorder classification.

Keywords: Deep Learning · 3D Convolutional Neural Networks · Neuroimaging · Psychiatric disorders

1 Introduction

It is shown in the literature [16,18] that psychiatric disorders, such as Schizophrenia (SZ) and Bipolar Disorder (BP), may affect the human's structure of the brain. The effect of these diseases can be observed by means of the Structural Neuroimages (sMRI). Neuroimages, also know as *brain scans*, are 3D images of the brain, used to diagnose neurodegenerative pathologies or to identify deformations of the brain.

Nowadays, the standard of brain based pathologies diagnosis is still based on the clinical assessment of patients which is based on the subjective experience

L. Oneto et al. (Eds.): INNSBDDL 2019, INNS 1, pp. 48–57, 2020.
https://doi.org/10.1007/978-3-030-16841-4_6

that is reported by the patient and on the observation of signs and non-verbal behavior. However, a psychiatric disorder is not always easy to diagnose, as aspecific symptoms might be present in different disorders and as psychiatric symptoms might be transiently observed in healthy individuals.

Usually, imaging techniques are used to confirm the diagnosis. One of these techniques is the well-known Voxel Based Morphometry (VBM), whose purpose is to investigate focal differences in brain anatomy, by using a statistical approach [14].

Recently, automatic mechanisms based on Machine Learning techniques have been used to alleviate the effort on the manual classification of neuroimaging [2,10]. The majority of these methods are based on classical Machine Learning algorithms, such as the well-known Support Vector Machine (SVM). In these approaches, the feature vector of a neuroimage is composed of the intensity value of each voxel of the brain. A voxel (Volumetric Picture Element) is a volumetric measure, and it represents a data point on a brain. Generally, a single voxel defines a cubic region of few millimeters.

Anyway, a serious drawback of these methods is that they consider only the intensity value of each voxel, without taking into account their position, with a consequent loss of useful information.

On the other hand, Deep Learning (DL) methods can be used to catch the spatiality information. DL is a large family of Machine Learning algorithms able to learn the best representation from raw data among a hierarchy of non-linear processing. Recently DL, and Convolutional Neural Networks (CNNs) reached considerable attention on visual recognition tasks. Standard 2D CNNs has been also widely used to solve neuroimaging tasks [17], including the analysis of Alzheimer's disease (AD) [6,13], Mild Cognitive Impairment (MCI) [11] and Attention Deficit Hyperactivity Disorder (ADHD) [11] pathologies classification.

Nevertheless, standard CNNs are not perfectly suitable in these tasks. Indeed, they are not able to handle the whole 3D structure of the brain. In order to overcome this issue, novel 3D CNN architectures have been recently proposed and applied on neuroimaging tasks, showing very promising results with respect to the other methods.

The main contribution of this work is an extensive analysis of 3D CNNs, comparing them against classical 2D architectures and the shallow SVM. These methods have been evaluated on 4 real-world neuroimaging tasks, including Schizophrenia and Bipolar Disorder classification.

The paper is organized as follows: the Sect. 2 discusses Convolutional Neural Networks and their application on neuroimaging tasks. Then, the Sect. 3 describes the experimental assessment in detail, including an exhaustive explanation of the preprocessing pipeline and the methodology. Results are shown in the Sect. 3.4. Finally, Sect. 4 concludes the article and discusses the results, the methodology and future works.

2   Background and Related Work

Convolutional Neural Networks (CNNs) use a variation of the Multi-Layered Perceptron algorithm, and they are designed to require a minimal preprocessing [7]. CNNs are inspired by the biological processes of the visual cortex of the brain [8], where the interconnections between the neurons emulate the restricted regions of the visual field, known as *receptive fields*. Each neuron partially overlaps another one until the entire visual field is covered. This way of grouping the neurons allows creating a network more effective and easier to train than classical artificial neural networks on visual recognition tasks.

Despite the good performances, CNNs showed their generalization capability and empirical effectiveness mainly when applied on flat 2D images.

Also in the Biomedical domain, Machine learning techniques have been mainly used with flat images, like Electronic Microscope scans. However, things get more complicated when dealing with 3D structures. In these cases a preprocessing phase is generally performed, where the original structure is divided into many different 2D slices. Then each slice is analyzed separately. However, slicing of the initial images may have a negative impact on the final results.

The extension of the Convolutional Networks from 2D to 3D opened to the possibility to analyze full 3D structures without slicing them. Concerning the biomedical images, this innovation has affected several research topics and areas. In terms of sMRI indeed, this means that exists the possibility of processing the whole brains volume at the same time with the consequential improvements of performances in pathologies detection.

2.1   Related Work

The classical Machine Learning approach to solve neuroimaging tasks is based on the usage of the linear Support Vector Machine (SVM). The SVM has been used on a wide range of studies conducted on several pathologies, as in the case of Alzheimer (AD), Mild Cognitive Impairment (MCI) [5], Attention Deficit Hyperactivity Disorder (ADHD) [4], Schizophrenia (SZ) and Bipolar disorder (BP) [15]. In those studies, the usage of SVM is based on the intensity values of the voxels. In other words, this algorithm, figure out the best hyperplayne to separate the healthy patient from the pathologic ones by using the intensity of the voxels as features. Since the usage of those techniques is widely biased due to the "a priori" preprocessing, nowadays, the research in neuroimages field is moving to the usage of Deep learning techniques [17].

This field appears to be indicated for usage of those techniques under different points of view. Indeed, Deep learning techniques are able to figure out hidden patterns that are not visible to scientists, and in some cases, they prevent and avoid the usage of a priori preprocessing techniques that may denature the structure of the image with a consequent biasing on the results. Moreover, due to their nature, 3D neuroimages appears to be indicated for the usage of 3D Convolutional Neural Networks. In some studies [11] carried out on diagnosis of

Alzheimer's disease (AD) and its prodromal stage and Mild Cognitive Impairment (MCI), it has been demonstrated that usage of 2D technique performed worse than 3D solution. This is potentially due to the fact that 3D images, as opposed to the 2D, contain more useful patterns.

3   Experimental Assessment

An intensive set of experiments have been performed on two neuroimaging datasets, showing the potentiality of the 3D CNN against 2D architectures and shallow methods.

3.1   Datasets Description

Two multi-class datasets of neuroimages have been used to evaluate the effectiveness of the mentioned approaches. These datasets contain neuroimages from Bipolar (BP), Schizophrenic (SZ) and Control (HC) patients. For each dataset, two different tasks have been analyzed separately. These tasks consist of the binary classification of SZ patients against HC and BP against HC. Hence, there are 4 different tasks. The Table 1 contains the numerosity information of each dataset. The name and the source of the datasets have been obscured to prevent privacy issues.

Unfortunately, as you can see these datasets are not suitable for the application of complex deep learning techniques due to the numerosity of examples. Despite these premises, the application of 3D CNNs in this field shown very promising results.

Table 1. Patient distribution over the datasets

Dataset	Control (HC)	Schizophrenic (SZ)	Bipolar (BP)	Total
Dataset-A	55	46	28	129
Dataset-B	122	54	49	225

3.2   Baselines and Neural Architectures

Several Neural Network architectures have been used in this work. The simplest one is a classical 2D CNN. This network is composed of 6 convolutional layers alternated with 3 maxpooling and 3 upsampling layers. At the top of the network, there are 4 dense layers with 1024, 512, 256 and 2 neurons for the classification. The hyperparameters and the architecture of this network have been chosen during a preliminary experimental phase. The input of the net consists of the middle axial slice of the brains.

Three different 3D architectures have been considered and compared. They are the VNet [9], the UNet [12] and the LeNet [7].

VNet and UNet architectures are already been used in the literature to solve similar tasks on 3D and 2D biomedical images. LeNet instead has been originally developed for number and character recognition.

Broadly speaking, the UNet and the VNet consist of a combination of 9 stages organized in two consecutive subnets, dubbed *left* and *right*. The left part contains convolutional layers whose aim is to learn a residual function. The right part operates the feature extraction and the expansion of the input by means of de-convolutional layers. PReLu and ReLu activation functions are used respectively in the VNet and UNet.

The name of these nets intrinsically represents the convolution and de-convolution mechanisms, which define a U and V shape. To get more details on these architectures, see the respective papers.

The LeNet instead, is more simple than the others. It contains two convolutional layers, each one followed by a max-pooling. At the top of the net, there are two dense layers. Relu activation functions have been applied through the whole network. These networks operate on the whole 3D brain.

As a further baseline, a classical SVM has been used on the whole 3D brain. The SVM uses a linear kernel, and a fixed cost $C = 1000$, selected from preliminary results.

Finally, the representation computed on the last layer of each neural network has been also used to feed a linear SVM.

3.3   Preprocessing Pipeline

Neuroimages have been preprocessed according to a standard pipeline based on the *Voxel-Based Morphometry* [3] (VBM) technique. VBM is a statistical approach widely used to analyze neuroimages, based on the investigation of the anatomical differences between the brains of a dataset. This technique defines a specific pipeline to preprocess and handle neuroimages. The preprocessing consist of several steps, where:

- Images have been converted from the *Digital Imaging and COmmunications in Medicine* (DICOM) original format to the *Neuroimaging Informatics Technology Initiative* (NIfTI) [1] format.
- Images have been aligned to synchronize the viewpoint of the brains. This procedure is essential for positioning all the brains in the same direction, making them easily comparable. This operation has been manually performed by expert neuro-scientists.
- Consequently, images have been *normalized* by using the DARTEL template to reduce the side-effect due to the different Magnetic Resonance machinery. This step aligns images by warping them to the *standard stereostatic space* through several non-linear transformations. These transformations include translations, rotations, scaling and shearing.

- Then, the *Segmentation* procedure has been applied. This step aims to remove useless structures from images, such as the skull residual. Gray and white matter have been separated.
- The *Modulation* is an optional step which performs a spatial transformation of regions of the brain to make the volume comparable. This process is useful to highlight differences in brain concentration and density, by keeping the total amount of Gray matter constant. However, the modulation might modify and denature the original volume of the brain in the image. This phenomenon is mainly visible especially when the data contains false positive samples.
- Finally, the latter stage is the *Smoothing*, which consists of the application of an isotropic Gaussian filter on the images. The aim of this process is to reduce issues related to the involuntary movements of the patient and to equally distributing the information. Typically, the width of the smoothing window is between 4 mm and 12 mm. Three different smoothing windows have been used in this work, which are: 4 mm, 8 mm and 12 mm.

The effect of some preprocessing steps are depicted in the Fig. 1

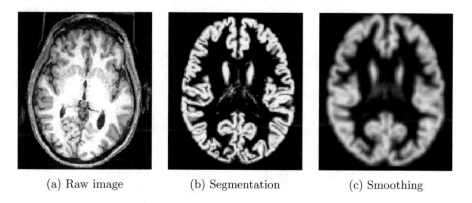

   (a) Raw image   (b) Segmentation   (c) Smoothing

Fig. 1. The effect of some preprocessing steps on the original brain.

The discussed pipeline has been used to create 8 different images from the original datasets, which are:

- swc1-x: the preprocessing includes smoothing with $x \in \{4, 8, 12\}$ millimeters;
- smwc1-x: the preprocessing includes smoothing with $x \in \{4, 8, 12\}$ millimeters and the modulation;
- wc1: images without smoothing nor modulation;
- mwc1: images with modulation.

The learning algorithms have been applied to each of these input representations.

3.4   Evaluation

Shallow SVM, 2D, and 3D CNNs have been compared by measuring their capability of dichotomize pathologic and healthy patients.

A 5-fold cross-validation procedure has been applied to evaluate each algorithm, and the average of AUC scores computed on each fold has been considered. To increase the stability of the results, the cross-validation procedure has been repeated 10 times with 10 different splits. The average and standard deviation of AUC scores reached on each run have been computed. The used splits are the same for each algorithm.

The complete procedure has been applied to each of the 8 image preprocessing formats. The Table 2 shows, for each algorithm and each task, the highest AUC score achieved by the application of the algorithm on each of the 8 formats. The average rank and the name of the best format is also exposed.

Table 2. Best AUC scores achieved on the 8 formats in each task of Dataset-A and Dataset-B

	Dataset-A		Dataset-B		Average rank
	SZ	BP	SZ	BP	
SVM	$82.66_{\pm6.69}$ wc1	$58.35_{\pm8.35}$ swc1-12	$65.57_{\pm9.60}$ wc1	$71.79_{\pm13.40}$ swc1-4	5.5
LeNet	$79.28_{\pm10.26}$ wc1	$62.04_{\pm9.19}$ wc1	$56.33_{\pm15.81}$ swc1-4	$65.78_{\pm11.64}$ wc1	7.0
VNet	$83.13_{\pm9.05}$ wc1	$63.63_{\pm10.87}$ swc1-4	$\mathbf{71.63}_{\pm12.87}$ wc1	$\mathbf{75.52}_{\pm13.71}$ wc1	**3.5**
UNet	$73.26_{\pm8.70}$ wc1	$\mathbf{66.43}_{\pm12.15}$ wc1	$58.75_{\pm11.65}$ mwc1	$71.81_{\pm13.23}$ swc1-4	5.0
2D CNN	$72.71_{\pm8.91}$ swc1-4	$61.91_{\pm13.72}$ swc1-12	$57.63_{\pm7.45}$ swc1-8	$72.00_{\pm4.29}$ swc1-8	6.25
LeNet + SVM	$84.52_{\pm6.72}$ wc1	$63.13_{\pm8.99}$ wc1	$68.17_{\pm11.64}$ wc1	$71.03_{\pm10.98}$ wc1	3.75
VNet + SVM	$\mathbf{86.30}_{\pm9.35}$ swc1-4	$56.56_{\pm9.75}$ swc1-4	$70.21_{\pm10.28}$ swc1-4	$70.39_{\pm10.93}$ wc1	3.75
UNet + SVM	$75.52_{\pm8.86}$ swc1-12	$62.77_{\pm9.02}$ mwc1	$67.93_{\pm12.41}$ mwc1	$71.84_{\pm12.35}$ swc1-8	5.0
2D CNN + SVM	$72.13_{\pm6.51}$ swc1-4	$57.14_{\pm11.18}$ swc1-12	$60.63_{\pm9.27}$ swc1-8	$72.09_{\pm6.33}$ swc1-8	6.25

The Fig. 2 shows the influence of different formats when using the VNet, that is the model which achieves the highest average rank on the 4 tasks. The application of the SVM on the representation computed by the VNet improves

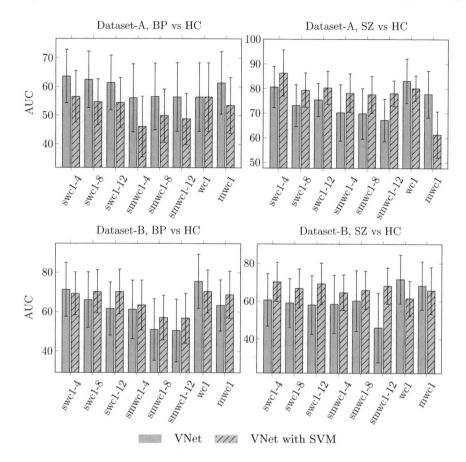

Fig. 2. AUC results of VNet and VNet with SVM for each image format.

the results on 3 tasks. However, due to the high standard deviation, it is difficult to assert which format is the most effective.

Despite the good results compared to the baselines, this methodology is strongly limited by the number of examples for each task. In order to overcome this issue, an augmentation procedure has been applied to observe empirically the effectiveness on larger datasets.

In the augmentation-experiment, the same learning pipeline has been used. Anyhow, during the 5-fold CV procedure, 200 positive and 200 negative artificial examples have been included in the training. These examples are computed as a convex combination of training ones. Figure 3 shows the empirical effectiveness of the training augmentation. On average, the augmentation has a positive impact on the AUC score. However, the high standard deviation makes the results not statistically significant.

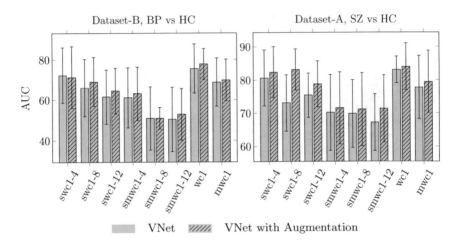

Fig. 3. Effectiveness of the augmentation procedure with the VNet.

4   Conclusions

The paper investigates the effectiveness of classical Support Vector Machines (SVMs), 2D and 3D Convolutional Neural Networks (CNNs) applied on 4 neuroimaging tasks, including the Schizophrenia and Bipolar disorder classification. 3D CNNs have been compared against 2D architectures and the linear SVM, showing that 3D models empirically outperform the baselines on average.

Results reached by the SVM are not surprising. It performs better than the 2D network, which works only on a slice.

These results emphasize two points. The first is that working on the whole 3D structure of the brain may improve the overall performances, and the information needed to solve these tasks is spread on the whole brain. The second point is that the spatial information about the position of each voxel is important and could further improve the performance.

SVM, Voxel-Based Morphometry and 3D CNN work on the whole 3D structure, and they use all the voxels, but only the 3D CNN is able to catch the spatial information correctly. As it is described in the paper, the SVM considers each voxel as a separate feature, without taking into account the neighborhood.

On one hand, the augmentation procedure increases on average the AUC score of the VNet on all of the tested configurations, with the exception on the task BP-vs-HP, Dataset-B, swc1-4. On the other hand, the training procedure becomes significantly more expensive.

However, the high values of the standard deviation suggest that the application of these methodologies is strongly limited by the dimension of the training set.

In the future, we will further investigate the application of these methodologies applied to more pathologies, larger datasets, and using pre-training procedures.

References

1. Statistical Parametric Mapping. http://www.fil.ion.ucl.ac.uk/spm/
2. Abraham, A., Pedregosa, F., Eickenberg, M., Gervais, P., Mueller, A., Kossaifi, J., Gramfort, A., Thirion, B., Varoquaux, G.: Machine learning for neuroimaging with scikit-learn. Front. Neuroinform. **8**, 14 (2014)
3. Ashburner, J., Friston, K.J.: Voxel-based morphometry–the methods. Neuroimage **11**(6), 805–821 (2000)
4. Bledsoe, J.C., Xiao, D., Chaovalitwongse, A., Mehta, S., Grabowski, T.J., Semrud-Clikeman, M., Pliszka, S., Breiger, D.: Diagnostic classification of ADHD versus control: support vector machine classification using brief neuropsychological assessment. J. Atten. Disord. 1087054716649666 (2016)
5. Fan, Y., Resnick, S.M., Wu, X., Davatzikos, C.: Structural and functional biomarkers of prodromal Alzheimer's disease: a high-dimensional pattern classification study. Neuroimage **41**(2), 277–285 (2008)
6. Gao, X.W., Hui, R.: A deep learning based approach to classification of CT brain images. In: SAI Computing Conference (SAI), pp. 28–31. IEEE (2016)
7. LeCun, Y., et al.: LeNet-5, Convolutional Neural Networks, p. 20 (2015). http://yann.lecun.com/exdb/lenet
8. Matsugu, M., Mori, K., Mitari, Y., Kaneda, Y.: Subject independent facial expression recognition with robust face detection using a convolutional neural network. Neural Netw. **16**(5–6), 555–559 (2003)
9. Milletari, F., Navab, N., Ahmadi, S.A.: V-Net: fully convolutional neural networks for volumetric medical image segmentation. In: 2016 Fourth International Conference on 3D Vision (3DV), pp. 565–571. IEEE (2016)
10. Orru, G., Pettersson-Yeo, W., Marquand, A.F., Sartori, G., Mechelli, A.: Using support vector machine to identify imaging biomarkers of neurological and psychiatric disease: a critical review. Neurosci. Biobehav. Rev. **36**(4), 1140–1152 (2012)
11. Payan, A., Montana, G.: Predicting Alzheimer's disease: a neuroimaging study with 3d convolutional neural networks. arXiv preprint arXiv:1502.02506 (2015)
12. Ronneberger, O., Fischer, P., Brox, T.: U-Net: convolutional networks for biomedical image segmentation. In: International Conference on Medical Image Computing and Computer-Assisted Intervention, pp. 234–241. Springer (2015)
13. Sarraf, S., Tofighi, G.: Classification of Alzheimer's disease using fMRI data and deep learning convolutional neural networks. arXiv preprint arXiv:1603.08631 (2016)
14. Scarpazza, C., De Simone, M.S.: Voxel-based morphometry: current perspectives. Neurosci. Neuroecon. **5**, 19–35 (2016)
15. Schnack, H.G., Nieuwenhuis, M., van Haren, N.E., Abramovic, L., Scheewe, T.W., Brouwer, R.M., Pol, H.E.H., Kahn, R.S.: Can structural mri aid in clinical classification? A machine learning study in two independent samples of patients with schizophrenia, bipolar disorder and healthy subjects. Neuroimage **84**, 299–306 (2014)
16. Shioya, A., Saito, Y., Arima, K., Kakuta, Y., Yuzuriha, T., Tanaka, N., Murayama, S., Tamaoka, A.: Neurodegenerative changes in patients with clinical history of bipolar disorders. Neuropathology **35**(3), 245–253 (2015)
17. Vieira, S., Pinaya, W.H., Mechelli, A.: Using deep learning to investigate the neuroimaging correlates of psychiatric and neurological disorders: methods and applications. Neurosci. Biobehav. Rev. **74**, 58–75 (2017)
18. Zipursky, R.B., Reilly, T.J., Murray, R.M.: The myth of schizophrenia as a progressive brain disease. Schizophr. Bull. **39**(6), 1363–1372 (2012)

Perturbed Proximal Descent to Escape Saddle Points for Non-convex and Non-smooth Objective Functions

Zhishen Huang$^{(\boxtimes)}$ and Stephen Becker(ID)

Department of Applied Mathematics, University of Colorado, Boulder, USA
{zhishen.huang,stephen.becker}@colorado.edu

Abstract. We consider the problem of finding local minimizers in non-convex and non-smooth optimization. Under the assumption of strict saddle points, positive results have been derived for first-order methods. We present the first known results for the non-smooth case, which requires different analysis and a different algorithm. *This is the extended version of the paper that contains the proofs.*

Keywords: Saddle-points · Proximal gradient descent · Non-smooth optimization

1 Introduction

We consider the problem of finding approximate local minimizers of the problem

$$\text{minimize}_{\mathbf{x} \in \mathbb{R}^d} \left(\Phi(\mathbf{x}) := f(\mathbf{x}) + g(\mathbf{x}) \right) \tag{1}$$

where $f(\mathbf{x})$ is not convex but smooth (and with full domain), and $g(\mathbf{x})$ is convex but not smooth. Many optimization problems in engineering, signal processing and machine learning can be cast in this framework, where f is a smooth loss function, and g is a non-smooth regularizer such as a norm. For example, our model captures regularized neural networks [11], where the regularization can induce sparsity as an alternative to dropout. In this paper, for simplicity we restrict our discussion to $g(\mathbf{x}) = \lambda \|\mathbf{x}\|_1$, where $\lambda \geq 0$ is a constant, but many of the results apply to more general choices of g. The first-order condition is $0 \in \nabla f(\mathbf{x}) + \partial g(\mathbf{x})$, and any $\mathbf{x}$ satisfying this condition is called a "stationary point" (see [2] for background on the subdifferential ∂g). All local minimizers are stationary points, but not vice-versa. We define a "saddle point" to be any stationary point where the Hessian is indefinite (and therefore not a local minimizer). This paper extends a recent line of work [13] to analyze when we can expect to find a local minimizer. It has been argued that in many machine learning problems, finding any local minimizer is often enough for good performance, but finding a saddle point is not useful [9].

The fact that g is non-smooth is crucially important, and it does more than just complicate the analysis, as it also requires a new algorithm. In the smooth

© Springer Nature Switzerland AG 2020
L. Oneto et al. (Eds.): INNSBDDL 2019, INNS 1, pp. 58–77, 2020.
https://doi.org/10.1007/978-3-030-16841-4_7

case, f is often minimized using gradient descent or an accelerated variant [16] with a fixed stepsize. Naïvely extending gradient descent to apply to (1) leads to subgradient descent with fixed-stepsize. Unfortunately, this method fails to converge as the example $d = 1, \lambda = 1$ and $f = 0$ shows [18] since for a generic choice of the initial point, the sequence is not Cauchy.

Instead of gradient descent, we use a perturbed version of proximal gradient descent. For a real-valued convex lower semi-continuous function g, define the "proximity" operator (or "prox" for short) as the map $\text{prox}_g(\mathbf{y}) = \text{argmin}_{\mathbf{x}} g(\mathbf{y}) + \frac{1}{2}\|\mathbf{x} - \mathbf{y}\|^2$ (throughout the paper, for vectors we use $\| \cdot \|$ to denote the Euclidean norm). Equivalently, $\text{prox}_g = (I + \partial g)^{-1}$, and thus the first-order condition is equivalent to $\mathbf{x} = \text{prox}_{\eta g}[\mathbf{x} - \eta \nabla f(\mathbf{x})]$ for any $\eta > 0$. Proximal gradient descent is the iteration $\mathbf{x}_{t+1} = \text{prox}_{\eta g}[\mathbf{x}_t - \eta \nabla f(\mathbf{x}_t)]$, so it immediately follows that if the sequence converges, it converges to a stationary point. Convergence of the sequence is known to follow from mild assumptions on f and g, the stepsize η, and boundedness of the sequence $\{\mathbf{x}_t\}$ [1].

We define a *second-order stationary point* to be a first-order stationary point $\mathbf{x}$ that additionally satisfies $\nabla^2 f(\mathbf{x}) \succ 0$, which is a sufficient condition for $\mathbf{x}$ to be a local minimizer. Our main contribution is showing that under suitable assumptions, a perturbed version of proximal gradient descent will generate a sequence that converges to an approximate second-order stationary point. We make assumptions on the second-order behavior of f, similar to assumptions under which it is known that gradient descent will always converge to a second-order stationary point except for adversarially chosen starting points [14]—in contrast to Newton's method, which is attracted to all stationary points. However, even in the smooth case when the sequence converges, gradient descent converges arbitrarily slowly [10] in the presence of a saddle point, so perturbation is necessary. In the non-smooth case, perturbation is even more important due to the proximal nature of the algorithm.

A Toy Example: Gaussian Bump. Consider the function $\Phi : \mathbb{R}^2 \to \mathbb{R}, x \mapsto \frac{1}{2}(x^2 - y^2)e^{-\frac{x^2+y^2}{5}} + \frac{1}{100}h_{100}(\mathbf{x})$ where $h_{100}(\mathbf{x})$ is the Huber function with parameter 100 [3]. The choice of this combination of Huber parameter and the magnitude of Huber function ensures that the origin is a saddle point. The Huber function approximates the ℓ_1 norm. The plot is show in Fig. 1.

This function has two local minima and a saddle point at $(0,0)$. Because the Huber function is both smooth and it has a known proximity operator, we can treat it as either part of the smooth f component or the non-smooth g component, and therefore run either gradient descent or proximal gradient descent. We experiment with both algorithms, randomly picking initial points at $\mathbf{x}_0 = (0.3, 0.01) + \boldsymbol{\xi}$ where $\boldsymbol{\xi}$ is sampled uniformly from $\mathbb{B}_0(\frac{1}{10}\|\mathbf{x}_0\|)$, and varying the stepsize η, with fixed maximum iteration 1000. Figure 2 shows the empirical success rate of finding a local minimizer (as opposed to converging to the saddle point at $(0,0)$).

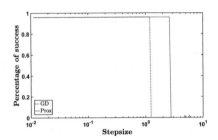

Fig. 1. Graph of function $\Phi(\mathbf{x})$

Fig. 2. The comparison between gradient descent (GD) and proximal gradient descent (Prox) on the percentage of success finding the correct local minima, as a function of the stepsize η

We observe that the range of stable step size for the proximal descent algorithm is wider than gradient descent, and the success rate of proximal descent is as high as the gradient descent. This example motivates us to adopt proximal descent over gradient descent in real application for better stability and equivalent, if not better, accuracy.

A Coincidence. In this toy example, the saddle point at $(0,0)$ happens to be a fixed point of proximal operator of $\eta\lambda\|\mathbf{x}\|_1$. Soft thresholding, as the proximal operator of $\lambda\|\mathbf{x}\|_1$ is known [7], has an attracting region that sets nearby points to 0. The radius of the attracting region (per dimension) is $\eta\lambda$, thus if $\|\mathbf{x}_{t_0} - \eta\nabla f(\mathbf{x}_{t_0})\|_\infty \leq \eta\lambda$ for some iteration t_0, then $\mathbf{x}_t = 0$ for all $t > t_0$. Proximal gradient descent performs even better when the saddle point is not in the attracting region.

Structure of the Paper. Section 2 states the algorithm, followed by Sect. 3 where the theoretical guarantee is presented with proof. Section 4 shows numerical experiments.

1.1 Related Literature

Second Order Methods for Smooth Objectives. Some recent second order methods, mainly based on either cubic-regularized Newton methods as in [17] or based on trust-region methods (as in Curtis et al. [8]), have been shown to converge to ε-approximate local minimizers of smooth non-convex objective functions in $\mathcal{O}(\varepsilon^{-1.5})$ iterations. See [6,13,21] for a more thorough review of these methods. We do not consider these methods further due to the high-cost of solving for the Newton step in large dimensions.

First Order Methods for Smooth Objectives. We focus on first order methods because each step is cheaper and these methods are more frequently adopted by the deep learning community. Xu et al. in [20] and Allen-Zhu et al. in [21] develop Negative-Curvature (NC) search algorithms, which find descent direction corresponding to negative eigenvalues of Hessian matrix. The NC search routines avoid using either Hessian or Hessian-vector information directly, and it can be applied in both online and deterministic scenarios. In the online setting, combining NC search routine with first-order stochastic methods will give algorithms NEON-$\mathcal{A}$ [20] and NEON2+SGD [21] with iteration cost $\mathcal{O}(\frac{d}{\varepsilon^{3.5}})$ and $\mathcal{O}(\varepsilon^{-3.5})$ respectively (the latter still depends on dimension, whose induced complexity is at least $\ln^2(d)$), and these methods generate a sequence that converges to an approximate local minimum with high probability. In the offline setting, Jin et al. in [13] provide a stochastic first order method that finds an approximate local minimizer with high probability at computational cost $\mathcal{O}(\frac{\ln^4(d)}{\varepsilon^2})$. Combining NEON2 with gradient descent or SVRG, the cost to find an approximate local minimum is $\mathcal{O}(\varepsilon^{-2})$, whose dependence on dimension is not specified but at least $\ln^2(d)$. These methods make Lipschitz continuity assumptions about the gradient and Hessian, so they do not apply to non-smooth optimization.

A recent preprint [15] approaches the problem of finding local minima using the forward-backward envelope technique developed in [19], where the assumption about the smoothness of objective function is weakened to local smoothness instead of global smoothness.

Non-smooth Objectives. In the offline settings, Boţ et al. propose a proximal algorithm for minimizing non-convex and non-smooth objective functions in [5]. They show the convergence to KKT points instead of approximate second-order stationary points. Other work [1,4] relies on the Kurdya-Lojasiewicz inequality and shows convergence to stationary points in the sense of the limiting subdifferential, which is not the same as a local minimizer or approximate second-order stationary point. In the online setting, Reddi et al. demonstrated in [12] that the proximal descent with variance reduction technique (proxSVRG) has linear convergence to a first-order stationary point, but not to a local minimizer.

2   Algorithm

The algorithm takes as input a starting vector $\mathbf{x}_0$, the gradient Lipschitz constant L, the Hessian Lipschitz constant ρ, the second-order stationary point tolerance ε, a positive constant c, a failure probability δ, and estimated function value gap Δ_ϕ. The key parameter for Algorithm 1 is the constant c. It should be made large enough so that the effect of perturbation will be significant enough for escaping saddle points, and at the same time not too large so that the iteration stepsize is of reasonable magnitude and the iteration will not go wild. The output of the algorithm is an ε-second-order stationary point (see Definition 3).

Algorithm 1. Perturbed Proximal Descent: input$(\mathbf{x}_0, L, \rho, \varepsilon, c, \delta, \Delta_\Phi)$

$\chi \leftarrow 3\max\{\ln(\frac{dL\Delta_\Phi}{c\varepsilon^2\delta}), 4\}, \ \eta \leftarrow \frac{c}{L}, \ r \leftarrow \frac{\sqrt{c}}{\chi^2} \cdot \frac{\varepsilon}{L}, \ g_{\text{thres}} \leftarrow \frac{\sqrt{c}}{\chi^2} \cdot \varepsilon, \ \Phi_{\text{thres}} \leftarrow \frac{c}{\chi^3} \cdot$
$\sqrt{\frac{\varepsilon^3}{\rho}}, \ t_{\text{thres}} \leftarrow \frac{\chi}{c^2} \cdot \frac{L}{\sqrt{\rho\varepsilon}}$
$t_{\text{noise}} \leftarrow -t_{\text{thres}} - 1$
for $t = 0, 1, \ldots$ **do**
    if $\|\mathbf{x} - \text{prox}_{\eta g}[\mathbf{x} - \eta\nabla f(\mathbf{x})]\| < g_{\text{thres}}$ and $t - t_{\text{noise}} > t_{\text{thres}}$ **then**
        $\tilde{\mathbf{x}}_t \leftarrow \mathbf{x}_t, \quad t_{\text{noise}} \leftarrow t$
        $\mathbf{x}_t \leftarrow \tilde{\mathbf{x}}_t + \boldsymbol{\xi}_t, \quad \boldsymbol{\xi}_t$ uniformly $\sim \mathbb{B}_0(r)$
    if $t - t_{\text{noise}} = t_{\text{thres}}$ and $\Phi(\mathbf{x}_t) - \Phi(\tilde{\mathbf{x}}_{t_{\text{noise}}}) > -\Phi_{\text{thres}}$ **then**
        return $\tilde{\mathbf{x}}_{t_{\text{noise}}}$
    $\mathbf{x}_{t+1} \leftarrow \text{prox}_{\eta g}[\mathbf{x}_t - \eta\nabla f(\mathbf{x}_t)]$

---

3 Escaping Saddle Points Through Perturbed Proximal Descent

The main step in the algorithm is a proximal gradient descent step applied to $f + g$, defined as

$$\mathbf{x}_{t+1} = \underset{\mathbf{y}}{\text{argmin}} \ f(\mathbf{x}_t) + \langle\nabla f(\mathbf{x}_t), \mathbf{y} - \mathbf{x}_t\rangle + \frac{\eta^{-1}}{2}\|\mathbf{y} - \mathbf{x}_t\|^2 + g(\mathbf{y})$$
$$= \text{prox}_{\eta g} \circ (I - \eta\nabla f)(\mathbf{x}_t) \tag{2}$$

One motivation of preferring proximal descent to gradient descent, as shown in Fig. 2, is the stability of the algorithm with respect to stepsize change. The proximal step is similar to the implicit/backward Euler scheme, as Eq. (2) can be written as $\mathbf{x}_{t+1} = \mathbf{x}_t - \eta(\nabla f(\mathbf{x}_t) + \partial g(\mathbf{x}_{t+1}))$. From this perspective, we expect that proximal descent will demonstrate at least the same convergence speed as gradient descent and stronger stability with respect to hyperparameter setting.

Definition 1 (Gradient Mapping). *Consider a function* $\Phi(\mathbf{x}) = f(\mathbf{x}) + g(\mathbf{x})$. *The gradient mapping is defined as* $G_\eta^{f,g}(\mathbf{x}) := \mathbf{x} - \text{prox}_{\eta g}[\mathbf{x} - \eta\nabla f(\mathbf{x})]$

In the rest of this paper, the super- and subscript of the gradient mapping are not specified, as it is always clear that f represents the smooth nonconvex part of Φ, g represents $\lambda\|\mathbf{x}\|_1$, and η is the stepsize used in the algorithm. Observe that the gradient map is just the gradient of f if $g \equiv 0$.

Definition 2 (First order stationary points). *For a function* $\Phi(\mathbf{x})$, *define first order stationary points as the points which satisfy* $G(\mathbf{x}) = 0$.

Definition 3 (ε-second-order stationary point). *Consider a function* $\Phi(\mathbf{x}) = f(\mathbf{x}) + g(\mathbf{x})$. *A point* $\mathbf{x}$ *is an* ε-*second-order stationary point if*

$$\|G(\mathbf{x})\| \leq \varepsilon \quad and \quad \lambda(\nabla^2 f(\mathbf{x}))_{\min} \geq -\sqrt{\rho\varepsilon} \tag{3}$$

where $\lambda(\cdot)_{\min}$ *is the smallest eigenvalue.*

The first Lipschitz assumption below is standard [3], and the assumption on the Hessian was used in [13] (for example, it is true if f is quadratic).

Assumption 1 (Lipschitz Properties). ∇f is L-Lipschitz continuous and $\nabla^2 f$ is ρ Lipschitz continuous. We write $\mathcal{H}$ as shorthand for $\nabla^2 f(\mathbf{x})$ when $\mathbf{x}$ is clear from context.

Assumption 2 (Moderate Nonsmooth Term). The magnitude of $\|\mathbf{x}\|_1$ term, which is denoted by λ, satisfies inequalities (7) and (9).

Theorem 1 (Main). There exists an absolute constant $c_{\max}$ such that if $f(\cdot)$ satisfies Assumptions 1 and 2, then for any $\delta > 0, \varepsilon \leq \frac{L^2}{\rho}, \Delta_\Phi \geq \Phi(\mathbf{x}_0) - \Phi^\star$, and constant $c \leq c_{\max}$, with probability $1 - \delta$, the output of $PPD(\mathbf{x}_0, L, \rho, \varepsilon, c, \delta, \Delta_f)$ will be a ε-second order stationary point, and terminate in iterations:

$$\mathcal{O}\left(\frac{L(\Phi(\mathbf{x}_0) - \Phi^\star)}{\varepsilon^2} \ln^4\left(\frac{dL\Delta_\Phi}{\varepsilon^2 \delta}\right)\right)$$

Remark. Assuming $\varepsilon \leq \frac{L^2}{\rho}$ does not lead to loss of generality. Recall the second order condition is specified as $\lambda\left(\nabla^2 f(\mathbf{x}^\star)\right)_{\min} \geq -\sqrt{\rho\varepsilon}$, since when $\varepsilon \geq \frac{L^2}{\rho}$, we always have $-\sqrt{\rho\varepsilon} \leq -L \leq \lambda\left(\nabla^2 f(\mathbf{x}^\star)\right)_{\min}$, where the second inequality follows from the fact that the Lipschitz constant is the upper bound for $\lambda(\nabla^2 f(\mathbf{x}))$ in norm. Consequently, when $\varepsilon \geq \frac{L^2}{\rho}$, every ε-second-order stationary point is automatically a first order stationary point.

For the proof of the main theorem, we introduce some notation and units for the simplicity of proof statement.

For matrices we use $\|\cdot\|$ to denote spectral norm. The operator $\mathcal{P}_\mathcal{S}(\cdot)$ denotes projection onto set $\mathcal{S}$. Define the local approximation of the smooth part of the objective function by

$$\tilde{f}_{\mathbf{x}}(\mathbf{y}) := f(\mathbf{x}) + \nabla^T f(\mathbf{x})(\mathbf{y} - \mathbf{x}) + \frac{1}{2}(\mathbf{y} - \mathbf{z})^T \mathcal{H}(\mathbf{y} - \mathbf{z}) \qquad (4)$$

Units. With the conditional number of the Hessian matrix $\kappa := \frac{L}{\gamma} \geq 1$, we define the following units for the convenience of proof statement:

$$\mathscr{F} := \eta L \frac{\gamma^3}{\rho^2} \cdot \ln^{-3}\left(\frac{d\kappa}{\delta}\right), \qquad \mathscr{G} := \sqrt{\eta L} \frac{\gamma^2}{\rho} \cdot \ln^{-2}\left(\frac{d\kappa}{\delta}\right)$$

$$\mathscr{S} := \sqrt{\eta L} \frac{\gamma}{\rho} \cdot \ln^{-1}\left(\frac{d\kappa}{\delta}\right), \qquad \mathscr{T} := \frac{\ln\left(\frac{d\kappa}{\delta}\right)}{\eta\gamma}$$

3.1   Lemma: Iterates Remain Bounded if Stuck Near a Saddle Point

Lemma 1. *For any constant $\hat{c} \geq 3$, there exists absolute constant $c_{\max}$: for any $\delta \in (0, \frac{d\kappa}{e}]$, let $f(\cdot), \tilde{\mathbf{x}}$ satisfies the condition in Lemma 6, for any initial point $\mathbf{u}_0$ with $\|\mathbf{u}_0 - \tilde{\mathbf{x}}\| \leq 2\mathscr{S}/(\kappa \cdot \ln(\frac{d\kappa}{\delta}))$, define:*

$$T = \min \left\{ \, \inf_t \left\{ t \mid \tilde{f}_{\mathbf{u}_0}(\mathbf{u}_t) - f(\mathbf{u}_0) + g(\mathbf{u}_t) - g(\mathbf{u}_0) \leq -3\mathscr{F} \right\}, \hat{c}\mathscr{T} \, \right\}$$

then, for any $\eta \leq c_{\max}/L$, we have for all $t < T$ that $\|\mathbf{u}_t - \tilde{\mathbf{x}}\| \leq 100(\mathscr{S} \cdot \hat{c})$.

Proof. We show if the function value did not decrease, then all the iteration updates must be constrained in a small ball. The proximal descent updates the solution as

$$\tilde{\mathbf{u}}_{t+1} = \mathbf{u}_t - \nabla f(\mathbf{u}_t) = (I - \nabla f)(\mathbf{u}_t)$$
$$\mathbf{u}_{t+1} = \text{prox}_{\eta g}(\tilde{\mathbf{u}}_{t+1}) = \text{prox}_{\eta g} \circ (I - \nabla f)(\mathbf{u}_t)$$

Without losing of generality, set $\mathbf{u}_0 = 0$ to be the origin. For any $t \in \mathbb{N}$,

$$\|\mathbf{u}_t - \mathbf{u}_0\| = \|\mathbf{u}_t - 0\| = \|\text{prox}_{\eta g}(\tilde{\mathbf{u}}_t) - \text{prox}_{\eta g}(0)\| \leq \|\tilde{\mathbf{u}}_t - 0\| = \|\tilde{\mathbf{u}}_t\|$$

Jin et al. prove in [13] by induction that if $\|\mathbf{u}_t\| \leq 100(\mathscr{S} \cdot \hat{c})$, then $\|\tilde{\mathbf{u}}_{t+1}\| \leq 100(\mathscr{S} \cdot \hat{c})$. Consequently, $\|\mathbf{u}_{t+1}\| \leq 100(\mathscr{S} \cdot \hat{c})$.

We point out that it is implicitly assumed that $\frac{2\mathscr{S}}{\kappa \cdot \ln(\frac{d\kappa}{\delta})} \ll \hat{c}$, so that for all $t < T$, $\|\tilde{\mathbf{x}}\| \ll \|\mathbf{u}_t\|$, and the relation $\|\mathbf{u}_t - \tilde{\mathbf{x}}\| \leq \|\mathbf{u}_t\| + \|\tilde{\mathbf{x}}\| \leq 100(\mathscr{S} \cdot \hat{c})$ holds.

3.2   Preparation for Building Pillars

Lemma 2 (Existence of lower bound for the difference sequence $\{\mathbf{v}_t\}_{t=1}^T$). *For iteration sequences $\{\mathbf{w}_t\}$ and $\{\mathbf{u}_t\}$ defined in Lemma 4, define the difference sequence as*

$$\mathbf{v}_t = \mathbf{w}_t - \mathbf{u}_t$$

There exists a positive lower bound for $\{\mathbf{v}_t\}$ when $t < \hat{c}\mathscr{T}$.

Proof. To show that the lower bound for iteration difference $\{\mathbf{v}_t\}_{t=1}^T$ exists, we consider bounding the iteration sequence $\tilde{\mathbf{v}}_{t+1}$ first. Define the difference between the proximal of l_1 penalty term and its coimage as $\mathcal{D}_g[\mathbf{x}] = \text{prox}_g[\mathbf{x}] - \mathbf{x} = \min\{\lambda\mathbb{1}, |\mathbf{x}|\} \otimes \text{sgn}(-\mathbf{x})$, where $\otimes$ is Hadamard product and the minimum is taken elementwise. We notice that $\|\mathcal{D}_{\eta\lambda\|\cdot\|_1}[\mathbf{x}]\| \leq \eta\lambda\sqrt{d}$. Thus, $\|\mathbf{w}_k - \mathbf{u}_k\| = \|\tilde{\mathbf{w}}_k - \tilde{\mathbf{v}}_k - \lambda(\mathcal{D}_{\eta g}[\tilde{\mathbf{w}}_k] - \mathcal{D}_{\eta g}[\tilde{\mathbf{u}}_k])\| \geq \|\tilde{\mathbf{w}}_k - \tilde{\mathbf{v}}_k\| - 2\eta\lambda\sqrt{d}$.

$$\begin{aligned}
\|\tilde{\mathbf{v}}_{t+1}\| &= \|\tilde{\mathbf{w}}_{t+1} - \tilde{\mathbf{u}}_{t+1}\| \\
&= \|(I - \eta\nabla f) \circ \text{prox}_{\eta g}(\tilde{\mathbf{w}}_k) - (I - \eta\nabla f) \circ \text{prox}_{\eta g}(\tilde{\mathbf{u}}_k)\| \\
&= \|\mathbf{w}_k - \mathbf{u}_k - \eta(\nabla f(\mathbf{w}_k) - \nabla f(\mathbf{u}_k))\| \\
&\geq \|\mathbf{w}_k - \mathbf{u}_k\| - \eta L\|\mathbf{w}_k - \mathbf{u}_k\| = (1 - \eta L)\|\mathbf{w}_k - \mathbf{u}_k\| \\
&\geq (1 - \eta L)(\|\tilde{\mathbf{w}}_k - \tilde{\mathbf{u}}_k\| - 2\eta\lambda\sqrt{d}) = (1 - \eta L)(\|\tilde{\mathbf{v}}_k\| - 2\eta\lambda\sqrt{d}) \\
&\geq (1 - \eta L)^t\|\tilde{\mathbf{v}}_1\| - 2\eta\lambda\sqrt{d}\sum_{i=1}^{t}(1 - \eta L)^i \\
&= (1 - \eta L)^t\|\tilde{\mathbf{v}}_1\| - 2\lambda\sqrt{d}\frac{(1 - \eta L)\bigl(1 - (1 - \eta L)^t\bigr)}{L}
\end{aligned}$$

As $\tilde{\mathbf{v}}_1 = (I - \eta\nabla f)\mathbf{v}_0 = (I - \eta\nabla f)\mu r\mathbf{e}_1 = \mu r(\mathbf{e}_1 - \eta\nabla^2 f(\boldsymbol{\xi})\theta\mathbf{e}_1) = \mu r(1 + \eta\gamma\theta)\mathbf{e}_1$, where $\theta \in (0, 1)$, we have

$$\|\tilde{\mathbf{v}}_{t+1}\| \geq (1 - \eta L)^t\mu r(1 + \eta\gamma\theta) - 2\lambda\sqrt{d}\frac{(1 - \eta L)(1 - (1 - \eta L)^t)}{\eta L} \tag{5}$$

To compare $\|\mathbf{v}_t\|$ and $\|\tilde{\mathbf{v}}_t\|$,

$$\|\mathbf{v}_{t+1}\| \geq \|\tilde{\mathbf{v}}_{t+1}\| - 2\eta\lambda\sqrt{d} \geq (1 - \eta L)^t\mu r(1 + \eta\gamma\theta) - 2\lambda\sqrt{d}\frac{(1 - \eta L)(1 - (1 - \eta L)^t) + \eta L}{L} \tag{6}$$

Therefore, as long as

$$\lambda < \frac{(1 - \eta L)^{\hat{c}\mathscr{T}}\mu\frac{1}{\kappa(\ln\frac{d\kappa}{\delta})^2}\sqrt{\eta}L^{\frac{3}{2}}\frac{\gamma}{\rho}(1 + \eta\gamma\theta)}{2\sqrt{d}[(1 - \eta L)(1 - (1 - \eta L)^{\hat{c}\mathscr{T}}) + \eta L]} \tag{7}$$

the difference sequence $\{\|\mathbf{v}_t\|\}$ has a positive lower bound on its norm.

Lemma 3 (Preservation of subspace projection monotonicity after prox of l_1 in rotated coordinate with small λ).

Denote the subspace of $\mathbb{R}^n$ spanned by $\{\mathbf{e}_1\}$ as $\mathbb{E}$, while the complement subspace spanned by $\{\mathbf{e}_2, \cdots, \mathbf{e}_n\}$ as $\mathbb{E}^\perp$. For a given vector $\mathbf{x}$ chosen from a lower bounded set $\mathcal{X}$, i.e. $\forall \mathbf{x} \in \mathcal{X}$, $\|\mathbf{x}\| \geq C$ for some constant $C > 0$, assume $\|\mathcal{P}_{\mathbb{E}^\perp}\mathbf{x}\| \leq K\|\mathcal{P}_{\mathbb{E}}\mathbf{x}\|$, where $0 < K \leq 1$ is a constant. If the parameter λ for the l_1 penalty term is small enough, then

$$\|\mathcal{P}_{\mathbb{E}^\perp}\text{prox}_{\eta g}(\mathbf{x})\| \leq K\|\mathcal{P}_{\mathbb{E}}\text{prox}_{\eta g}(\mathbf{x})\|$$

Proof. We want to find a constraint on λ such that when λ is small enough, if the projection in the original coordinate demonstrates the monotonicity relation $\|\mathcal{P}_{\mathbb{E}}\mathbf{x}\| \leq \|\mathcal{P}_{\mathbb{E}^\perp}\mathbf{x}\|$, this monotonicity relation will be preserved after proximal operator of l_1 is applied on the input vector.

Naturally there exists a normal vector, denoted as $\hat{\boldsymbol{n}}_{\text{boundary}} \equiv \hat{\boldsymbol{n}}$, for the boundary hyperplane on which $\|\mathcal{P}_{\mathbb{E}}\mathbf{x}\| = K\|\mathcal{P}_{\mathbb{E}^{\perp}}\mathbf{x}\|$. By moving along $\hat{\boldsymbol{n}}$, a point approaches the boundary most efficiently. Any vector inside the hyperplane is perpendicular to $\hat{\boldsymbol{n}}$, which we denote as $\hat{\boldsymbol{n}}^{\perp}$.

Define

$$\hat{\mathbf{v}}_{\text{move}}(\mathbf{x}) = \begin{cases} -\eta\lambda \cdot \text{sgn}(x_i) & \text{if } |x_i| \geq \eta\lambda \\ -x_i & \text{if } |x_i| < \eta\lambda \end{cases} = \min\{|x|, \eta\lambda\mathbb{1}\} \otimes \text{sgn}(-\mathbf{x}) \quad (8)$$

where $\otimes$ is the Hadamard product, and the minimum is taken elementwise. Because $\text{prox}_{\eta g}(\mathbf{x}) = \mathbf{x} + \hat{\mathbf{v}}_{\text{move}}$, a sufficient condition to be imposed on λ to guarantee the preservation of projection monotonicity $\|\mathcal{P}_{\mathbb{E}^{\perp}}\text{prox}_{\eta g}(\mathbf{x})\| \leq K\|\mathcal{P}_{\mathbb{E}}\text{prox}_{\eta g}(\mathbf{x})\|$ is that

$$\lambda < \left\| \frac{\text{Proj}_n \mathbf{x}}{\hat{\mathbf{v}}_{\text{move}} \cdot \hat{\boldsymbol{n}}} \right\| = \left\| \frac{\mathbf{x} \cdot \hat{\boldsymbol{n}}}{\hat{\mathbf{v}}_{\text{move}} \cdot \hat{\boldsymbol{n}}} \right\| \leq \frac{\|\mathbf{x}\|}{\|\hat{\mathbf{v}}_{\text{move}} \cdot \hat{\boldsymbol{n}}\|}$$

which means the moving distance caused by applying the l_1 proximal operator (soft shrinkage) projected on the direction of $\hat{\boldsymbol{n}}$ is less that the distance between $\mathbf{x}$ to the boundary hyperplane, hence rendering the vector stay on the same side of the boundary after moving.

Therefore, as long as

$$\lambda < \frac{C}{\|\hat{\mathbf{v}}_{\text{move}} \cdot \hat{\boldsymbol{n}}\|} \quad (9)$$

the monotonicity of projection onto subspaces can be preserved.

Remark 1 for Lemma 3. As an examples in $\mathbb{R}^2$, set $K = 1$, we visualise the shift caused by proximal operator and the boundary of projection-monotonicity preserving region. Assume $\mathbf{e}_{1,2}$ are orthonormal basis of Cartesian coordinate in the standard position. The directional vector for region division boundary is $\hat{\mathbf{e}}_{\text{boundary}} = \hat{\boldsymbol{n}}^{\perp} = \frac{\pm\hat{\mathbf{e}}_1 \pm \hat{\mathbf{e}}_2}{\sqrt{2}}$, and $\hat{\mathbf{e}}_{\text{boundary}}^{\perp} = \hat{\boldsymbol{n}}$ is the corresponding perpendicular directional vector. For l_1 norm, $\hat{\mathbf{v}}_{\text{move}}$ is $(\pm 1, \pm 1)$.

Remark 2 for Lemma 3. We point out that the upper bound for the parameter λ is related to the alignment of the eigenspace of $\mathcal{H}$. If the eigenspace of $\mathcal{H}$ is aligned with canonical orthonormal basis of $\mathbb{R}^d$, then $\lambda \in (0, \infty)$. The most stringent restriction on the upper bound of λ applies when $\hat{\mathbf{v}}_{\text{move}}$ is parallel to $\hat{\boldsymbol{n}}$.

3.3   Lemma: Perturbed Iterates Will Escape the Saddle Point

Lemma 4. *There exists absolute constant $c_{\max}, \hat{c}$ such that: for any $\delta \in (0, \frac{d\kappa}{e}]$, let $f(\cdot), \tilde{\mathbf{x}}$ satisfies the condition in Lemma 6, and sequences $\{\mathbf{u}_t\}, \{\mathbf{w}_t\}$ satisfy the conditions in Lemma 6, define:*

$$T = \min\left\{ \inf_t \left\{ t | \tilde{f}_{\mathbf{w}_0}(\mathbf{w}_t) + g(\mathbf{w}_t) - f(\mathbf{w}_0) - g(\mathbf{w}_0) \leq -3\mathscr{F} \right\}, \hat{c}\mathscr{T} \right\}$$

then, for any $\eta \leq c_{\max}/L$, *if* $\|\mathbf{u}_t - \tilde{\mathbf{x}}\| \leq 100(\mathscr{S} \cdot \hat{c})$ *for all* $t < T$, *we will have* $T < \hat{c}\mathscr{T}$.

Proof. We show that if the iterate sequence before time T starting from $\mathbf{u}_0$ does not provide sufficient function value decrease, the other iterate sequence, which starts from $\mathbf{w}_0$, will be able to achieve the function value decrease purpose. Ultimately, we will prove $T < \hat{c}\mathscr{T}$. We establish the inequality about T by considering the difference between $\mathbf{w}_t$ and $\mathbf{u}_t$. Define $\mathbf{v}_t = \mathbf{w}_t - \mathbf{u}_t$. The assumption of the Lemma 4, $\mathbf{v}_0 = \mu[\mathscr{S}/(\kappa \cdot \ln(\frac{d\kappa}{\delta}))]\mathbf{e}_1$, $\mu \in [\delta/(2\sqrt{d}), 1]$.

We bound $\|\mathbf{v}_t\|$ from both sides for all $t < T$ to obtain an inequality about T. Recall that the proximal descent updates the solution as

$$\tilde{\mathbf{u}}_{t+1} = \mathbf{u}_t - \nabla f(\mathbf{u}_t) = (I - \eta \nabla f)(\mathbf{u}_t)$$

$$\mathbf{u}_{t+1} = \text{prox}_{\eta g}(\tilde{\mathbf{u}}_{t+1}) = \text{prox}_{\eta g} \circ (I - \eta \nabla f)(\mathbf{u}_t)$$

Simple algebraic computation gives

$$\tilde{\mathbf{v}}_{t+1} = (I - \eta \mathcal{H} - \eta \Delta'_t)\mathbf{v}_t \tag{10}$$

where $\Delta'_t = \int_0^1 \nabla^2 f(\mathbf{u}_t + \theta \mathbf{v}_t)\, d\theta - \mathcal{H}$, and $\tilde{\mathbf{v}}_t = \tilde{\mathbf{w}}_t - \tilde{\mathbf{u}}_t$.

Consider $\|\tilde{\mathbf{u}}_t\|$ and $\|\tilde{\mathbf{w}}_t\|$. Because $\mathbf{v}_0 = \tilde{\mathbf{v}}_0$, we have $\|\tilde{\mathbf{w}}_0 - \tilde{\mathbf{x}}\| \leq \|\tilde{\mathbf{u}}_0 - \tilde{\mathbf{x}}\| + \|\tilde{\mathbf{v}}_0\| \leq 2\mathscr{S}/(\kappa \cdot \ln(\frac{d\kappa}{\delta}))$. With same logic in the proof for Lemma 1, we see $\|\tilde{\mathbf{u}}_t\| \leq 100(\mathscr{S} \cdot \hat{c})$, and $\|\tilde{\mathbf{w}}_t\| \leq 100(\mathscr{S} \cdot \hat{c})$. (Same relation hold for $\|\mathbf{u}_t\|$ and $\|\mathbf{w}_t\|$ respectively.) As a result, $\|\tilde{\mathbf{v}}_t\| \leq \|\tilde{\mathbf{w}}_t\| + \|\tilde{\mathbf{u}}_t\| \leq 200(\mathscr{S} \cdot \hat{c})$ for all $t < T$. Also,

$$\|\mathbf{v}_t\| \leq 200(\mathscr{S} \cdot \hat{c}) \tag{11}$$

Equation (11) and Hessian Lipschitz gives for $t < T$, $\|\Delta'_t\| \leq \rho(\|\mathbf{u}_t\| + \|\mathbf{v}_t\| + \|\tilde{\mathbf{x}}\|) \leq \rho\mathscr{S}(300\hat{c} + 1) = \frac{\zeta}{\eta}$, where $\zeta = \eta\rho\mathscr{S}(300\hat{c} + 1)$.

Denote ψ_t be the norm of $\mathbf{v}_t$ projected onto $\mathbf{e}_1$ direction $(\mathcal{S})$, and φ_t be the norm of $\mathbf{v}_t$ projected onto the remaining subspace $(\mathcal{S}^c)$, while $\tilde{\psi}_t$ be the norm of $\tilde{\mathbf{v}}_t$ projected onto $\mathcal{S}$, and $\tilde{\varphi}_t$ be the norm of $\tilde{\mathbf{v}}_t$ projected onto $\mathcal{S}^c$.

Equation (10) gives

$$\tilde{\psi}_{t+1} \geq (1 + \gamma\eta)\psi_t - \zeta\sqrt{\psi_t^2 + \varphi_t^2} \tag{12}$$

$$\tilde{\varphi}_{t+1} \leq (1 + \gamma\eta)\varphi_t + \zeta\sqrt{\psi_t^2 + \varphi_t^2} \tag{13}$$

To obtain the lower bound of $\|\mathbf{v}_t\|$, we prove the following relation as preparation:

$$\text{for all } t < T, \quad \varphi_t \leq 4\zeta t \cdot \psi_t \tag{14}$$

By hypothesis of Lemma 4, we know $\varphi_0 = 0$, thus the base case of induction holds. Assume Eq. (14) is true for $\tau \leq t$, for $t + 1 \leq T$, we have

$$\tilde{\varphi}_{t+1} \leq 4\zeta t(1 + \gamma\eta)\psi_t + \zeta\sqrt{\psi_t^2 + \varphi_t^2}$$

$$4\zeta(t + 1)\left[(1 + \gamma\eta)\psi_t - \zeta\sqrt{\psi_t^2 + \varphi_t^2}\right] \leq 4\zeta(t + 1)\tilde{\psi}_{t+1} \tag{15}$$

By choosing $\sqrt{c_{\max}} \leq \frac{1}{300\hat{c}+1}\min\{\frac{1}{2\sqrt{2}}, \frac{1}{4\hat{c}}\}$, and $\eta \leq \frac{c_{\max}}{L}$, we have $4\zeta(t+1) \leq 4\zeta T \leq 4\eta\rho\mathscr{S}(300\hat{c}+1)\hat{c}\mathscr{T} = 4\sqrt{\eta L}(300\hat{c}+1)\hat{c} \leq 1$. This gives $4(1+\gamma\eta)\psi_t \geq 4\psi_t \geq (1+1)\sqrt{2\psi_t^2} \geq (1+4\zeta(t+1))\sqrt{\psi_t^2 + \varphi_t^2}$. i.e.

$$(1 + 4\zeta(t+1))\sqrt{\psi_t^2 + \varphi_t^2} \leq 4\psi_t \tag{16}$$

Connecting two parts of Eq. (15), we obtain

$$\tilde{\varphi}_{t+1} \leq 4\zeta(t+1)\tilde{\psi}_{t+1} \tag{17}$$

Now we switch our focus to the eigenspace of Hessian $\mathcal{H}$. Assume the orthonormal basis for the eigensapce of $\mathcal{H}$ is $\{\mathbf{e}_1, \mathbf{e}_2, \cdots, \mathbf{e}_d\}$. The order of dimension aligns with the increasing order of the corresponding eigenvalues. This coordinate transformation does not lead to loss of generality, as it is unitary.

By Lemma 2, we know the iteration difference sequence $\mathbf{v}_t$ has a positive lower bound in terms of 2-norm. Therefore, by Lemma 3, with the virtue of Eq. (17) $\sqrt{\sum_{i=2}^d (\mathbf{e}_i^T \tilde{\mathbf{v}}_{t+1})^2} \leq 4\zeta(t+1)\|\mathbf{e}_1^T \tilde{\mathbf{v}}_{t+1}\|$, we still have the projection monotonicity on the subspace of eigenspace of $\mathcal{H}$, i.e.

$$\varphi_{t+1} = \sqrt{\sum_{i=2}^d (\mathbf{e}_i^T \mathrm{prox}_g(\tilde{\mathbf{v}}_{t+1}))^2} \leq 4\zeta(t+1)\|\mathbf{e}_1^T \mathrm{prox}_g(\tilde{\mathbf{v}}_{t+1})\| = 4\zeta(t+1)\psi_{t+1}$$

Until here we finish the induction.

Recall that $4\zeta(t+1) \leq 1$, we thus have $\varphi_t \leq 4\zeta t\psi_t \leq \psi_t$, which gives

$$\psi_{t+1} \geq (1+\gamma\eta)\psi_t - \sqrt{2}\zeta\psi_t \geq \left(1 + \frac{\gamma\eta}{2}\right)\psi_t \tag{18}$$

where the last inequality follows from $\zeta = \eta\rho\mathscr{S}(300\hat{c}+1) \leq \sqrt{c_{\max}}(300\hat{c}+1)\gamma\eta \cdot \ln^{-1}(\frac{d\kappa}{\delta}) \leq \frac{\gamma\eta}{2\sqrt{2}}$.

Finally, combining (11) and (18), we have for all $t < T$:

$$200(\mathscr{S} \cdot \hat{c}) \geq \|\mathbf{v}_t\| \geq \psi_t \geq \left(1 + \frac{\gamma\eta}{2}\right)^t \psi_0 = \left(1 + \frac{\gamma\eta}{2}\right)^t c_0 \frac{\mathscr{S}}{\kappa}\ln^{-1}\left(\frac{d\kappa}{\delta}\right)$$

$$\geq \left(1 + \frac{\gamma\eta}{2}\right)^t \frac{\delta}{2\sqrt{d}}\frac{\mathscr{S}}{\kappa}\ln^{-1}\left(\frac{d\kappa}{\delta}\right)$$

This implies

$$T < \frac{1}{2}\frac{\ln[400\frac{\kappa\sqrt{d}}{\delta} \cdot \hat{c}\ln(\frac{d\kappa}{\delta})]}{\ln(1 + \frac{\gamma\eta}{2})} \leq \frac{\ln[400\frac{\kappa\sqrt{d}}{\delta} \cdot \hat{c}\ln(\frac{d\kappa}{\delta})]}{\gamma\eta} \leq (2 + \ln(400\hat{c}))\mathscr{T}$$

The last inequality is due to $\delta \in (0, \frac{d\kappa}{e}]$, we have $\ln(\frac{d\kappa}{\delta}) \geq 1$. By choosing the constant $\hat{c}$ to be large enough to satisfy $2 + \ln(400\hat{c}) \leq \hat{c}$, we will have $T < \hat{c}\mathscr{T}$, which finishes the proof.

3.4   Combining Previous Results

Lemma 5. *There exists a universal constant $c_{\max}$, for any $\delta \in (0, \frac{d\kappa}{e}]$, let $f(\cdot), \tilde{\mathbf{x}}$ satisfies the conditions in Lemma 6, and without loss of generality let $\mathbf{e}_1$ be the minimum eigenvector of $\nabla^2 f(\tilde{\mathbf{x}})$. Consider two gradient descent sequences $\{\mathbf{u}_t\}, \{\mathbf{w}_t\}$ with initial points $\mathbf{u}_0, \mathbf{w}_0$ satisfying: (denote radius $r = \mathscr{S}/(\kappa \cdot \ln(\frac{d\kappa}{\delta})))$*

$$\|\mathbf{u}_0 - \tilde{\mathbf{x}}\| \leq r, \quad \mathbf{w}_0 = \mathbf{u}_0 + \mu \cdot r \cdot \mathbf{e}_1, \quad \mu \in [\delta/(2\sqrt{d}), 1]$$

Then, for any stepsize $\eta \leq c_{\max}/L$, and any $T \geq \frac{1}{c_{\max}}\mathscr{T}$, we have:

$$\min\{f(\mathbf{u}_T) + g(\mathbf{u}_T) - f(\mathbf{u}_0) - g(\mathbf{u}_0), f(\mathbf{w}_T) + g(\mathbf{w}_T) - f(\mathbf{w}_0) - g(\mathbf{w}_0)\} \leq -2.7\mathscr{F}$$

Proof. Without losing generality, let $\tilde{\mathbf{x}} = 0$ be the origin. Let $(c_{\max}^{(2)}, \hat{c})$ be the absolute constant so that Lemma 4 holds, also let $c_{\max}^{(1)}$ be the absolute constant to make Lemma 1 holds based on our current choice of $\hat{c}$. We choose $c_{\max} \leq \min\{c_{\max}^{(1)}, c_{\max}^{(2)}\}$ so that our learning rate $\eta \leq c_{\max}/L$ is small enough which make both Lemmas 1 and 4 hold. Let $T^* := \hat{c}\mathscr{T}$ and define:

$$T' = \inf_t \left\{ t | \tilde{f}_{\mathbf{u}_0}(\mathbf{u}_t) + g(\mathbf{u}_t) - f(\mathbf{u}_0) - g(\mathbf{u}_0) \leq -3\mathscr{F} \right\}$$

Let's consider following two cases:

Case $T' \leq T^*$: In this case, by Lemma 1, we know $\|\mathbf{u}_{T'-1}\| \leq O(\mathscr{S})$, and therefore

$$\|\mathbf{u}_{T'}\| \leq \|\mathbf{u}_{T'-1}\| + \eta\|\nabla f(\mathbf{u}_{T'-1})\| \leq \|\mathbf{u}_{T'-1}\| + \eta\|\nabla f(\tilde{\mathbf{x}})\| + \eta L\|\mathbf{u}_{T'-1}\| \leq O(\mathscr{S})$$

By choosing $c_{\max}$ small enough and $\eta \leq c_{\max}/L$, this gives:

$$
\begin{aligned}
&f(\mathbf{u}_{T'}) + g(\mathbf{u}_{T'}) - f(\mathbf{u}_0) - g(\mathbf{u}_0) \\
&\leq \nabla f(\mathbf{u}_0)^\top(\mathbf{u}_{T'} - \mathbf{u}_0) + \frac{1}{2}(\mathbf{u}_{T'} - \mathbf{u}_0)^\top \nabla^2 f(\mathbf{u}_0)(\mathbf{u}_{T'} - \mathbf{u}_0) + \frac{\rho}{6}\|\mathbf{u}_{T'} - \mathbf{u}_0\|^3 + g(\mathbf{u}_{T'}) - g(\mathbf{u}_0) \\
&\leq \tilde{f}_{\mathbf{u}_0}(\mathbf{u}_{T'}) - f(\mathbf{u}_0) + g(\mathbf{u}_{T'}) - g(\mathbf{u}_0) + \frac{\rho}{2}\|\mathbf{u}_0 - \tilde{\mathbf{x}}\|\|\mathbf{u}_{T'} - \mathbf{u}_0\|^2 + \frac{\rho}{6}\|\mathbf{u}_{T'} - \mathbf{u}_0\|^3 \\
&\leq -3\mathscr{F} + O(\rho\mathscr{S}^3) = -3\mathscr{F} + O(\sqrt{\eta L} \cdot \mathscr{F}) \leq -2.7\mathscr{F}
\end{aligned}
$$

The first and second inequality exploit Hessian Lipschitz property of smooth function f, and $\|\mathbf{u}_0 - \tilde{\mathbf{x}}\| \leq O(\mathscr{S})$, $\|\mathbf{u}_{T'} - \mathbf{u}_0\| \leq O(\mathscr{S})$. By choose $c_{\max} \leq \min\{1, \frac{1}{\hat{c}}\}$. We know $\eta < \frac{1}{L}$, by *sufficient decrease lemma* for proximal descent, we know each proximal descent iteration decreases function value. Therefore, for any $T \geq \frac{1}{c_{\max}}\mathscr{T} \geq \hat{c}\mathscr{T} = T^* \geq T'$, we have:

$$\Phi(\mathbf{u}_T) - \Phi(\mathbf{u}_0) \leq \Phi(\mathbf{u}_{T^*}) - \Phi(\mathbf{u}_0) \leq \Phi(\mathbf{u}_{T'}) - \Phi(\mathbf{u}_0) \leq -2.7\mathscr{F}$$

Case $T' > T^*$: In this case, by Lemma 1, we know $\|\mathbf{u}_t\| \leq O(\mathscr{S})$ for all $t \leq T^*$. Define

$$T'' = \inf_t \left\{ t | \tilde{f}_{\mathbf{w}_0}(\mathbf{w}_t) + g(\mathbf{w}_t) - f(\mathbf{w}_0) - g(\mathbf{w}_0) \leq -3\mathscr{F} \right\}$$

By Lemma 4, we immediately have $T'' \leq T^*$. Apply same argument as in the case $T' \leq T^*$, we have for all $T \geq \frac{1}{c_{\max}}\mathscr{T}$ that $f(\mathbf{w}_T) + g(\mathbf{w}_T) - f(\mathbf{w}_0) - g(\mathbf{w}_0) \leq f(\mathbf{w}_{T^*}) + g(\mathbf{w}_{T^*}) - f(\mathbf{w}_0) - g(\mathbf{w}_0) \leq -2.7\mathscr{F}$.

3.5   Main Lemma

Lemma 6 (Main Lemma). *There exists universal constant $c_{\max}$, for $f(\cdot)$ satisfies Assumption 1, for any $\delta \in (0, \frac{d\kappa}{e}]$, suppose we start with point $\tilde{\mathbf{x}}$ satisfying following conditions:*

$$\|G(\tilde{\mathbf{x}})\| = \left\| L\left(\tilde{\mathbf{x}} - \mathrm{prox}_{\frac{1}{L}g}\left(\tilde{\mathbf{x}} - \frac{1}{L}\nabla f(\tilde{\mathbf{x}})\right)\right)\right\| \le \mathscr{G} \quad and \quad \lambda_{\min}(\nabla^2 f(\tilde{\mathbf{x}})) \le -\gamma$$

Let $\mathbf{x}_0 = \tilde{\mathbf{x}} + \boldsymbol{\xi}$ where $\boldsymbol{\xi}$ come from the uniform distribution over ball with radius $\mathscr{S}/(\kappa \cdot \ln(\frac{d\kappa}{\delta}))$, and let $\mathbf{x}_t$ be the iterates of gradient descent from $\mathbf{x}_0$. Then, when stepsize $\eta \le c_{\max}/L$, with at least probability $1 - \delta$, we have following for any $T \ge \frac{1}{c_{\max}}\mathscr{T}$:

$$f(\mathbf{x}_T) + g(\mathbf{x}_T) - f(\tilde{\mathbf{x}}) - g(\tilde{\mathbf{x}}) \le -\mathscr{F}$$

Proof. Denote $T_{\frac{1}{L}}(\mathbf{x}) = \mathrm{prox}_{\frac{1}{L}g}[\mathbf{x} - \frac{1}{L}\nabla f(\mathbf{x})]$. The first order stationary condition is equivalent to $\|\tilde{\mathbf{x}} - T_{\frac{1}{L}}(\tilde{\mathbf{x}})\| = \|\nabla f(\tilde{\mathbf{x}}) + \partial g(T_{\frac{1}{L}}(\tilde{\mathbf{x}}))\| \le \mathscr{G}$, where ∂g is the subgradient of the function g.

As $g(\mathbf{x}) = \lambda\|\mathbf{x}\|_1$ has Lipschitz constant λ, we have

$$f(\mathbf{x}_0) + g(\mathbf{x}_0) \le f(\tilde{\mathbf{x}}) + \langle\nabla f(\tilde{\mathbf{x}}), \boldsymbol{\xi}\rangle + \frac{L}{2}\|\boldsymbol{\xi}\|^2 + g(\tilde{\mathbf{x}}) + \langle\partial g(\tilde{\mathbf{x}}), \boldsymbol{\xi}\rangle + \frac{\lambda}{2}\|\boldsymbol{\xi}\|^2$$

Notice

$$\|\nabla f(\tilde{\mathbf{x}}) + \partial g(\tilde{\mathbf{x}})\| = \|\nabla f(\tilde{\mathbf{x}}) + \partial g(T_{\frac{1}{L}}(\mathbf{x})) - (\partial g(T_{\frac{1}{L}}(\mathbf{x})) - \partial g(\tilde{\mathbf{x}}))\|$$
$$\le \mathscr{G} + \lambda\mathscr{G}$$

By adding perturbation, in worst case we increase function value by:

$$f(\mathbf{x}_0) - f(\tilde{\mathbf{x}}) + g(\mathbf{x}_0) - g(\tilde{\mathbf{x}}) \le \|\nabla f(\tilde{\mathbf{x}}) + \partial g(\tilde{\mathbf{x}})\|\|\boldsymbol{\xi}\| + \frac{L + \lambda}{2}\|\boldsymbol{\xi}\|^2$$
$$\le (1 + \lambda)\mathscr{G}\left(\frac{\mathscr{S}}{\kappa \cdot \ln(\frac{d\kappa}{\delta})}\right) + \frac{1}{2}(L + \lambda)\left(\frac{\mathscr{S}}{\kappa \cdot \ln(\frac{d\kappa}{\delta})}\right)^2$$
$$\le \left(\frac{3}{2} + \frac{1}{5}\right)\mathscr{F}$$

where the last inequality follows from the fact that $\lambda \ll \min\{1, l\}$ per Eq. (7).

On the other hand, let radius $r = \frac{\mathscr{S}}{\kappa \cdot \ln(\frac{d\kappa}{\delta})}$. We know $\mathbf{x}_0$ come froms uniform distribution over $\mathbb{B}_{\tilde{\mathbf{x}}}(r)$. Let $\mathcal{X}_{\mathrm{stuck}} \subset \mathbb{B}_{\tilde{\mathbf{x}}}(r)$ denote the set of bad starting points so that if $\mathbf{x}_0 \in \mathcal{X}_{\mathrm{stuck}}$, then $\Phi(\mathbf{x}_T) - \Phi(\mathbf{x}_0) > -2.7\mathscr{F}$ (thus stuck at a saddle point); otherwise if $\mathbf{x}_0 \in B_{\tilde{\mathbf{x}}}(r) - \mathcal{X}_{\mathrm{stuck}}$, we have $\Phi(\mathbf{x}_T) - \Phi(\mathbf{x}_0) \le -2.7\mathscr{F}$.

By applying Lemma 5, we know for any $\mathbf{x}_0 \in \mathcal{X}_{\mathrm{stuck}}$, it is guaranteed that $(\mathbf{x}_0 \pm \mu r \mathbf{e}_1) \notin \mathcal{X}_{\mathrm{stuck}}$ where $\mu \in [\frac{\delta}{2\sqrt{d}}, 1]$. Denote $I_{\mathcal{X}_{\mathrm{stuck}}}(\cdot)$ be the indicator function of being inside set $\mathcal{X}_{\mathrm{stuck}}$; and vector $\mathbf{x} = (x^{(1)}, \mathbf{x}^{(-1)})$, where $x^{(1)}$ is the component along $\mathbf{e}_1$ direction, and $\mathbf{x}^{(-1)}$ is the remaining $d-1$ dimensional vector. Recall $\mathbb{B}^{(d)}(r)$ be d-dimensional ball with radius r; By calculus, this gives an upper bound on the volume of $\mathcal{X}_{\text{stuck}}$:

$$
\begin{aligned}
\text{Vol}(\mathcal{X}_{\text{stuck}}) &= \int_{\mathbb{B}_{\tilde{\mathbf{x}}}^{(d)}(r)} d\mathbf{x} \cdot I_{\mathcal{X}_{\text{stuck}}}(\mathbf{x}) \\
&= \int_{\mathbb{B}_{\tilde{\mathbf{x}}}^{(d-1)}(r)} d\mathbf{x}^{(-1)} \int_{\tilde{x}^{(1)}-\sqrt{r^2-\|\tilde{\mathbf{x}}^{(-1)}-\mathbf{x}^{(-1)}\|^2}}^{\tilde{x}^{(1)}+\sqrt{r^2-\|\tilde{\mathbf{x}}^{(-1)}-\mathbf{x}^{(-1)}\|^2}} dx^{(1)} \cdot I_{\mathcal{X}_{\text{stuck}}}(\mathbf{x}) \\
&\leq \int_{\mathbb{B}_{\tilde{\mathbf{x}}}^{(d-1)}(r)} d\mathbf{x}^{(-1)} \cdot \left(2 \cdot \frac{\delta}{2\sqrt{d}} r\right) = \text{Vol}(\mathbb{B}_0^{(d-1)}(r)) \times \frac{\delta r}{\sqrt{d}}
\end{aligned}
$$

Then, we immediately have the ratio:

$$
\frac{\text{Vol}(\mathcal{X}_{\text{stuck}})}{\text{Vol}(\mathbb{B}_{\tilde{\mathbf{x}}}^{(d)}(r))} \leq \frac{\frac{\delta r}{\sqrt{d}} \times \text{Vol}(\mathbb{B}_0^{(d-1)}(r))}{\text{Vol}(\mathbb{B}_0^{(d)}(r))} = \frac{\delta}{\sqrt{\pi d}} \frac{\Gamma(\frac{d}{2}+1)}{\Gamma(\frac{d}{2}+\frac{1}{2})} \leq \frac{\delta}{\sqrt{\pi d}} \cdot \sqrt{\frac{d}{2}+\frac{1}{2}} \leq \delta
$$

The second last inequality is by the property of Gamma function that $\frac{\Gamma(x+1)}{\Gamma(x+1/2)} < \sqrt{x+\frac{1}{2}}$ as long as $x \geq 0$. Therefore, with at least probability $1-\delta$, $\mathbf{x}_0 \notin \mathcal{X}_{\text{stuck}}$. In this case, we have:

$$
\begin{aligned}
\Phi(\mathbf{x}_T) - \Phi(\tilde{\mathbf{x}}) &= \Phi(\mathbf{x}_T) - \Phi(\mathbf{x}_0) + \Phi(\mathbf{x}_0) - \Phi(\tilde{\mathbf{x}}) \\
&\leq -2.7\mathscr{F} + 1.7\mathscr{F} \leq -\mathscr{F}
\end{aligned}
$$

which finishes the proof.

3.6 Main Theorem, and Its Proof

Lemma 7 (Sufficient Decrease Lemma for Proximal Descent, [3]).
Assume the function f is real-valued and lower semi-continuous. Then for any $L \in (\frac{L}{2}, \infty)$ where $\eta = \frac{1}{L}$, we have $\Phi(\mathbf{x}_t) - \Phi(\mathbf{x}_{t+1}) \geq \frac{L-\frac{L}{2}}{L^2}\|G_{\frac{1}{L}}(\mathbf{x}_t)\|$.

Proof of the Main Theorem

Proof. Denote $\tilde{c}_{\max}$ to be the absolute constant allowed in Lemma 6 when it is given following parameters $\eta = \frac{c}{L}$, $\gamma = \sqrt{\rho\varepsilon}$, and $\delta = \frac{dL}{\sqrt{\rho\varepsilon}}e^{-\chi}$. In this theorem, we let $c_{\max} = \min\{\tilde{c}_{\max}, 1/2\}$, and choose any constant $c \leq c_{\max}$.

In this proof, we will actually achieve some point satisfying following condition:

$$
\|G(\mathbf{x})\| \leq g_{\text{thres}} \equiv \frac{\sqrt{c}}{\chi^2} \cdot \varepsilon, \qquad \lambda_{\min}(\nabla^2 f(\mathbf{x})) \geq -\sqrt{\rho\varepsilon} \qquad (19)
$$

Since $c \leq 1$, $\chi \geq 1$, we have $\frac{\sqrt{c}}{\chi^2} \leq 1$, which implies any $\mathbf{x}$ satisfy Eq. (19) is also a ε-second-order stationary point.

Starting from $\mathbf{x}_0$, we know if $\mathbf{x}_0$ does not satisfy Eq. (19), there are only two possibilities:

1. $\|G(\mathbf{x}_0)\| > g_{\text{thres}}$: In this case, Algorithm 1 will not add perturbation. By Lemma 7:

$$\Phi(\mathbf{x}_1) - \Phi(\mathbf{x}_0) \leq -\frac{\eta}{2} \cdot g_{\text{thres}}^2 = -\frac{c^2}{2\chi^4} \cdot \frac{\varepsilon^2}{L}$$

2. $\|G(\mathbf{x}_0)\| \leq g_{\text{thres}}$: In this case, Algorithm 1 will add a perturbation of radius r, and will perform proximal gradient descent (without perturbations) for the next t_{thres} steps. Algorithm 1 will then check termination condition. If the condition is not met, we must have:

$$\Phi(\mathbf{x}_{t_{\text{thres}}}) - \Phi(\mathbf{x}_0) \leq -\Phi_{\text{thres}} = -\frac{c}{\chi^3} \cdot \sqrt{\frac{\varepsilon^3}{\rho}}$$

This means on average every step decreases the function value by

$$\frac{\Phi(\mathbf{x}_{t_{\text{thres}}}) - \Phi(\mathbf{x}_0)}{t_{\text{thres}}} \leq -\frac{c^3}{\chi^4} \cdot \frac{\varepsilon^2}{L}$$

In case 1, we can repeat this argument for $t = 1$ and in case 2, we can repeat this argument for $t = t_{\text{thres}}$. Hence, we can conclude as long as Algorithm 1 has not terminated yet, on average, every step decrease function value by at least $\frac{c^3}{\chi^4} \cdot \frac{\varepsilon^2}{L}$. However, we clearly can not decrease function value by more than $\Phi(\mathbf{x}_0) - \Phi^\star$, where $\Phi^\star$ is the function value of global minima. This means Algorithm 1 must terminate within the following number of iterations:

$$\frac{\Phi(\mathbf{x}_0) - \Phi^\star}{\frac{c^3}{\chi^4} \cdot \frac{\varepsilon^2}{L}} = \frac{\chi^4}{c^3} \cdot \frac{L(\Phi(\mathbf{x}_0) - \Phi^\star)}{\varepsilon^2} = O\left(\frac{L(\Phi(\mathbf{x}_0) - \Phi^\star)}{\varepsilon^2} \ln^4\left(\frac{dL\Delta_\Phi}{\varepsilon^2\delta}\right)\right)$$

Finally, we would like to ensure when Algorithm 1 terminates, the point it finds is actually an ε-second-order stationary point. The algorithm can only terminate when the gradient mapping is small, and the function value does not decrease after a perturbation and t_{thres} iterations. We shall show every time when we add perturbation to iterate $\tilde{\mathbf{x}}_t$, if $\lambda_{\min}(\nabla^2 f(\tilde{\mathbf{x}}_t)) < -\sqrt{\rho\varepsilon}$, then we will have $\Phi(\mathbf{x}_{t+t_{\text{thres}}}) - \Phi(\tilde{\mathbf{x}}_t) \leq -\Phi_{\text{thres}}$. Thus, whenever the current point is not an ε-second-order stationary point, the algorithm cannot terminate.

According to Algorithm 1, we immediately know $\|G(\tilde{\mathbf{x}}_t)\| \leq g_{\text{thres}}$ (otherwise we will not add perturbation at time t). By Lemma 6, we know this event happens with probability at least $1 - \frac{dL}{\sqrt{\rho\varepsilon}}e^{-\chi}$ each time. On the other hand, during one entire run of Algorithm 1, the number of times we add perturbations is at most:

$$\frac{1}{t_{\text{thres}}} \cdot \frac{\chi^4}{c^3} \cdot \frac{L(\Phi(\mathbf{x}_0) - \Phi^\star)}{\varepsilon^2} = \frac{\chi^3}{c} \cdot \frac{\sqrt{\rho\varepsilon}(\Phi(\mathbf{x}_0) - \Phi^\star)}{\varepsilon^2}$$

By the union bound, for all these perturbations, with high probability Lemma 6 is satisfied. As a result Algorithm 1 works correctly. The probability of that is at least

$$1 - \frac{dL}{\sqrt{\rho\varepsilon}}e^{-\chi} \cdot \frac{\chi^3}{c} \cdot \frac{\sqrt{\rho\varepsilon}(\Phi(\mathbf{x}_0) - \Phi^\star)}{\varepsilon^2} = 1 - \frac{\chi^3 e^{-\chi}}{c} \cdot \frac{dL(\Phi(\mathbf{x}_0) - \Phi^\star)}{\varepsilon^2}$$

Perturbed Proximal Descent   73

Recall our choice of $\chi = 3\max\{\ln(\frac{dL\Delta_f}{c\varepsilon^2\delta}), 4\}$. Since $\chi \geq 12$, we have $\chi^3 e^{-\chi} \leq e^{-\chi/3}$, this gives:

$$\frac{\chi^3 e^{-\chi}}{c} \cdot \frac{dL(\Phi(\mathbf{x}_0) - \Phi^\star)}{\varepsilon^2} \leq e^{-\chi/3}\frac{dL(\Phi(\mathbf{x}_0) - \Phi^\star)}{c\varepsilon^2} \leq \delta$$

which finishes the proof.

Remarks on Large λ. We point out that when λ is large enough so that the g term alters the local landscape of the objective function $\Phi(\mathbf{x})$, it is inevitable that new local minima will be introduced to the landscape of the objective function, and potentially change the stability of saddle points. We hypothesize that perturbed proximal descent will still converge to an ε-second-order stationary point regardless of the magnitude of λ.

An example for the new local minima introduced by large λ is Fig. 3(b). We see new wrinkles are introduced to the four legs of the octopus function as λ increases from 1 to 10. If an iteration starts in the neighborhood of creases, it can converge to the bottom of the creases. Figure 3(c) is an extreme scenario where the original landscape of the octopus function is completely altered to conform to the behavior of ℓ_1 penalty term.

3.7 From ε-Second-Order Stationary Point to Local Minimizers

Assumption 3 (Nondegenerate Saddle). *For all stationary points* $\mathbf{x}_c$, $\exists m > 0$ *such that* $\min_{i=1,2,\cdots,d} |\lambda_i(\nabla^2 f(\mathbf{x}_c))| > m > 0$, *where* λ_i *are the eigenvalues (not to be confused with the parameter* λ*).*

With this nondegenerate saddle assumption, the main theorem can be strengthened to the following corollary, whose proof is immediate as one sets the ε value in the main theorem as m^2/ρ and realizes that there is no eigenvalue of $\nabla^2 f$ existing between $-\sqrt{\rho\varepsilon}$ and the first positive eigenvalue.

Corollary 1. *There exists an absolute constant* $c_{\max}$ *such that if* $f(\cdot)$ *satisfies Assumptions 1, 2 and 3, then for any* $\delta > 0$, $\Delta_\Phi \geq \Phi(\mathbf{x}_0) - \Phi^\star$, *constant* $c \leq c_{\max}$, *and* $\varepsilon = \frac{m^2}{\rho}$, *with probability* $1 - \delta$, *the output of* $PPD(\mathbf{x}_0, L, \rho, \varepsilon, c, \delta, \Delta_f)$ *will be a local minimizer of* $f + \lambda\|\mathbf{x}\|_1$, *and terminate in iterations:*

$$\mathcal{O}\left(\frac{L(\Phi(\mathbf{x}_0) - \Phi^\star)}{\varepsilon^2}\ln^4\left(\frac{dL\Delta_\Phi}{\varepsilon^2\delta}\right)\right)$$

4 Numerical Experiment

We set f to be the "octopus" function described in [10] and use perturbed proximal descent to minimize the objective function $\Phi(\mathbf{x}) = f(\mathbf{x}) + \lambda\|\mathbf{x}\|_1$. Plots of octopus function defined in $\mathbb{R}^2$ for various λ are shown in Fig. 3.

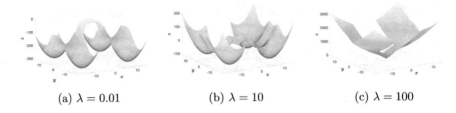

(a) $\lambda = 0.01$ (b) $\lambda = 10$ (c) $\lambda = 100$

Fig. 3. The octopus function with different λ values

The "octopus" family of functions is parameterized by τ, which controls the width of the "legs," and M and γ which characterize how sharp each side is surrounding a saddle point, related to the Lipschitz constant. The example illustrated in Fig. 3 uses parameters $M = \mathrm{e}, \gamma = 1, \tau = \mathrm{e}$.

We are interested in the octopus family of functions because it can be generalized to any dimension d, and it has $d-1$ saddle points (not counting the origin) which are known to slow down standard gradient descent algorithms. The usual minimization iteration sequence, if starting at the maximum value of the octopus function, will successively go through *each* saddle point before reaching the global minimum, thus rendering the iteration progress easy to track and visualize.

Specifics of Octopus Function. We define octopus function in first quadrant of $\mathbb{R}^d$. And then, by even function reflection, the octopus can be continued to all other quadrants.

Define the *auxiliary gluing functions* as

$$\mathcal{G}_1(x_i) = -\gamma x_i^2 + \frac{-14L + 10\gamma}{3\tau}(x_i - \tau)^3 + \frac{5L - 3\gamma}{2\tau}(x_i - \tau)^4$$

$$\mathcal{G}_2(x_i) = -\gamma - \frac{10(L+\gamma)}{\tau^3}(x_i - 2\tau)^3 - \frac{15(L+\gamma)}{\tau^4}(x_i - 2\tau)^4 - \frac{6(L+\gamma)}{\tau^5}(x_i - 2\tau)^5$$

Define the *gluing function* and *gluing balance constant* respectively as

$$\mathcal{G}(x_i, x_{i+1}) = \mathcal{G}_1(x_i) + \mathcal{G}_2(x_i)x_{i+1}^2$$

$$\nu = -\mathcal{G}_1(2\tau) + 4L\tau^2 = \frac{26L + 2\gamma}{3}\tau^2 + \frac{-5L + 3\gamma}{2}\tau^3$$

For a given $i = 1, \cdots, d-1$, when $6\tau \geq x_1, \cdots, x_{i-1} \geq 2\tau, \tau \geq x_i \geq 0, \tau \geq x_{i+1}, \cdots, x_d \geq 0$

$$f(\mathbf{x}) = \sum_{j=1}^{i-1} L(x_j - 4\tau)^2 - \gamma x_i^2 + \sum_{j=i+1}^{d} Lx_j^2 - (i-1)\nu \equiv f_{i,1}(\mathbf{x}) \qquad (20)$$

and if $6\tau \geq x_1, \cdots, x_{i-1} \geq 2\tau, 2\tau \geq x_i \geq \tau, \tau \geq x_{i+1}, \cdots, x_d \geq 0$, we have

$$f(\mathbf{x}) = \sum_{j=1}^{i-1} L(x_j - 4\tau)^2 + \mathcal{G}(x_i, x_{i+1}) + \sum_{j=i+2}^{d} Lx_j^2 - (i-1)\nu \equiv f_{i,2}(\mathbf{x}) \quad (21)$$

and for $i = d$, if $6\tau \geq x_1, \cdots, x_{d-1} \geq 2\tau, \tau \geq x_d \geq 0$

$$f(\mathbf{x}) = \sum_{j=1}^{d-1} L(x_j - 4\tau)^2 - \gamma x_d^2 - (d-1)\nu \equiv f_{d,1}(\mathbf{x}) \quad (22)$$

and if $6\tau \geq x_1, \cdots, x_{d-1} \geq 2\tau, 2\tau \geq x_d \geq \tau$

$$f(\mathbf{x}) = \sum_{j=1}^{d-1} L(x_j - 4\tau)^2 + \mathcal{G}_1(x_d) - (d-1)\nu \equiv f_{d,2}(\mathbf{x}) \quad (23)$$

and if $6\tau \geq x_1, \cdots, x_d \geq 2\tau$,

$$f(\mathbf{x}) = \sum_{j=1}^{d} L(x_j - 4\tau)^2 - d\nu \equiv f_{d+1,1}(\mathbf{x}) \quad (24)$$

Remark. All saddle points happen at $(\pm 4\tau, \pm 4\tau, \cdots, \pm 4\tau, 0, 0, \cdots, 0)$, and the global minimum is at $(\pm 4\tau, \cdots, \pm 4\tau)$. Regions in the form of $[2\tau, 6\tau] \times \cdots \times [2\tau, 6\tau] \times [\tau, 2\tau] \times [0, \tau] \times \cdots \times [0, \tau]$ are transition zones described by the gluing functions which connect separate pieces to make f a continuous function. The octopus function can be constructed first in the first quadrant, and then using even function reflection to define it in all other quadrants. A typical descent algorithm applied to the octopus generates iterations that take multiple turns like walking down a spiral staircase, each staircase leading to a new dimension.

4.1   Results

We apply the perturbed proximal descent (PPD) on the octopus function plus $0.01\|\mathbf{x}\|_1$ when the dimension varies between $d = 2, 5, 10, 20$. We set the constant $c = 3$. For comparison, we apply perturbed gradient descent (PGD) as well since $\|\mathbf{x}\|_1$ is differentiable almost everywhere; for both algorithms, the norm of the perturbation $\boldsymbol{\xi}$ is 0.1.

We see that PPD successfully finds the local minimum in the first three cases within 1000 iterations, and in the case of $d = 20$, PPD almost finds the local minimum within 1000 iterations. In contrast, unperturbed proximal descent (PD), gradient descent (GD), and perturbed gradient descent (PGD) sequences are trapped near saddle points (Fig. 4).

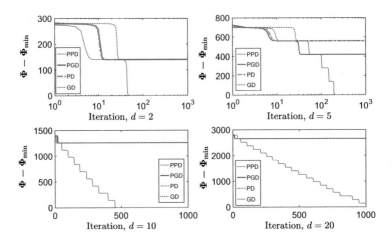

Fig. 4. Performance of our proposed PPD algorithm on the octopus function with $\lambda = 0.01$

5 Conclusion

This paper provides an algorithm to minimize a non-convex function plus a ℓ_1 penalty of small magnitude, with a probabilistic guarantee that the returned result is an approximate second-order stationary point, and hence for a large class of functions, a local minimum instead of a saddle point. The complexity is of $\mathcal{O}(\varepsilon^{-2})$ and the result depends on dimension in $\mathcal{O}(\ln^4 d)$.

The deficiency of the result is that the magnitude of ℓ_1 penalty needs to be small to let our theoretical result hold. Meanwhile, we also notice that a large λ will lead to creation of new local minima to the objective function altering the original landscape. Our future work will address the case of large λ in the iteration process.

References

1. Attouch, H., Bolte, J., Svaiter, B.F.: Convergence of descent methods for semi-algebraic and tame problems: proximal algorithms, forward-backward splitting, and regularized Gauss-Seidel methods. Math. Program. **137**, 1–39 (2011)
2. Bauschke, H.H., Combettes, P.L.: Convex Analysis and Monotone Operator Theory in Hilbert Spaces, 2nd edn. Springer, New York (2017)
3. Beck, A.: First-Order Methods in Optimization: MOS-SIAM Series on Optimization. Society for Industrial and Applied Mathematics, Philadelphia (2017)
4. Bolte, J., Sabach, S., Teboulle, M.: Proximal alternating linearized minimization for nonconvex and nonsmooth problems. Math. Prog. **146**(1–2), 459–494 (2014)
5. Bot, R.I., Csetnek, E.R., Nguyen, D.-K.: A proximal minimization algorithm for structured nonconvex and nonsmooth problems. arXiv preprint arXiv:1805.11056v1 [math.OC] (2018)
6. Carmon, Y., Duchi, J., Hinder, O., Sidford, A.: Accelerated methods for nonconvex optimization. SIAM J. Optim. **28**(2), 1751–1772 (2018)

7. Combettes, P.L., Wajs, V.R.: Signal recovery by proximal forward-backward splitting. SIAM Multiscale Model. Simul. **4**(4), 1168–1200 (2005)
8. Curtis, F.E., Robinson, D.P., Samadi, M.: A trust region algorithm with a worst-case iteration complexity of $\mathcal{O}(\epsilon^{\frac{3}{2}})$ for nonconvex optimization. Math. Program. **162**(1), 1–32 (2017)
9. Dauphin, Y.N., Pascanu, R., Gulcehre, C., Cho, K., Ganguli, S., Bengio, Y.: Identifying and attacking the saddle point problem in high-dimensional non-convex optimization. In: Advances in Neural Information Processing Systems, pp. 2933–2941 (2014)
10. Du, S.S., Jin, C., Lee, J.D., Jordan, M.I., Singh, A., Poczos, B.: Gradient descent can take exponential time to escape saddle points. In: Advances in Neural Information Processing Systems, pp. 1067–1077 (2017)
11. Girosi, F., Jones, M., Poggio, T.: Regularization theory and neural networks architectures. Neural Comput. **7**(2), 219–269 (1995)
12. Reddi, S.J., Sra, S., Poczos, B., Smola, A.J.: Proximal stochastic methods for nonsmooth nonconvex finite-sum optimization. Adv. Neural Inf. Process. Syst. **29**, 1145–1153 (2016)
13. Jin, C., Ge, R., Netrapalli, P., Kakade, S.M., Jordan, M.I.: How to escape saddle points efficiently. In: ICML (2017)
14. Lee, J.D., Simchowitz, M., Jordan, M.I., Recht, B.: Gradient descent only converges to minimizers. In: Conference on Learning Theory, pp. 1246–1257 (2016)
15. Liu, Y., Yin, W.: An envelope for Davis-Yin splitting and strict saddle point avoidance. arXiv preprint arXiv:1804.08739 (2018)
16. Nesterov, Y.: A method for unconstrained convex minimization problem with the rate of convergence $\mathcal{O}(1/k^2)$. In: Doklady AN SSSR (translated as Soviet Math. Docl.), vol. 269, pp. 543–547 (1983)
17. Nesterov, Y., Polyak, B.T.: Cubic regularization of Newton method and its global performance. Math. Program. **108**, 177–205 (2006)
18. Shor, N.Z.: An application of the method of gradient descent to the solution of the network transportation problem. Materialy Naucnovo Seminara po Teoret i Priklad. Voprosam Kibernet. i Issted. Operacii, Nucnyi Sov. po Kibernet, Akad. Nauk Ukrain. SSSR, vyp **1**, 9–17 (1962)
19. Stella, L., Themelis, A., Patrinos, P.: Forward-backward Quasi-Newton methods for nonsmooth optimization problems. Comput. Optim. Appl. **67**(3), 443–487 (2017)
20. Xu, Y., Jin, R., Yang, T.: First-order stochastic algorithms for escaping from saddle points in almost linear time. arXiv preprint (2018). arXiv:1711.01944v3 [math.OC]
21. Zhu, Z., Li, Y.: Neon2: finding local minima via first-order oracles. arXiv preprint (2018). arXiv:1711.06673 [cs.LG]

Deep-Learning Domain Adaptation Techniques for Credit Cards Fraud Detection

Bertrand Lebichot[1]([✉])[ID], Yann-Aël Le Borgne[1][ID], Liyun He-Guelton[2],
Frédéric Oblé[2], and Gianluca Bontempi[1][ID]

[1] Machine Learning Group, Computer Science Department, Faculty of Sciences,
ULB, Université Libre de Bruxelles, City of Brussels, Belgium
bertrand.lebichot@ulb.ac.be
[2] High Processing & Volume, R&D, Worldline, Lyon, France
http://mlg.ulb.ac.be

Abstract. Although the incidence of credit card fraud is limited to a small percentage of transactions, the related financial losses may be huge. This demands the design of automatic Fraud Detection Systems (FDS) able to detect fraudulent transactions with high precision and deal with the heterogeneous nature of the fraudster behavior. Indeed, the nature of the fraud behavior may strongly differ according to the payment system (e.g. e-commerce or shop terminal), the country and the population segment. Given the high cost of designing data-driven FDSs, it is more and more important for transactional companies to reuse existing pipelines and adapt them to different domains and contexts: this boils down to a well-known problem of transfer learning.

This paper deals with deep transfer learning approaches for credit card fraud detection and focuses on transferring classification models learned on a specific category of transactions (e-commerce) to another (face-to-face). In particular we present and discuss two domain adaptation techniques in a deep neural network setting: the first one is an original domain adaptation strategy relying on the creation of additional features to transfer properties of the source domain to the target one and the second is an extension of a recent work of Ganin *et al.*

The two methods are assessed, together with three state-of-the-art benchmarks, on a five-months dataset (more than 80 million e-commerce and face-to face transactions) provided by a major card issuer.

Keywords: Fraud detection · Domain adaptation · Transfer learning

This work was supported by the DefeatFraud project funded by Innoviris (2017-R-49a). We thank this institution for giving us the opportunity to conduct both fundamental and applied research. B. Lebichot also thanks LouRIM, Université catholique de Louvain, Belgium for their support.

L. Oneto et al. (Eds.): INNSBDDL 2019, INNS 1, pp. 78–88, 2020.
https://doi.org/10.1007/978-3-030-16841-4_8

1   Introduction

Global card fraud losses amounted to 22.8 billion US dollar in 2017 and is foreseen to continue to grow [20]. In recent years, machine learning techniques appeared as an essential component of any detection approach dealing automatically with massive amounts of transactions [12]. The existing work showed however that a detection strategy needs to take into account some peculiarities of the fraud phenomenon [10,11]: unbalancedness (frauds are less than 1% of all transactions), concept drift (typically due to seasonal aspects and fraudster strategies) and the big data and streaming nature [4]. Disregarding those aspects might lead to high false alert rate, low detection accuracy or slow detection (see [1] for more details). As a result the design of an accurate Fraud Detection System (FDS) goes beyond the integration of some conventional off-the-shelf learning libraries and requires a deep understanding of the fraud phenomenon. This means that the reuse of existing FDS in new settings, like a new market (e.g. with a different fraud ratio) or a new payment system, is neither immediate nor straightforward.

This paper depicts transfer learning strategies [23] in the adaptation of existing and effective models to new domains. In particular we focus on the heterogeneous nature of the credit-card transactions, related to the physical presence of the card holder, which distinguishes between face-to-face (F2F) and e-commerce (EC) settings. Face-to-face transactions occur when the buyer and merchant have to physically meet in order to complete a purchase. In e-commerce (e.g. exchange of goods or services through a computer network, like internet) transactions can take place when the card holder is not physically with the merchant.

E-commerce fraud detection settings have been more studied in literature [3, 12,15,22,29] than face-to-face ones [21]. Though F2F frauds are typically less frequent because personal identification number (PIN) is often required, their impact is not negligible and worthy of consideration. Well-known examples of F2F frauds are due to *skimming* i.e. the criminal action retrieving the card holder information from the card magnetic strip.

The differences between the EC and F2F setting, notably in terms of different ratios of genuine vs. fraudulent transactions and different attitudes of the fraudsters, are reflected in different statistical properties of the two detection tasks. At the same time much of the detection process is similar, for instance in terms of feature representation of the transactions. It is therefore important for card issuer companies to understand how much of the modelling and design effort done in setting up an accurate detection system for EC (source domain) can be reused and transferred to F2F (target domain).

We first review the topic of transfer learning and presents some state-of-the art *domain adaptation* methods which can be used to transfer the knowledge learned during EC fraud detection to enhance the detection of F2F frauds. Then, it presents two original contributions: (i) the proposal of an original transfer strategy relying on the creation of ad-hoc features (related to the marginal and the conditional distribution) to improve the transfer from the source to the target domain, (ii) the customization of an existing adversarial deep-learning strategy (typically used in image recognition task) to the specific task of credit card fraud detection (e.g. taking into account issues of unbalancedness and concept drift).

To the best of our knowledge this is the first paper assessing the quality of transfer learning techniques for credit card fraud detection. Another specificity of the paper is the extensive assessment procedure carried out on a five-months real-life dataset obtained from the major credit card issuer in Belgium.

The rest of this paper is structured as follows: Sect. 2 introduces background and notation. Section 3 reviews related work. Section 4 details the methodological contributions of the paper. Experimental comparisons are presented and analyzed in Sect. 5 and Sect. 6 discusses the results. Section 7 concludes this paper.

2 Background and Notation

Transfer learning (TL) is a crucial aspect of real-world learning: for instance, learning to recognize apples might help to recognize pears, or knowing to play piano might help learning electric organ [23]. Suppose also that you trained a learning machine to label a website on the basis of existing websites. Transfer learning could help to adapt the learner to deal with brand new websites [25]. The rationale of transfer learning is to fill the gap between two learning tasks and to reuse as much as possible what was learned from the first task to better solve the second one. More precisely, given a source domain D_s and learning task T_s, a target domain D_t and learning task T_t, transfer learning aims to improve the learning of the target predictive function $f_t(\cdot)$ in D_t using knowledge in D_s and T_s where $D_s \neq D_t$ or $T_s \neq T_t$. Transfer learning requires at least a change in domains or in tasks [23,26]:

- A *domain* D is defined as a tuple $(\mathbf{X}, P_{\mathbf{X}}(\cdot))$, where $\mathbf{X}$ denotes the multivariate input and $P_{\mathbf{X}}(\cdot)$ its marginal probability distribution.
- Given a specific domain, a *task* T is defined as a tuple $(\mathbf{Y}, f(\cdot))$ where $\mathbf{Y}$ is the label space and $f(\cdot)$ summarizes the conditional dependency (either the regression function or conditional distribution).

Domain adaptation (DA) is a sub-field of transfer learning: there is a change in the domain $D_s \neq D_t$ but the task remains the same $T_s = T_t$. In this paper, the task is the same, fraud detection, but there is a change of domain. In this context, by *source* (with subscript s) we denote the original domain (e-commerce) while *target* (with subscript t) refers to the new domain (face-to-face).

The transaction dataset is a collection of n vectors $\mathbf{x}_i$ of size m where m is the number of features (or attributes). The features are identical in the source and the target domain but the marginal distribution changes between them. Finally we define $\mathbf{y}$ as the column vector containing the class labels (fraudulent or genuine) of the n transactions.

3   Related Work

Transfer learning and domain adaptation have been widely studied in the last 20 years but none, to our knowledge, were devoted to FDS. There are four main classes of DA techniques in literature [23–25] which are described below.

Instance weighting for covariate shift: the *covariate shift* scenario refers to a non-stationary environment, that is a setting where the input distribution changes while the conditional distribution remains the same [27]). This scenario may occur when the training data has been biased toward one region of the input space or is selected in a non-I.I.D. manner. Instance weighting methods aim to weight samples in the source domain to match the target domain. For example the paper [23] proposes to estimate weights of the source domain, trying to make the weighted distributions of both domains as similar as possible.

Self-labeling methods: they include unlabeled target domain samples in the training process, initialize their labels and then iteratively refine them. This is often done using Expectation-Maximization (EM) algorithms (for example TrAdaBoost [9]). Hard versions add samples with specific labels while others [28] assign label confidences when fitting the model. While efficient, the EM procedure can be really computationally intensive, especially with large datasets.

Feature representation methods: they aim to find a new feature representation for the data and belong to two categories. Distribution similarity approaches aim explicitly to make the source and target domain sample distributions similar, either by penalizing or removing features whose statistics vary between domains or by learning a feature space projection in which a distribution divergence statistic is minimized [16,17]. On the other hand, latent feature approaches aim to construct new features using (often unlabeled) source and target domain data or, more widely, to find an abstracted representation for domain-specific features [13,17]. We discuss an easy way [13] to do so in Sect. 4.

Cluster-based learning methods: these methods construct a graph where the labeled and unlabeled samples are nodes. Edge weights, between samples, are based on their similarity. Labels are then propagated according to the graphs (e.g. by means of graph-based classification). The main assumption is that samples connected by high-density paths are likely to have the same label if there is a high density path between them [17]. Also those methods may be highly computationally intensive, especially when working with large graphs.

Furthermore, deep neural networks methods (DNN) have been widely used for TL and DA: their multi-layer nature can capture the intricate non-linear representations of data, and provide useful level features for transfer learning [23]. Multitask learning [2] can be easily implemented by DNN, by training two or more related tasks with a network sharing inputs and hidden layers but having separate output layers. As far as domain adaptation is concerned, hidden layers trained by the source task can be reused on a different target task. For the target task model, only the last classification layer needs to be retrained, though any layer of the new model could be fine-tuned if needed [23]. In other configurations, the hidden parameters related to the source task can be used to initialize the target model [8]. Autoencoders can also be used to gradually changing the training distribution: In [7], a sequence of deep autoencoders are trained layer-by-layer, while gradually replacing source-domain samples with target-domain samples. In [18] the authors simply trains a single deep autoencoder for both domains. Finally, Ganin and al. used DNN in an adversarial way (we will discuss more extensively this approach in Sect. 4) to tackle domain adaptation [16].

Table 1. This table summarizes the five considered methods of this paper. More details about the strategy and parameters can be found in Sect. 4.

Acronym	Strategy	Train set	Test set	Selected parameter
BDNN	Baseline	F2F	F2F	-
NDNN	Naive	EC	F2F	-
FEDADNN	Imputation	EC + F2F	F2F	-
AugDNN	Add EC-related features	Extended F2F	Extended F2F	$\lambda = 10$
AdvDNN	Adversarial	EC + F2F	F2F	$n_{PC} = 2$

4   Transfer Learning Strategies for Fraud Detection

As discussed in the previous section, several approaches may be taken into account to transfer information related to a source task to a target one. Though in a realistic situation none or very few labeled training samples could be available in the target domain, for the sake of assessment, we consider here an experimental setting where training samples are available for both the source and the target tasks. The target dataset is splitted in a training and test portion. The test portion makes possible a sound paired assessment of all the considered strategies while the training portion enables us to assess how much improvement may derive from integrating the source dataset with (part of) the target dataset.

Table 1 lists the alternative strategies which differentiate in terms of training set (different combinations of source and target) and transfer methodology. To avoid biases related to the learning machine, all strategies share the same DNN topology composed of two fully connected hidden layers and implemented on Keras [6]. Based of preliminary results (not reported here), we set the number of neurons in the hidden layers n_h to 1.5 times the number of features (here, 37).

The five considered methods are (code on https://github.com/B-Lebichot):

- *BDNN:* this is our baseline DNN classifier ($n_h = 55$). The train and test data are composed of target (F2F) samples: no transfer learning in this baseline.
- *NDNN:* this is the naive strategy which simply consists in training the same DNN ($n_h = 55$) as BDNN on the source (EC) dataset and test it on the target (F2F) testset. This approach is also often considered in literature as a baseline [13] to assess the added value of a transfer learning strategy.

– *FEDADNN*: this is a basic feature representation method (Sect. 3) which uses three versions of the original feature set: a general version, a source-specific version and a target-specific version [13].
  Each source column-feature $\mathbf{x_s}$ is simply replaced by $\phi^s(\mathbf{x_s}) = \langle \mathbf{x_s}, \mathbf{x_s}, \mathbf{0} \rangle$ and each target column-feature $\mathbf{x_t}$ is replaced by $\phi^t(\mathbf{x_t}) = \langle \mathbf{x_t}, \mathbf{0}, \mathbf{x_t} \rangle$. Where $\mathbf{0}$ is a vector full of zeros. Where $\phi^s(\mathbf{x_s})$ is the source mapping and $\phi^s(\mathbf{x_t})$ is the target mapping (roughly this is $\langle$general, source-specific, target-specific$\rangle$).
  This strategy allows to express both domains in an extended feature space, imputing missing values. The augmented source data therefore contains only general and source-specific versions while the augmented target data contains both general and target-specific versions. Finally, ϕ^t is used to obtain the test set from the original target data. As the number of features tripled compared to BDNN/NDNN, we set $n_h = 165$.
– *AugDNN*: this is an original technique whose rationale is to use information from the source domain (e.g. conditional distribution, marginal input distribution) as additional features of the classifier (for both training set and test set). This strategy allows the classifier to learn from data how the relatedness [5] between source and target samples is associated to the classification output. The main difficulty is that such information is not explicitly available but can only be estimated. This is the reason why we fit both a classifier and a dimensionality reduction to extract from the source domain information about the conditional distribution and the input distribution (see Fig. 1(a) for an illustration). Other variants were studied (Gaussian mixture models, ...) but we select the more informative subset of features. As a result we add three new features (denoted *Aug1* to *Aug3*) to the original feature set (therefore $n_h = 60$) obtained as follows:
  • We train a regular DNN source classifier. The first feature (*Aug1*) is then the predicted activation value (i.e. estimated conditional probability) of the output neuron for each F2F sample. [13] used a similar idea but used the binary predicted value instead.
  • We build a principal component analysis (e.g. PCA [19]) on the source training set. The projections of the F2F transactions on the first two PCs ($n_{PC} = 2$) return *Aug2 and Aug3*.
  The aim of adding such features is to encode in the training set the relatedness between the source and target distributions, both from a marginal and conditionally dependent perspective.
– *AdvDNN*: this is an adaptation of Ganin *et al.* [16] approach to the fraud detection setting. The rationale is that the prediction model must use features that cannot discriminate between the source and target domains. The original approach was used for image recognition and combines a labeled source domain and unlabeled target domain while here both source and target are labeled. The method learns domain invariant features by jointly optimizing the feature layer from the label predictor (here genuine versus fraudulent) and the domain label (here F2F versus EC) predictor. The domain classifier

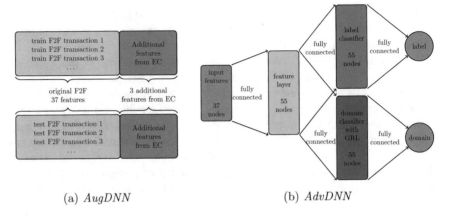

(a) *AugDNN*   (b) *AdvDNN*

Fig. 1. Illustration of methods *AugDNN* and *AdvDNN*. For (b), $n_h = 55$ and all successive layers are fully connected. Notice that removing the domain classifier reduces the network to the BDNN and NDNN baselines.

uses a gradient reversal layer (GRL) and a few fully connected layers [16]. The effect of the GRL is to multiply all domain-related gradients by a negative constant λ during back-propagation.

During the training, the feature layer is optimized to both minimize the label classifier loss and to maximize (due to GRL) the domain classifier loss. This approach promotes the emergence of features that are discriminative for the main learning task on the source domain and non-discriminative with respect to the domain tag [16]. A network illustration can be found on Fig. 1(b).

5 Experimental Comparisons

In this section, the different methods of Table 1 are compared on a real-life credit card transaction dataset obtained from our industrial partner.

The source database is composed of 37,882,367 e-commerce transactions (138 days: 85 training, 3 validation days and 50 test days) and 37 features. Validation days are used to tune Λ. The fraud ratio is 0.366%. The target database is composed of 47,619,852 face-to-face transactions (the same days and features as the source database). The fraud ratio is 0.033%.

The accuracy indicator is Precision@100 (Pr@100) (for details on such measure see [12]) which reports the number of true compromised cards among 100 investigated. The number 100 is chosen since this is compliant with the daily effort of the team of human investigators who manually check the transactions. For a deeper discussion on accuracy indicators for FDS, see [4,10,12,22]. Transaction-based precision is sometimes used instead of card-based precision: We obtained similar conclusions using transaction-based precision.

6   Discussion

Figures 2 and 3 compare methods described on Table 1 through a Friedman/ Nemenyi test [14] for cards-based Pr@100. We adopt a sliding window approach (see Subsect. 5). Friedman null hypothesis is rejected with $\alpha = 0.05$ and Nemenyi critical difference is equal to 0.837. A method is considered as significantly better than another if its mean rank is larger by more than this amount.

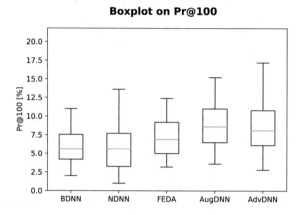

Fig. 2. Boxplots representing the card-based precision@100 for each method (50 days with one precision score per day). Notice that the a priori fraud probability is only 0.033%. The largest relative mean increase (NDNN versus AugDNN) is +59%.

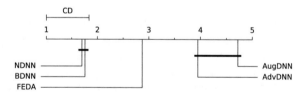

Fig. 3. Mean rank (the higher, the better) and critical difference (CD) of the Friedman/Nemenyi test for the five methods. A method is considered as significantly better than another if its mean rank is larger by more than the critical difference $CD = 0.838$.

From Fig. 2, it appears that the F2F detection accuracy is much lower than the EC one. In previous works [4, 12] we showed that the EC Pr@100 can attain 50% when no domain adaptation is involved. This is explained by the fact that the F2F a priori fraud probability is ten times lower than for EC.

We also observe that BDNN and NDNN (the two baselines) are not such different, showing that both tasks are actually related (this is confirmed by the Friedman/Nemenyi test on Fig. 3). However, the three considered domain adaptation methods increase the accuracy with respect to the baselines.

From Fig. 3, the FEDADNN algorithm is clearly less accurate than the AugDNN and AdvDNN approach. Those two last methods cannot be significantly discriminated on our data by the Friedman/Nemenyi test. However, a Wilcoxon test between AugDNN and AdvDNN indicates that AugDNN significantly outperforms with $\alpha < 0.05$. Overall, AugDNN could be considered as the best method emerging from our experimental assessment.

All experiments were carried on a server with 10 cores, 256 GB RAM and three Asus GTX 1080 TI. These are the related execution times (feature manipulation and classification only): BDNN, NDNN, FEDADNN, AugDNN, and AdvDNN runs in 12, 16, 23, 41, and 46 min, respectively. Note that accuracy comes at the price of increase execution time and that AugDNN is slightly superior to AdvDNN both in terms of accuracy and run time.

7   Conclusion

The paper studied the use of domain adaptation strategies in transaction-based fraud detection systems. In particular, we considered e-commerce transactions as the source domain and face-to-face transactions as the target domain.

We introduced two original methods. The first adds e-commerce related features (both predictive and distribution-related) to the face-to-face transactions to improve predictions. The second is an adaptation of the work from [16]. This method learns domain invariant features by jointly optimizing the underlying feature layer from the fraud tag predictor (here genuine versus fraudulent) and the domain tag (face-to-face versus e-commerce).

Those two methods, and three others, were tested on a five-months (more than 80 millions of transactions) real-life e-commerce and face-to-face credit card transaction dataset obtained from a large credit card issuer in Belgium. The second proposed method outperforms all the considered approaches and the first proposed method comes second (in terms of performance and run time). Those results are shown to be significant using statistical tests.

Future work will focus on scenarios characterized by a low ratio of labeled transactions in the target domain. Better density estimation (using for example Gaussians mixture models) for the source distribution of our first method (AugDNN) will also be studied.

References

1. Abdallah, A., Maarof, M.A., Zainal, A.: Fraud detection system. J. Netw. Comput. Appl. **68**, 90–113 (2016)
2. Ahmed, A., Yu, K., Xu, W., Gong, Y., Xing, E.: Training hierarchical feed-forward visual recognition models using transfer learning from pseudo-tasks. In: ECCV (3), pp. 69–82 (2008)
3. Bolton, R., Hand, D.: Statistical fraud detection: a review. Stat. Sci. **17**, 235–249 (2002)

4. Carcillo, F., Dal Pozzolo, A., Le Borgne, Y.A., Caelen, O., Mazzer, Y., Bontempi, G.: SCARFF: a scalable framework for streaming credit card fraud detection with spark. Inf. Fusion **41**(C), 182–194 (2018)
5. Caruana, R.: Multitask learning. Mach. Learn. **28**(1), 41–75 (1997)
6. Chollet, F., et al.: Keras (2015). https://keras.io
7. Chopra, S., Balakrishnan, S., Gopalan, R.: DLID: deep learning for domain adaptation by interpolating between domains. In: ICML Workshop on Challenges in Representation Learning (2013)
8. Ciresan, D.C., Meier, U., Schmidhuber, J.: Transfer learning for Latin and Chinese characters with deep neural networks. In: IJCNN, pp. 1–6. IEEE (2012)
9. Dai, W., Yang, Q., Xue, G.R., Yu, Y.: Boosting for transfer learning. In: Proceedings of the 24th International Conference on Machine Learning, ICML 2007, pp. 193–200. ACM (2007)
10. Dal Pozzolo, A., Boracchi, G., Caelen, O., Alippi, C., Bontempi, G.: Credit card fraud detection and concept-drift adaptation with delayed supervised information. In: Proceedings of the International Joint Conference on Neural Networks, pp. 1–8. IEEE (2015)
11. Dal Pozzolo, A., Boracchi, G., Caelen, O., Alippi, C., Bontempi, G.: Credit card fraud detection: a realistic modeling and a novel learning strategy. IEEE Trans. Neural Netw. Learn. Syst. **29**(8), 3784–3797 (2018)
12. Dal Pozzolo, A., Caelen, O., Le Borgne, Y.A., Waterschoot, S., Bontempi, G.: Learned lessons in credit card fraud detection from a practitioner perspective. Expert. Syst. Appl. **10**(41), 4915–4928 (2014)
13. Daume III, H.: Frustratingly easy domain adaptation. In: Proceedings of the 45th Annual Meeting of the Association of Computational Linguistics, pp. 256–263. Association for Computational Linguistics, Prague, Czech Republic, June 2007
14. Demsar, J.: Statistical comparaison of classifiers over multiple data sets. J. Mach. Learn. Res. **7**, 1–30 (2006)
15. Fawcett, T., Provost, F.: Adaptive fraud detection. Data Min. Knowl. Discov. **1**, 291–316 (1997)
16. Ganin, Y., Ustinova, E., Ajakan, H., Germain, P., Larochelle, H., Laviolette, F., Marchand, M., Lempitsky, V.: Domain-adversarial training of neural networks. J. Mach. Learn. Res. **17**(1), 2096–2030 (2016)
17. Gao, J., Fan, W., Jiang, J., Han, J.: Knowledge transfer via multiple model local structure mapping. In: Proceedings of the 14th ACM SIGKDD International Conference on Knowledge Discovery and Data Mining, KDD 2008, pp. 283–291. ACM, New York (2008)
18. Glorot, X., Bordes, A., Bengio, Y.: Domain adaptation for large-scale sentiment classification: a deep learning approach. In: Proceedings of the Twenty-eight International Conference on Machine Learning, ICML (2011)
19. Hastie, T., Tibshirani, R., Friedman, J.: The Elements of Statistical Learning, 2nd edn. Springer, New York (2009)
20. HSN Consultants, Inc.: The Nilson report (consulted on 2018-10-23) (2017). https://nilsonreport.com/upload/content_promo/The_Nilson_Report_Issue_1118.pdf
21. Jurgovsky, J., Granitzer, M., Ziegler, K., Calabretto, S., Portier, P.E., He, L., Caelen, O.: Sequence classification for credit-card fraud detection. Expert. Syst. Appl. **100**, 234–245 (2018)
22. Lebichot, B., Braun, F., Caelen, O., Saerens, M.: A graph-based, semi-supervised, credit card fraud detection system, pp. 721–733. Springer, Cham (2017)

23. Lu, J., Behbood, V., Hao, P., Zuo, H., Xue, S., Zhang, G.: Transfer learning using computational intelligence: a survey. Knowl. Based Syst. **80**, 14–23 (2015)
24. Margolis, A.: A literature review of domain adaptation with unlabeled data. Technical report, University of Washington (2011)
25. Pan, S.J., Yang, Q.: A survey on transfer learning. IEEE Trans. Knowl. Data Eng. **22**(10), 1345–1359 (2010)
26. Pan, S.J., Yang, Q., et al.: A survey on transfer learning. IEEE Trans. Knowl. Data Eng. **22**(10), 1345–1359 (2010)
27. Sugiyama, M., Kawanabe, M.: Machine Learning in Non-Stationary Environments: Introduction to Covariate Shift Adaptation. The MIT Press, Cambridge (2012)
28. Tan, S., Cheng, X., Wang, Y., Xu, H.: Adapting Naive Bayes to domain adaptation for sentiment analysis. In: Proceedings of the 31th European Conference on IR Research on Advances in Information Retrieval, ICML 2009, pp. 337–349. Springer (2009)
29. Van Vlasselaer, V., Bravo, C., Caelen, O., Eliassi-Rad, T., Akoglu, L., Snoeck, M., Baesens, B.: APATE: a novel approach for automated credit card transaction fraud detection using network-based extensions. Decis. Support Syst. **75**, 38–48 (2015)

Selective Information Extraction Strategies for Cancer Pathology Reports with Convolutional Neural Networks

Hong-Jun Yoon[✉], John X. Qiu, J. Blair Christian, Jacob Hinkle,
Folami Alamudun, and Georgia Tourassi

Biomedical Sciences, Engineering and Computing Group,
Health Data Sciences Institute, Oak Ridge National Laboratory,
Oak Ridge, TN 37831, USA
yoonh@ornl.gov

Abstract. To trust model predictions, it is important to ensure new data scored by the model comes from the same population used for model training. If the model is used to score new data different than the model's training data, then predictions and model performance metrics cannot be trusted. Identifying and excluding these anomalous data points is an important task when using models in the real world. Traditional machine learning algorithms and classifiers don't have the capability to abstain in this case. Here we propose a data-novelty detection algorithm for the Convolutional Neural Network classifier, yielding a rejection score for each new data point scored. It is a post-modeling procedure which examines the distribution of convolution filters to determine if the prediction should be trusted. We apply this algorithm to an information extraction model for a natural language text corpus. We evaluated the algorithm performance using a primary cancer site classification model applied to cancer pathology reports. Results demonstrate that the algorithm is an effective way to exclude cancer pathology reports from model scoring when they do not contain the expected information necessary to accurately classify the primary cancer type.

Keywords: Novelty detection · Uncertainty determination · Cancer pathology reports · Information extraction · Natural language processing · Convolutional neural networks

This manuscript has been authored by UT-Battelle, LLC under Contract No. DE-AC05-00OR22725 with the U.S. Department of Energy. The United States Government retains and the publisher, by accepting the article for publication, acknowledges that the United States Government retains a non-exclusive, paid-up, irrevocable, world-wide license to publish or reproduce the published form of the manuscript, or allow others to do so, for United States Government purposes. The Department of Energy will provide public access to these results of federally sponsored research in accordance with the DOE Public Access Plan (http://energy.gov/downloads/doe-public-access-plan).

© Springer Nature Switzerland AG 2020
L. Oneto et al. (Eds.): INNSBDDL 2019, INNS 1, pp. 89–98, 2020.
https://doi.org/10.1007/978-3-030-16841-4_9

1 Introduction

In 2018, the estimated cancer prevalence in the United States was 1,735,350 cases [19]. Much of the data used to report cancer prevalence is in the form of individualized and highly detailed natural language text reports. Using humans to read and annotate country-level volumes of data is very costly and can be error-prone due to subjective interpretation or even human fatigue. Thus, algorithmic approaches to automate information extraction from cancer pathology reports have been researched to provide a consistent, fast, and scalable approach to process large-scale data to support cancer reporting and epidemiology research. Much research into algorithmic information extraction has been conducted [8,13,14], and we contributed a Deep Learning (DL)-based approach [16].

Conventionally, neural networks for classification tasks terminate with a softmax layer which first estimates a log-likelihood value for a set of predetermined classes, then normalize all predictions into probability space [10]. An implicit assumption is that conventional metrics such as F-1 score applied to validation set predictions should generalize to a real-world production environment [2,17]. To evaluate this assumption, Statistical Machine Learning have developed techniques ranging from conformal prediction for providing error bounds per case [15,17] to uncertainty quantification for assessing predictions as true probability distributions [18].

Deep learning improves upon prior statistical approaches by automatically learning to transform data into latent knowledge representations [11,16]. Although powerful and robust, such representations lack an explicit interpretation to the original data, thereby resulting in difficulty assessing the appropriateness of new prediction data. For example, submitting a cat image to a dog-breed classification model would return a corresponding breed despite the model's inappropriateness. Though the uncertainty quantification literature has explored the downstream impact of feature variance, such efforts were limited to fully parametric statistical models and do not generalize to latent neural features [5,6]. Although prior DL research has attempted to address latent feature variation [4,7,20], there has yet to be any evaluation of representations from out-of-scope cases.

There are two situations in clinical data in particular that impact classification performance. The first situation is the frequent presence of under-represented classes with relatively little training data. This imbalance often occurs in real-world observational data; for example, the American Cancer Association [19], estimated there would be 234,030 new cases of lung and bronchus cancers in 2018, 3,540 cases of eye and eye orbit cancers. Though larger datasets enable the training of models with many more parameters and increased learning capacity, under-represented classes may not be fully learned by the network, resulting in an inability to identify these rare cases in real-world applications. The second situation that may impact classification performance is that individual reports may be incomplete or lack the informative content for a reasonable classification, such as documents consisting of only an addendum without a report body. Population-scale data means incomplete or inappropriate data is inevitable, so developing a systematic method to recognize such data is necessary to trust model predictions.

In this paper, we develop an algorithm to determine when DL models are exposed to novel cases by assessing the network's intermediate layer responses and comparing them to prior uncertainty quantification techniques. This algorithm can be utilized to flag these novel cases for human annotation. Specifically, we extend the Convolutional Neural Network for information extraction from pathology reports to quantify the novelty of input data relative to the training set. We performed the studies with a cancer pathology report corpus provided by the CT, HI, KY, NM, and Seattle cancer registries. In Sect. 2, we summarize the characteristics of the corpus, DL models, and algorithms utilized. In Sect. 3, we detail our experimental results, and in Sect. 4 we provide our discussion and conclusion.

2 Methods

2.1 Cancer Pathology Data Corpus

The corpus in this study consists of 2,505 deidentified cancer pathology reports provided from five different Surveillance, Epidemiology, and End Results (SEER) cancer registries (CT, HI, KY, NM, Seattle) with the proper IRB-approved protocol. From 1,197 human-annotated pathology reports, we used 942 reports labeled with one of 12 primary cancer subsites as detailed in the third edition of the International Classification of Diseases for Oncology (ICD-O-3) [12]. These labels correspond to seven breast and five lung subsites, and comprise the model's training data.

Gold-standard subsite labels were constructed by cancer registry experts who manually annotated the labeled set using SEER coding guidelines. For consistency, our training set used only pathology reports with a single topography code in the Final Diagnosis section of the report to minimize variation in our training data. Since pathology report sections vary across pathology labs and registries, we aggregated the text content of every section in the pre-processing phase.

To assess our algorithm's performance on data novel to the model's training data, we used 1,308 pathology reports outside the training data; 472 were colon and rectal cancers, 409 were prostate cancer, and 427 had unknown cancer types. A summary of the numbers and types of pathology is listed in Table 1.

2.2 Convolutional Neural Networks for Natural Text Data

Convolutional Neural Networks (CNN) for natural language processing [9] use learned latent representations of words for document representation known as word vectors. We tokenized each document on the word and non-alphanumeric symbol, and formed an $l \times k$ document matrix A, where l is the number of documents and k is the length of the word vector. We applied a convolution

Table 1. Label names, label descriptions, and counts of cases annotated with the label in the dataset.

Site	Description	# cases
C34.0	Main bronchus	26
C34.1	Upper lobe of lung	139
C34.2	Middle lobe of lung	11
C34.3	Lower lobe of lung	78
C34.9	Lung, NOS	191
C50.1	Central portion of breast	13
C50.2	Upper-Inner quadrant of breast	36
C50.3	Lower-Inner quadrant of breast	10
C50.4	Upper-Outer quadrant of breast	63
C50.5	Lower-Outer quadrant of breast	21
C50.8	Overlapping lesion of breast	62
C50.9	Breast, NOS	292
Novel	Cancer site other than breast or lung	1,308

filter w of size $k \times h$, where h is the length of convolution filter, to the document matrix followed by the Rectified Linear Unit (ReLU) activation function. We then applied global max-pooling, resulting in a scalar value corresponding to the maximum filter response given a particular document matrix,

$$a_n = \text{GlobalMaxPooling}(\text{ReLU}(A \cdot w_n)), \text{ for } n = 1, \cdots, N, \tag{1}$$

where N is the number of convolution filters. The filter responses were then concatenated and fed through a penultimate fully-connected hidden layer and a softmax classification layer,

$$\sigma(y)_i = \frac{e^{y_i}}{\sum_{k=1}^{K} e^{y_k}}, \text{ for } i = 1, \cdots, K, \tag{2}$$

where K is the number of class labels.

The classification performance scores of the CNN applied to the breast and lung training data were Micro-$F_1 = 0.772$ and Macro-$F_1 = 0.551$, resulting from 10-fold cross-validation. However, using the whole dataset consisting of both the breast and lung training data combined with the novel cancer reports, the scores decreased to Micro-$F_1 = 0.324$ and Macro-$F_1 = 0.407$, because the CNN classifier had no labels for cancer types novel to the training data, thus incorrectly classifying all cancers not breast or lung.

2.3   Selective Classification Strategies

To study how the CNN's network behaves when exposed to data novel to its training data, we designed the following algorithm to identify pathology report

data for cancer types untrained by the CNN model using two selective classification strategies. First, we used an uncertainty quantification approach based on
the model's softmax classification scores. Second, we used our proposed metric
based on the response of our network's convolutional filters.

Softmax-Based Uncertainty Determination. Networks utilized for classification tasks train a final layer to first predict class-specific logits, then applying
the softmax function to normalize predictions into a probability. This results
in predictions for class i as $\sigma_i \in (0,1)$ where $\sum_{i=1}^{K} \sigma_i = 1$. We can determine
the confidence of the classifier's prediction by ensuring the max class prediction
$s = \max(\sigma_i)$, is greater than a given acceptance threshold θ_s.

Novelty Detection. Convolutional filters in the CNN are trained to encode
features from words and phrases relevant to the training corpus [9]. Our approach
explores whether convolution filters produce a systematically different response
if the input is novel to the model's training data. We define a measurement of
relevance v as the number of convolution responses, a_n, greater than a threshold
θ_a, such that $v = \sum_{n=1}^{N} [a_n > \theta_a]$. Values of the statistic v for the training data
and training+novel data are in Fig. 1. Our novelty detection algorithm classifies
input data as "novel" if v is smaller than a given acceptance threshold θ_v.

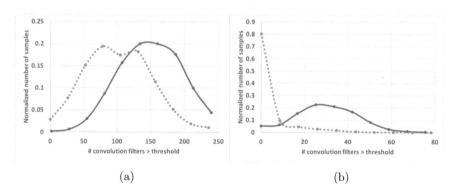

(a)   (b)

Fig. 1. Plots of the distribution of the number of convolution filters exceeding a threshold for training data consisting of lung and breast cancers (blue solid line) and for
cancers other than breast and lung (orange dotted line). The left panel uses a filter
threshold of (a) $\theta_a = 0.2$ and the right panel uses a threshold of (b) $\theta_a = 1.0$.

2.4   10-Fold Cross-Validation Study of the Novelty Detection Algorithm

In this study, there are two metrics of interest, (1) the ability of the algorithm
to identify novel data, and (2) after excluding novel data, the ability of the
model to correctly predict class labels. To get unbiased class predictions for (2)

predicting class labels, we used 10-fold Cross-Validation (CV). In order to use the same model for both metrics we partitioned the reports other than breast or lung cancer into 10 folds of approximately equal size and added them to the testing sets.

We determined our experimental thresholds θ_s and θ_u by identifying the 10th, 20th, 30th, 40th, and 50th percentiles of the softmax values s_k, and of the novelty metrics u_k, where $k \in \{1, \ldots, K\}$, where K is the number of lung and breast cancer subsites, 12 in this study. To determine thresholds for θ_a utilized in our novelty detection algorithm, we compare values of θ_a from 0.2 to 1.8, with a step size of 0.2.

To determine a threshold for the first metric of interest, identification of novel data, we use the correct and incorrect rejection ratio by the selective classification algorithms,

$$
R_s^{cor} = \sum_{k=1}^{N_u} [s_k^u < \theta_s]/N_u, \quad R_s^{inc} = \sum_{k=1}^{N_a} [s_k^a < \theta_s]/N_a,
$$
$$
R_v^{cor} = \sum_{k=1}^{N_u} [v_k^u < \theta_u]/N_u, \text{ and } R_v^{inc} = \sum_{k=1}^{N_a} [v_k^a < \theta_u]/N_a,
$$
(3)

where N_u is the number of pathology reports other than breast or lung cancer, and N_a is the number of breast and lung cancer annotated pathology reports in the corpus. To evaluate the second metric of interest, model prediction accuracy, we use both Micro- and Macro-averaged F_1 scores for the CNN classifier denoted F_1^{mic} and F_1^{mac}.

Intuitively, we must adjust the thresholds to balance the statistical type I and type II error rates. By increasing the thresholds θ_s and θ_v, we may increase correct rejection rates, R_s^{cor} and R_v^{cor}, meaning the algorithm correctly rejected more novel samples, but at the same time it would increase R_s^{inc} and R_v^{inc}, meaning we incorrectly reject more breast and lung cancer cases.

3   Results

We implemented the CNN model and additive functions for novelty detection with the Keras [3] and TensorFlow [1] backend on Python 3 environment. We applied the same 10-fold splits to both of the algorithms. Table 2 lists F_1 scores and the rejection rates R from the algorithms.

We observed that our algorithm performed optimally with a threshold of $\theta_a = 1.2$. We observed the 10th percentile of θ_u successfully rejected 95.9% of the reports other than breast or lung cancer, while incorrectly rejecting only 9.4% of the breast or lung cancer reports, which resulted in improved F_1 scores over our baseline results. Increasing θ_u further resulted in a sub-optimal trade-off of improved rejection of reports other than breast or lung cancer (up to 98.8%) with an increased percentage of incorrectly identified breast and lung reports (up to 48.3%).

Table 2. Clinical task performance scores (F_1^{mic} and F_1^{mac}), correct rejection rates (R_u^{cor}), and incorrect rejection rates (R_u^{inc}) from the novelty detection algorithm to the cancer pathology corpus, with respect to the convolution filter response threshold θ_a and the acceptance threshold θ_u.

$\theta_a = 0.2$				$\theta_a = 0.4$				$\theta_a = 0.6$			
θ_u F_1^{mic}	F_1^{mac}	R_u^{cor}	R_u^{inc}	F_1^{mic}	F_1^{mac}	R_u^{cor}	R_u^{inc}	F_1^{mic}	F_1^{mac}	R_u^{cor}	R_u^{inc}
10 0.447	0.443	0.530	0.116	0.542	0.465	0.718	0.116	0.653	0.498	0.873	0.125
20 0.481	0.455	0.651	0.203	0.581	0.484	0.815	0.223	0.681	0.503	0.923	0.227
30 0.503	0.454	0.741	0.300	0.600	0.484	0.874	0.324	0.687	0.500	0.947	0.322
40 0.525	0.455	0.812	0.403	0.632	0.485	0.926	0.413	0.684	0.489	0.966	0.418
50 0.550	0.448	0.865	0.487	0.656	0.467	0.956	0.505	0.701	0.493	0.979	0.517

$\theta_a = 0.8$				$\theta_a = 1.0$				$\theta_a = 1.2$			
θ_u F_1^{mic}	F_1^{mac}	R_u^{cor}	R_u^{inc}	F_1^{mic}	F_1^{mac}	R_u^{cor}	R_u^{inc}	F_1^{mic}	F_1^{mac}	R_u^{cor}	R_u^{inc}
10 0.724	0.510	0.946	0.116	0.734	0.507	0.955	0.101	0.735	0.501	0.959	0.094
20 0.728	0.512	0.961	0.195	0.744	0.509	0.975	0.193	0.748	0.512	0.979	0.192
30 0.721	0.513	0.969	0.310	0.744	0.513	0.981	0.287	0.753	0.522	0.985	0.288
40 0.711	0.501	0.979	0.416	0.733	0.508	0.985	0.412	0.736	0.509	0.985	0.377
50 0.722	0.510	0.987	0.515	0.735	0.516	0.989	0.519	0.741	0.523	0.988	0.483

$\theta_a = 1.4$				$\theta_a = 1.6$				$\theta_a = 1.8$			
θ_u F_1^{mic}	F_1^{mac}	R_u^{cor}	R_u^{inc}	F_1^{mic}	F_1^{mac}	R_u^{cor}	R_u^{inc}	F_1^{mic}	F_1^{mac}	R_u^{cor}	R_u^{inc}
10 0.641	0.479	0.853	0.074	0.361	0.421	0.192	0.019	0.324	0.407	0.000	0.000
20 0.749	0.513	0.977	0.199	0.744	0.515	0.969	0.159	0.341	0.409	0.097	0.012
30 0.749	0.513	0.984	0.271	0.750	0.508	0.980	0.247	0.452	0.442	0.587	0.160
40 0.742	0.498	0.989	0.386	0.748	0.504	0.987	0.358	0.740	0.468	0.983	0.333
50 0.745	0.478	0.990	0.439	0.731	0.454	0.991	0.485	0.750	0.461	0.991	0.456

We repeated the same 10-fold CV tests but with our identified softmax threshold, and the results are in Table 3. Note that there is only one threshold θ_s involved in the algorithm.

Compared to the optimal incorrect rejection rate results from Table 2 of $R_u^{inc} = 0.192$, we observed a similar incorrect rejection rate in Table 3 of $R_s^{inc} = 0.210$. Our novelty detection measure yielded a 97.9% correct rejection rate while the softmax method only correctly rejected 75.5%. We consistently observed that the novelty detection performed better than the softmax method.

However, comparing a situation where the softmax algorithm incorrectly rejects more, at a threshold of $\theta_s = 40$ which yields a $R_s^{inc} = 0.548$, than our algorithm, which yields $R_u^{inc} = 0.483$ at a threshold of $\theta_u = 50$, the softmax method scored substantially higher Micro- and Macro- F_1 scores than those from the novelty detection method (Micro-F_1: 0.879 versus 0.741, Macro-F_1: 0.578 versus 0.523). This shows that while the softmax algorithm may be classifying better than our algorithm, it is at the expense of incorrectly rejecting breast and lung cancer reports at a higher rate.

Table 3. Clinical task performance scores (F_1^{mic} and F_1^{mac}), correct rejection rates (R_s^{cor}), and incorrect rejection rates (R_s^{inc}) from the softmax-based algorithm to the cancer pathology corpus, with respect to the acceptance threshold θ_s.

θ_s	F_1^{mic}	F_1^{mac}	R_s^{cor}	R_s^{inc}
10	0.595	0.485	0.755	0.210
20	0.769	0.517	0.926	0.331
30	0.846	0.521	0.973	0.447
40	0.879	0.578	0.982	0.548
50	0.892	0.451	0.986	0.632

4   Discussion

In this study, we compared two selective classification strategies to identify data not similar to a model's training data and evaluated their performance using rejection rates as well as clinical task performance on information extraction from a cancer pathology report corpus. The results demonstrated that the algorithms successfully identified data samples not found in the training set and by excluding them from model scoring, were able to increase the classification performance score. However, some breast and lung cancer reports from the training set were incorrectly excluded.

The novelty detection measure has higher correct rejection rates than the softmax-based algorithm. However, the softmax-based method had substantially higher F_1 scores than the novelty detection method. We posit that the novelty detection method verifies that the data sample possesses relevant vocabularies and keywords, which itself does not guarantee higher accuracy from the CNN model. On the contrary, the softmax is the measure of the model's prediction confidence, but lower confidence does not always imply that the data is irrelevant to the task. It suggests that a combination of these two algorithms may boost prediction performance while excluding novel data entries effectively, which is a future research topic.

Acknowledgment. This work has been supported in part by the Joint Design of Advanced Computing Solutions for Cancer (JDACS4C) program established by the U.S. Department of Energy (DOE) and the National Cancer Institute (NCI) of the National Institutes of Health. This work was performed under the auspices of the U.S. Department of Energy by Argonne National Laboratory under Contract DE-AC02-06-CH11357, Lawrence Livermore National Laboratory under Contract DE-AC52-07NA27344, Los Alamos National Laboratory under Contract DE-AC5206NA25396, and Oak Ridge National Laboratory under Contract DE-AC05-00OR22725.

The authors wish to thank Valentina Petkov of the Surveillance Research Program from the National Cancer Institute and the SEER registry at Connecticut, Hawaii, Kentucky, New Mexico and Seattle for the pathology reports used in this investigation.

This research used resources of the Oak Ridge Leadership Computing Facility at the Oak Ridge National Laboratory, which is supported by the Office of Science of the U.S., Department of Energy under Contract No. DE-AC05-00OR22725.

References

1. Abadi, M., Barham, P., Chen, J., Chen, Z., Davis, A., Dean, J., Devin, M., Ghemawat, S., Irving, G., Isard, M., et al.: Tensorflow: a system for large-scale machine learning. OSDI **16**, 265–283 (2016)
2. Amodei, D., Olah, C., Steinhardt, J., Christiano, P., Schulman, J., Mané, D.: Concrete problems in AI safety. arXiv preprint arXiv:1606.06565 (2016)
3. Chollet, F., et al.: Keras (2015). https://github.com/fchollet/keras
4. Deng, Y., Bao, F., Deng, X., Wang, R., Kong, Y., Dai, Q.: Deep and structured robust information theoretic learning for image analysis. IEEE Trans. Image Process. **25**(9), 4209–4221 (2016)
5. Gelman, A., Stern, H.S., Carlin, J.B., Dunson, D.B., Vehtari, A., Rubin, D.B.: Bayesian Data Analysis. Chapman and Hall/CRC, Berkeley (2013)
6. Goodman, L.A.: On the exact variance of products. J. Am. Stat. Assoc. **55**(292), 708–713 (1960)
7. Goroshin, R., Mathieu, M.F., LeCun, Y.: Learning to linearize under uncertainty. In: Advances in Neural Information Processing Systems, pp. 1234–1242 (2015)
8. Kavuluru, R., Hands, I., Durbin, E.B., Witt, L.: Automatic extraction of ICD-O-3 primary sites from cancer pathology reports. In: AMIA Summits on Translational Science Proceedings 2013, p. 112 (2013)
9. Kim, Y.: Convolutional neural networks for sentence classification. arXiv preprint arXiv:1408.5882 (2014)
10. Krizhevsky, A., Sutskever, I., Hinton, G.E.: ImageNet classification with deep convolutional neural networks. In: Advances in Neural Information Processing Systems, pp. 1097–1105 (2012)
11. LeCun, Y., Bengio, Y., Hinton, G.: Deep learning. Nature **521**(7553), 436 (2015)
12. Louis, D.N., Ohgaki, H., Wiestler, O.D., Cavenee, W.K., Burger, P.C., Jouvet, A., Scheithauer, B.W., Kleihues, P.: The 2007 WHO classification of tumours of the central nervous system. Acta Neuropathol. **114**(2), 97–109 (2007)
13. Meystre, S.M., Savova, G.K., Kipper-Schuler, K.C., Hurdle, J.F.: Extracting information from textual documents in the electronic health record: a review of recent research. Yearb. Med. Inform. **17**(01), 128–144 (2008)
14. Nguyen, A., Moore, J., Lawley, M., Hansen, D., Colquist, S.: Automatic extraction of cancer characteristics from free-text pathology reports for cancer notifications. Stud. Health Technol. Inform. **168**, 117–124 (2011)
15. Papadopoulos, H.: Inductive conformal prediction: theory and application to neural networks. In: Tools in Artificial Intelligence. InTech (2008)
16. Qiu, J.X., Yoon, H.J., Fearn, P.A., Tourassi, G.D.: Deep learning for automated extraction of primary sites from cancer pathology reports. IEEE J. Biomed. Health Inform. **22**(1), 244–251 (2018)
17. Shafer, G., Vovk, V.: A tutorial on conformal prediction. J. Mach. Learn. Res. **9**, 371–421 (2008)

18. Smith, R.C.: Uncertainty Quantification: Theory, Implementation, and Applications, vol. 12. SIAM, Philadelphia (2013)
19. American Cancer Society: Cancer facts & figures. The Society (2018)
20. Yoon, H.J., Robinson, S., Christian, J.B., Qiu, J.X., Tourassi, G.D.: Filter pruning of convolutional neural networks for text classification: a case study of cancer pathology report comprehension. In: 2018 IEEE EMBS International Conference on Biomedical & Health Informatics (BHI), pp. 345–348. IEEE (2018)

An Information Theoretic Approach to the Autoencoder

Vincenzo Crescimanna$^{(\boxtimes)}$ and Bruce Graham

University of Stirling, Stirling, UK
`vincenzo.crescimanna1@stir.ac.uk`

Abstract. We present a variation of the Autoencoder (AE) that explicitly maximizes the mutual information between the input data and the hidden representation. The proposed model, the InfoMax Autoencoder (IMAE), by construction is able to learn a robust representation and good prototypes of the data. IMAE is compared both theoretically and then computationally with the state of the art models: the Denoising and Contractive Autoencoders in the one-hidden layer setting and the Variational Autoencoder in the multi-layer case. Computational experiments are performed with the MNIST and Fashion-MNIST datasets and demonstrate particularly the strong clusterization performance of IMAE.

Keywords: Infomax · Autoencoder · Representation learning

1 Introduction

Nowadays, Deep Neural Networks (DNNs) are considered the best machine learning structure for many different tasks. The principal property of this family is the ability to learn a representation of the input data in the hidden layers. Indeed, as underlined in [4], in order to understand the world, an intelligent system needs to learn a compact representation of the high dimensional input data.

In biological Neural Networks (NNs) processing sensory information, it is hypothesised that the first layers encode a task-independent representation that maximizes the mutual information with the input [2]. Given that it is desirable to learn a representation that is able to identify and disentangle the underlying explanatory hidden factors in input data, and inspired by the behaviour of real NNs, we propose an unsupervised Infomax Autoencoder (IMAE) NN that maximizes the mutual information between the encoded representation and the input. The results of computational experiments show that IMAE is both robust to noise and able to identify the principal features of the data.

The paper is structured as follows. In the next section we describe related work. Section 3 describes the derivation of the model, which is based on a classic approximation of differential entropy. Due to comparison with other AEs, we give also a geometric and a generative description of the model. The theoretical differences are confirmed in the fourth section by computational experiments, where

© Springer Nature Switzerland AG 2020
L. Oneto et al. (Eds.): INNSBDDL 2019, INNS 1, pp. 99–108, 2020.
https://doi.org/10.1007/978-3-030-16841-4_10

the clusterization (representation) and the robustness of IMAE are evaluated. The work is concluded with a brief comment on the results and a description of future work.

2   Related Work

The first neural network defined to explicitly maximize the information between the input and the hidden layer was proposed by Linsker [13], giving to the objective function the name, *Infomax principle*. This model is a linear model that actually maximizes only the entropy of the representation, performing Principal Component Analysis (PCA) in the case of Gaussian distributed data. The same objective function was applied by Bell and Sejnowsky [3], showing that a 1-layer neural network with sigmoid activation performs Independent Component Analysis (ICA).

Both these models are quite restrictive, indeed they work only under the assumption that the visible data is a linear combination of hidden features. A more general way to extract hidden features is given by the Autoencoder (AE), a NN that is a composition of an encoder and a decoder map, respectively W and V. This model can be seen as a generalization of the PCA model, because in the assumption W and V are linear maps, the space spanned by W is the same of the one spanned by principal components, see. e.g. [1].

An information theoretic description of an AE was given by Vincent et al. [18] where, with restrictive assumptions, they observed that reducing the reconstruction loss of an AE is related to maximizing the mutual information between the visible and hidden variables.

The IMAE, as we will see in the next section, is an AE that is able to learn a robust representation. In the literature there are many models in this family that are defined to learn a good representation, see e.g. [9] and references therein. For practical reasons, in the following section we compare the IMAE with the AEs that are recognized to learn the best features: Denoising, Contractive and Variational Autoencoders.

3   Infomax Autoencoder Model

Shannon [17] developed Information theory in order to define a measure of the efficiency and reliability of signal. In this theory a key role is assumed by Mutual Information I, i.e. a measure of the information shared between two variables, in the signal transmission case: the original message and the received one.

Formally, given two random variables: X and Y, the mutual information between these two variables is defined as

$$I(X,Y) = h(Y) - h(Y|X) \tag{1}$$

where $h(Y)$ is the (differential) entropy of Y, a measure of the average amount of information conveyed, and $h(Y|X)$ is the conditional entropy Y with respect to X, a measure of the average of the information conveyed by Y given X, i.e. the information about the processing noise, rather than about the system input X.

Given the assumption the hidden representation Y lives in a manifold embedded in a subspace of the visible space, by the definition above, the Infomax objective function finds the projection from the visible (higher dimension space) to the hidden manifold that preserves as much information as possible. Defining W as the projection map, the Infomax objective can be state as: $\max_W I(X, W(X))$, with the representation, $Y = W(X)$, that maximizes the mutual information with the system input X.

In general, to compute the mutual information between two variables is not trivial, indeed it is necessary to know the distribution of the input and of the hidden representation, and to compute an integral. For this reason, we propose an approximation of the mutual information $I(X, Y)$.

In the following, capital letters $X, \hat{X}, Y$ denote the random variables and lowercase letters $x, \hat{x} \in \mathbb{R}^d, y \in \mathbb{R}^l$ the respective realisations. The map between the input space $\mathbb{R}^d$ to the hidden representation $\mathbb{R}^l$ is $W, Y = W(X)$, and the map from $\mathbb{R}^l$ to the input space is $V, \hat{X} = V(Y)$. In the first step we assume that the map W is invertible, then both $x, y \in \mathbb{R}^d$.

Our approximation starts by considering the conditional entropy in Eq. (1). In order to estimate this quantity it is necessary to suppose there exits an unbiased efficient estimator $\hat{X} = V(Y)$ of the input X, in this way the conditional entropy $h(Y|X)$ can be approximated by $h(\hat{X}|X)$. Indeed, the conditional entropy is the information about the processing noise, and the noise added in the encoding process is proportional to the noise between the input and its reconstruction. Formally, see e.g. [15], p. 565, in the case of a bijective function the following holds

$$h(\hat{X}|X) \leq h(Y|X) + \int \log|J_V(y)|dp(y), \tag{2}$$

where the equality holds in the case that V is an orthogonal transformation, with J_V the Jacobian of the function V and $|\cdot| = |\det(\cdot)|$.

Let us start by assuming the second term on the right-hand-side (RHS) is bounded (in the end we show that this term is bounded). In order to have the inequality as tight as possible, the idea is to assume the conditional variable $\hat{X}|X$ is Gaussian distributed with mean X and variance $E[(X - \hat{X})(X - \hat{X})^T]$. Indeed, the Gaussian distribution has maximum entropy. Since $\hat{X}|X$ is Gaussian distributed, the conditional entropy $h(\hat{X}|X)$ is an (increasing) function of the covariance matrix $\Sigma_{\hat{x}|x}$, i.e. $h(\hat{X}|X) \propto \text{const} + \log(|\Sigma_{\hat{x}|x}|)$. By the symmetric definition of the covariance matrix and the well known relationship $log(|\Sigma|) = trace(\log(\Sigma))$, it is sufficient to consider as the loss function to minimize the trace of $\Sigma_{\hat{x}|x}$, i.e. the L_2-norm of the input with its reconstruction. In this way, we obtain formally that V should be near, in the L_2 space of functions, to the inverse of W, i.e. $V \circ W \approx Id$; then also the Jacobian $J_{V \circ W} \approx Id$.

Now let us focus on the first term of the RHS of (1), the differential entropy $h(Y)$. Defining the function W as a composition of two functions: $W = \sigma \circ W_0$, the entropy function can be written as follows:

$$h(Y) = \int \log |J_\sigma(Y_0)| p(y_0) + h(Y_0) \tag{3}$$

where $Y_0 = W_0(X)$. In our application, we consider the non-linearity σ as the logistic function $\sigma(x) = (1 + \exp(-x))^{-1}$; this allows to have the representation Y_0 distributed following the logistic distribution, see [3]. In order to compute the second term, $h(Y_0)$, following [7,8], this term can be seen as a sparse penalty in the Y_0 output, that, as observed in [12], encourages to have an independent distribution of the components of Y_0. Since the Y_0 components are almost independent, a good approximation for the determinant of the Jacobian of the sigmoid is the product of its diagonal elements. Then observing that the variance of Y_0 is controlled by the Jacobian of the sigmoid, the entropy of Y, can be approximated as:

$$h(Y) = \sum_i \sigma_i(Y_0)(1 - \sigma_i(Y_0)) - (\log(\cosh(Y_0)_i))^2, \tag{4}$$

where the first term in the RHS is associated to the log of the determinant sigmoid Jacobian, and the second term is the sparsity penalty, associated to $h(Y_0)$. The sparsity penalty in the equation above ensures W to be a bounded operator and, by the open mapping theorem, J_W and J_V have bounded determinants different from zero, thus the inequality (2) is well defined.

Generalization to Under- and Over-Complete Representations. The approximation of the conditional entropy as the L_2 reconstruction loss was derived under the condition that the input and the representation live in the same space. A way to generalize the derivation to the under- and over-complete setting is thinking of this structure as a noisy channel, where some units (components) in the hidden and input layers are masked. Indeed, since by definition V behaves as the inverse of the (corrupted) map W, it is not adding more, or even different, noise than that already added in the map W. Thus, we can conclude that the approximation $h(Y|X) \approx h(\hat{X}|X)$ holds both in the over- and under- complete setting. By construction, the approximation of the entropy term does not have a particular requirement, so can be described for a generic number of hidden units.

Properties. The difference of IMAE from the other models in the literature is given by the latent loss term i.e. the maximization of the latent entropy. Choosing the sigmoid activation in the latent layer, in accordance with the theory of Independent Component Analysis and processing signals [6], we suppose that the latent distribution is logistic and concentrated around the mean. In particular, the first term of the entropy (4), ensures that the distribution is peaked around the mean value, guaranteeing the learning of good prototypes of the input data. Instead, the sparseness penalty encourages the independence between the variables, guaranteeing a model robust to noise.

Relationship with Other Approaches. In order to describe the properties of IMAE from different perspectives, we compare it with the most common AE variants able to learn good representations. In all the cases that we consider the encoder map has the form $\sigma \circ W_0$.

Contractive Autoencoder (CAE). Starting from the idea that a low-derivative hidden representation is more robust, because it is less susceptible to small noise, Rifai et al. [16] suggested a model that explicitly reduces (contracts) the Jacobian of the encoder function. The proposed Contractive Autoencoder (CAE) has a structure similar to IMAE, with sigmoid activation in the hidden layer, but the latent loss is the Froebenius norm of the encoder Jacobian J_W. In the case W_0 is a linear matrix, the latent loss can be written as

$$\|J_W(x)\|_F^2 = \sum_i^l (\sigma_i(1 - \sigma_i))^2 \sum_j^d (W_0)_{ij}^2. \tag{5}$$

From Eq. (5), it is clear that CAE encourages a flat hidden distribution, the opposite of IMAE, that encourages a big derivative around the mean value; the consequences of such differences will be underlined and clarified in the computational experiment section, below. From an information theory perspective, observing that the Froebenius norm is an approximation of the absolute value of the determinant, the CAE representation can be described as a low entropy one. Indeed, by changing variables in the formula, in the case of a complete representation, the entropy of Y is a linear function of the log-determinant of the Jacobian of W.

Denoising Autoencoder (DAE). Assuming that robust reconstruction of the input data implies robust representation, Vincent et al. [18] proposed a variant of the AE trained with corrupted input, termed the Denoising Autoencoder (DAE). From a manifold learning perspective, Vincent et al. observed that the DAE is able to learn the latent manifold, because it learns the projection from the corrupted $\tilde{X}$ to the original X input. The problem, as highlighted in Sect. 4, is that there is no assumption on the distribution of the data around the hidden manifold i.e. DAE learns only a particular type of projection and then is robust to only one type of noise. Differently in IMAE, the preferred projection direction is suggested by the choice of the hidden activation.

Variational Autoencoder (VAE). The Variational Autoencoder (VAE) [11] is not a proper neural network, but actually a probabilistic model; differently from the other models described until now it does not discover a representation of the hidden layer, but rather the parameters λ that characterize a certain distribution chosen a-priori $q_\lambda(y_0|x)$ as a model of the unknown probability $p(y_0|x)$. In the case that q_λ is Gaussian, the parameters are only the mean and variance and VAE can be written as a classic AE with loss function:

$$\mathcal{L}_{VAE} = -E_q(\log p(x|y_0)) + D_{KL}(q_\lambda(y_0|x)\|p(y_0)). \tag{6}$$

The first term of $\mathcal{L}_{VAE}$ is approximately the reconstruction loss. The Kullback-Leibler (K-L) divergence term, D_{KL}, which is the divergence in the probability space between the a-priori and the real distributions, represents the latent loss of the neural network, approximated in our case as: $\sum_i \mu(y_0)_i^2 + \sigma(y_0)_i^2 - \log(\sigma(y_0)_i^2) - 1$.

From the IMAE perspective, we observe that the first term of the latent loss, $-h(Y)$, encourages the Y_0-distribution to be logistic with small variance, and the second term penalizes the dependencies between the Y_0 components. Then we can conclude that, from a probabilistic perspective, the IMAE latent loss works like a K-L divergence, with the target function a product of independent logistic distributions.

4   Computational Experiments

In this section we describe the robustness and the clusterization performance of IMAE, comparing it with the models described above and the classic AE. We carry out the tests with two different architectures: single- and multi-hidden layers. The first comparison is made with AE, CAE and DAE; in the second comparison, which is suitable for a small representation, we compare IMAE with VAE.

4.1   1-Hidden Layer

The first tests are carried using the MNIST dataset, a collection of 28×28 grey-scale handwritten digits, subdivided into ten classes, with 60 000 training and 10 000 test images. We consider two different AutoEncoder architectures, one under-complete with 200 hidden units and the other over-complete with 1000 hidden units. All the neural networks are trained using batch Gradient Descent with learning rate equal to 0.05 and batch size composed by 500 input patterns (images), trained for 2000 epochs. For all the models, the encoder and decoder functions have form: $W(x) = \sigma(W_0 x)$ and $V(x) = Vx$. The listed results are obtained with tied weights, $W_0 = V^T$, but equivalent results are obtained with W and V not related to each other. The DAE is trained both with Gaussian (DAE-g) and mask (DAE-b) noise. In DAE-g, each pixel of the training images is corrupted with zero-mean Gaussian noise with standard deviation $\sigma = 0.3$; in DAE-b, each pixel is set to zero with probability $p = 0.3$. In CAE and IMAE, the hyper-parameters associated with the latent loss are set equal to 0.1 and 1, respectively.

Robust Reconstruction. As suggested by [18], a way to evaluate the robustness of the representation is by evaluating the robustness of the reconstruction. Under this assumption, for each model we compute the L_2-loss between the reconstructed image and the original (clean) one. Results for this measure are listed in Table 1. From these results we can deduce that IMAE is equally robust in all the considered settings and, excluding the respective DAEs trained with the test noise, has the best reconstruction performance. In contrast, DAE is robust only to the noise with respect to which it was trained and CAE is only robust in the over-complete setting with respect to Gaussian noise. The classic AE is best in the zero-noise case.

Table 1. Reconstruction loss performance with MNIST input data corrupted with masking noise with probability $p = 0, 0.3, 0.5, 0.75$ (left column) or with zero-mean Gaussian noise with standard deviation $\sigma = 0.03, 0.015, 0.35, 0.45$ (right column), 1000 hidden units (above), 200 hidden units (below).

Mask noise, p					Gaussian noise, σ			
Model	0	0.3	0.5	0.75	Model	0.03	0.15	0.35	0.45
AE	**12.8**	110.2	145.7	183.1	AE	**44.3**	130.5	178.1	190.9
CAE	129.4	140	157.7	254.3	CAE	133.2	149.1	156.9	159.5
DAE-b	70.5	**68.9**	**97.6**	**151.7**	DAE-b	86.6	165.4	210	226.1
DAE-g	263.3	247.4	283.6	419.6	DAE-g	202.4	**78.6**	**77.5**	**92**
IMAE	53.7	106.5	132.9	161.8	IMAE	73.4	126.9	156.6	166.4

Model	0	0.3	0.5	0.75	Model	0.03	0.15	0.35	0.45
AE	**37.4**	97.4	133.	176.3	AE	**50.4**	122.2	163.6	176.6
CAE	141.1	148.3	157.7	206.4	CAE	143.4	151.6	158.5	160.7
DAE-b	71.3	**70.5**	**97.1**	**152**	DAE-b	86.6	154.6	193	205.3
DAE-g	158.2	148.7	167.2	226.4	DAE-g	133.5	**72.2**	**77.9**	**91.5**
IMAE	103.8	125	138.9	155.8	IMAE	112.5	135.7	148.1	152.7

Prototype Identification. A good, task-independent representation should include the hidden prototypes for each class. Assuming that the hidden representation of each element lies close to the respective prototype, to evaluate representation quality we measure the clusterization performance. In particular, we compute the Rand-index between the ten clusters identified by the K-means algorithm on the hidden space and the ones defined by ground-truth in the MNIST data set. Formally we compute the Rand-index R as follows:

$$R = \max_m \frac{\sum_{i=1}^{N} \mathbf{1}(m(c_i) = l_i)}{N} \tag{7}$$

where $c_i \in \{1, \ldots, k\}$ is the hidden representation cluster in which the current input, x_i, belongs, $l_i \in \{1, \ldots, k\}$ is the ground truth label associated with $x_i \in \{x_i\}_1^N$, m is a 1-1 map from cluster to label, and $\mathbf{1}$ is the indicator function, that is 1 if $m(c_i) = l_i$, 0 otherwise. In our case $k = 10$, the number of different classes, and $N = 1000$; the results are shown in Table 2.

IMAE has the best clusterization performance (high Rand index) and is also the more robust to noise. In particular, we observe that these performances are associated with the mean derivative of the non-linear hidden activation, where a high derivative is associated with good performance. This observation leads us

Table 2. Average over 50 iterations of the Rand indices obtained with clean, R, and noisy, R_ν ($\nu \sim \mathcal{N}(0, (0.2)^2)$), test data. The σ' row denotes the average derivative of the hidden representation (the order of the derivative is 10^{-2}; note all the CAE derivatives are around the order 10^{-5}).

	AE	CAE	DAE-b	DAE-g	IMAE
R	53.5	17.9	53.8	51.3	**54.4**
R_ν	53.5	17.8	54	50.2	**55.2**
σ'	1.9	-	1.5	1.11	5.9

200 hidden units

	AE	CAE	DAE-b	DAE-g	IMAE
R	55.7	17.3	55.6	51.4	**55.8**
R_ν	55.6	17.3	55.3	50.5	**55.6**
σ'	3.9	-	4.5	2.5	5.8

1000 hidden units

to conclude that the derivative value can be considered as an empirical measure of the clusterization performance.

Concerning both sets of results, we underline that reconstruction robustness is not associated with robust clusterization. For example, according to the dropout idea, DAE-b has a more robust representation than DAE-g also in the case of the Gaussian noise corruption on which it was not trained.

4.2 Multi-hidden Layers

In many applications a shallow network structure is not sufficient, because the considered data are more redundant than the MNIST data set and the desired representation is really small. For this reason it is important to consider a deeper structure. In this setting AE, CAE and DAE cannot naturally be extended, so we compare IMAE with VAE, the AE that, better than others, is able to learn a hidden representation. We now consider a network with hidden architecture of type: 1100-700-n_h-700-1100, with softplus activation in each hidden layer except for the n_h units, where a sigmoid nonlinearity is applied. In the following, we use hidden units to refer to the n_h-variables, where $n_h \in \{5, 10, 20\}$. The tests are carried out with MNIST and Fashion-MNIST (F-MNIST), a MNIST-like dataset where the digits are replaced by images of clothes, [19]. This dataset is suitable for this network structure since it is more redundant than MNIST but still low-dimensional. The network is trained as in the 1-hidden layer architecture, with learning rate equal to $5 \cdot 10^{-3}$.

Since we are interested to learn a small representation of the data, we compare the clusterization performance of VAE and IMAE. We do not consider the robustness of reconstruction because, by construction, the models are robust to really small noise.

Clusterization Performance. Table 3 lists the Rand-indices obtained in the hidden space, with possibly Gaussian-noise-corrupted input data. In the noisy case the input data is corrupted with zero mean Gaussian noise with standard deviation, respectively, $\sigma = 0.01, 0.1$ for MNIST and F-MNIST. For MNIST, we choose this relatively small noise because, since it is not particularly redundant, the deep structure is not robust.

Table 3. Average over 50 iterations of Rand-indices of VAE and IMAE with networks having 5, 10 or 20 hidden units.

Model	MNIST						F-MNIST					
	n_h			Noisy			n_h			Noisy		
	5	10	20	5	10	20	5	10	20	5	10	20
VAE	70.1	57.2	52.9	61	51.7	36	40.3	42.3	42.4	38.1	41.8	41.3
IMAE	**76.8**	**75.7**	**54.8**	**68.7**	**60.5**	**42.2**	**55.5**	**59.3**	**56**	**55**	**55.3**	**55.3**

The results highlight how IMAE is able to clusterize the data in a small hidden space and identify the prototypes for each class also in the case of noisy input data i.e. the hidden prototypes are robust. In order to clarify the robustness of the clusters, Fig. 1 visualises the hidden representations of the two models in the case of noisy input data. In the IMAE representation, despite the noise, it is possible to clearly distinguish the different clusters; instead in the VAE representation all the points are mixed, except largely the orange ones (bottom cluster), which represent the digit 1.

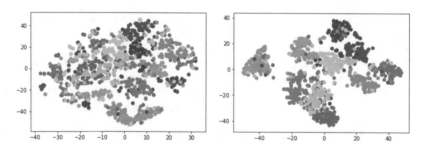

Fig. 1. Illustration of t-SNE encoding [14] of VAE (left) and IMAE (right) representation obtained with noisy MNIST, $n_h = 10$.

5   Conclusions and Future Work

In this paper, we derived formally an Autoencoder that explicitly maximizes the information between the input data and the hidden representation and introduces an empirical measure of the clusterization performance on the hidden space. The experiments show that IMAE is versatile, works both in shallow and deep structures, has good reconstruction performance in the case of noisy input data, and is able to learn good and robust prototypes in the hidden space. The latter point, that is also the principal innovation, is particularly useful for applications, such as case-based reasoning approaches [10], where it is necessary to identify clear prototypes to represent each case.

Acknowledgments. This research is funded by the University of Stirling CONTEXT research programme and by Bambu (B2B Robo advisor, Singapore).

References

1. Baldi, P., Hornik, K.: Neural networks and principal component analysis: learning from examples without local minima. Neural Netw. **2**(1), 53–58 (1989)
2. Barlow, H.: Possible principles underlying the transformation of sensory messages. In: Sensory Communication. MIT Press, Cambridge (1961)
3. Bell, A.J., Sejnowski, T.J.: An information-maximization approach to blind separation and blind deconvolution. Neural Comput. **7**(6), 1129–1159 (1995)
4. Bengio, Y., Courville, A., Vincent, P.: Representation learning: a review and new perspectives. IEEE Trans. Pattern Anal. Mach. Intell. **35**(8), 1798–1828 (2013)
5. Brunel, N., Nadal, J.P.: Mutual information, Fisher information, and population coding. Neural Comput. **10**(7), 1731–1757 (1998)
6. Hyvärinen, A., Hurri, J., Hoyer, P.O.: Natural Image Statistics: A Probabilistic Approach to Early Computational Vision. Springer, London (2009)
7. Hyvrinen, A.: New approximations of differential entropy for independent component analysis and projection pursuit. Neural Inf. Process. Syst. **10**, 273–279 (1998)
8. Hyvrinen, A., Oja, E.: Independent component analysis: algorithms and applications. Neural Netw. **13**(4–5), 411–430 (2000)
9. Karhunen, J., Raiko, T., Cho, K.: Unsupervised deep learning: a short review. In: Advances in Independent Component Analysis and Learning Machines, pp. 125–142 (2015)
10. Kim, B., Rudin, C., Shah, J.A.: The Bayesian case model: a generative approach for case-based reasoning and prototype classification. In: Advances in Neural Information Processing Systems, pp. 1952–1960 (2014)
11. Kingma, D.P., Welling, M.: Auto-encoding variational Bayes. In: Proceedings of the 2nd International Conference on Learning Representations (2014)
12. Lee, T.-W., Girolami, M., Bell, A.J., Sejnowski, T.J.: A unifying information-theoretic framework for independent component analysis. Comput. Math. Appl. **39**(11), 1–21 (2000)
13. Linsker, R.: Self-organization in a perceptual network. Computer **21**(3), 105–117 (1988)
14. Maaten, L.V.D., Hinton, G.: Visualizing data using t-SNE. J. Mach. Learn. Res. **9**, 2579–2605 (2008)
15. Papoulis, A., Pillai, S.U.: Probability, Random Variables, and Stochastic Processes. Tata McGraw-Hill Education, New Delhi (2002)
16. Rifai, S., Vincent, P., Muller, X., Glorot, X., Bengio, Y.: Contractive auto-encoders: explicit invariance during feature extraction. In: Proceedings of the 28th International Conference on International Conference on Machine Learning, pp. 833–840. Omnipress (2011)
17. Shannon, C.E.: A mathematical theory of communication. Bell Syst. Tech. J. **27**(3), 379–423 (1948)
18. Vincent, P., Larochelle, H., Bengio, Y., Manzagol, P.A.: Extracting and composing robust features with denoising autoencoders. In: Proceedings of the 25th International Conference on Machine Learning, July, pp. 1096–1103 (2008)
19. Xiao, H., Rasul, K., Vollgraf, R.: Fashion-MNIST: a novel image dataset for benchmarking machine learning algorithms. arXiv preprint arXiv:1708.07747 (2017)

Deep Regression Counting: Customized Datasets and Inter-Architecture Transfer Learning

Iam Palatnik de Sousa$^{(\boxtimes)}$ ⓘ, Marley Maria Bernardes Rebuzzi Vellasco, and Eduardo Costa da Silva

Pontifical Catholic University of Rio de Janeiro, Rio de Janeiro, RJ, Brazil
iam.palat@gmail.com

Abstract. The problem of regression counting is revisited and analyzed by generating custom data and simplified Residual Network architectures. The results provide three key insights: A deeper understanding of the inherent challenges to this problem with regards to the data characteristics; the influence of architecture depth on the regression counting performance; and ideas for a transfer learning strategy between dissimilar architectures that allow training deeper networks with knowledge gained from shallower ones. In a striking example, a network with 30 convolution layers is successfully initialized with the weights from a trained architecture containing only 7 convolutions, whereas convergence was previously unattainable with random initialization. The two datasets consist of 20,000 images containing 3 and 5 classes of shapes to be counted, respectively. The network architectures are simplified Residual Networks with varying depths. The images are made to be inexpensive computationally to train, allowing for easy future comparisons with the baseline set by this work.

Keywords: Regression counting · Residual Networks · Transfer learning

1 Introduction

The task of counting objects in an image has many applications, from counting cells in histopathology images [12], to estimating the number of individuals in crowds [6], to counting fruit, trees or other relevant objects in agriculture [7], to objects in satellite images [8], among others.

Several techniques have been employed for this task, both relying on unsupervised and supervised methods. Lempitsky and Zisserman [6] mention how the supervised methods have been shown to have better performance in general, and how they can usually be divided into detection counters and regression counters.

Aich and Stavness [1] mention how object counting is a less complex task than segmentation or object detection, with the latter ones requiring more laborious

© Springer Nature Switzerland AG 2020
L. Oneto et al. (Eds.): INNSBDDL 2019, INNS 1, pp. 109–119, 2020.
https://doi.org/10.1007/978-3-030-16841-4_11

annotation of datasets for training. This often limits the size and availability of large datasets for this purpose.

Indeed, often datasets used for object counting are composed by a small number of large high resolution images. These are often divided in smaller patches for training, depending on how they are annotated. This makes artificial datasets with simulated data especially interesting, since in this case the annotation, whether for segmentation, detection or regression, is automatic. It does require however that the dataset closely resembles the objects in the application of interest. One such example is the framework [5] created by Lehmussola et al. It simulates images from fluorescence microscopy with cell populations, allowing the user of the framework to control a series of parameters regarding cell number, shape and the image in general.

Furthermore, there are still aspects of this counting task that seem to not be fully described in literature, such as what characteristics of the counted objects or images might make the training process harder. This could correspond to asking whether a high variability in the size of the objects is more of an obstacle than a color variability; whether the presence of multiple types of shapes is a relevant factor; whether treating the problem as a multi-label task influences the performance of training, among others.

The answer to these questions could greatly influence the process of building and acquiring data to train such counting systems. Given this overview, the motivation arose to generate a computationally inexpensive and highly customizable dataset for quickly testing counting algorithms and related ideas.

The main concept is to use simple scripts to create an arbitrary number of images that consist on different colored shapes against a dark background. If at a given point the dataset becomes too easy to learn, new more complex versions can be created by simply increasing the number of shapes, or giving each shape more complex characteristics. For the purposes of this manuscript, two such datasets were generated. One with 3 classes of shapes, and one with 5. These were termed the Easy and Hard datasets, respectively.

Figure 1 shows an example of a large image generated with the methodology described in the next section. For brevity, the dataset will be called throughout the manuscript as the Shape Counting (ShaCo) Dataset.

Regarding the choice of network architecture, Convolutional Networks are the most commonly used for Computer Vision tasks. However, although a large number of pre-trained weights are available for transfer learning in classification and segmentation tasks, the same is not true for regression counting. As such, the intention was to train networks from scratch. This included the commonly used models such as the VGG architectures [9], Inception [10] and Residual Nets (ResNets) [3].

Nevertheless, as described in more detail on the following sections, preliminary tests using these models yielded unsatisfactory results. It seemed hard to train networks of many layers directly for this task.

A more meticulous architecture testing approach was then used, defining a shallow base network to which convolution layers were added, one at a time. The

Fig. 1. Two image examples from the Easy dataset (Panel A) and the Hard dataset (Panel B).

architecture was a simplified version of ResNet, containing the characteristic shortcut connections. These not only aid in training deeper models without suffering with the degeneration problem but also gives layers an ensemble-like behavior, as described by Veit et al. [11]. Furthermore, the transfer learning of learned weights from shallower to deeper models was tested.

2   Materials and Methods

2.1   Datasets

As previously mentioned, two datasets were generated. In the following subsections a more detailed description of each is provided.

Easy. ShaCo Easy is comprised of 20,000 3 channel RGB mode images. The dataset is divided into 16,000 training images, and 4,000 test images. The images have dimensions of 60 by 60 pixels. Then a random 20% of this remaining training set is used for validation. The three classes used in this case are blue circles, green squares and red squares.

The shapes are allowed to overlap, and are transparent so that even in the case of full overlaps it is still possible to discern how many are present. They are, however, defined so as not to cross the edge of the image, and are always fully inside the boundaries.

The data labels for each image are the quantities of each shape class on that particular image.

As for the properties of each type of shape (radius and side length dimensions are measured in pixels):

- Blue circles - Random amount between 1 and 5 per image. The radius is randomly assigned for each, and can be either 5 or 6.
- Green squares - Random amount between 1 and 4 per image. Side length is randomly assigned for each, and can be 5, 6 or 7.

- Red squares - Random amount between 0 and the number of Green Squares on that specific image. Side length is always 5.

The Easy training set contains 100,204 shapes in total: 48,261 blue circles, 39,966 green squares and 11,977 red squares. The test set contains 24,983 shapes in total: 12,071 blue circles, 9,912 green squares and 3,000 red squares.

The choice of 60 × 60 pixel images was intended allow for computationally. inexpensive training. The models used in this study were trained roughly with training epochs taking 5 to 8 s in a Tesla K80 GPU.

Hard. The hard version of the dataset includes the 3 previously cited types of shapes, with the same characteristics and parameters, but also includes two additional more challenging ones:

- Light Blue Ellipses - Random amount between 1 and 5 per image. Each ellipse has a shortest radius of size between 1 and 3 and largest radius of size between 4 and 7. The ellipses may be rotated about their center by a random angle between 0 and 360°.
- Blue squares - Random amount between 1 and 3 per image. Side length is randomly assigned for each, and can be any whole number between 3 and 29.

These new shapes have a much higher shape and size variability.

The Hard training set contains 180,443 shapes in total: 47,976 blue circles, 40,023 green squares, 12,134 red squares, 32,043 blue squares and 48,267 light blue ellipses. Finally, the Hard test set has within its images 45,050 shapes in total: 11,927 blue circles, 9,992 green squares, 3,038 red squares, 8,041 blue squares and 12,052 light blue ellipses.

2.2 Architecture

As briefly discussed in the Introduction, some of the most commonly used Convolutional Neural Network (CNN) models were tested with the datasets. The poor performance of almost 100% miscounts motivated the creation of custom models.

The neural networks were implemented in Keras [2]. They are inspired in the ResNet models, but greatly simplified. As such, for the sake of brevity, they will be referred to as SimpleResNets, or SiReNs, for the remainder of this manuscript.

These architectures start with the same convolution and max-pooling used in the original ResNets [3]. A base model with no additional convolutions was implemented and called model1. This model, besides the already described convolution and pooling, has a Keras flattening operation, a 0.2 Dropout layer, and two densely connected classification layers with 512 neurons and 3/5 neurons, respectively (3 for the Easy dataset, 5 for the Hard).

Moving from model1 to model2, two convolutions are added between the max-pooling and the flattening operations. They will serve as inputs for the shortcut connection on the next model. All convolutions besides the first one, in this network, share the same hyper parameters to facilitate the comparison of their effect on the overall result.

For the next iteration, model3 contains an additional convolution compared to model2. This new layer receives the shortcut connection formed from the two convolutions that precede it, in the same fashion as in the original ResNet implementation. The same process is repeated for models 4 through 30. Figure 2 shows this process in a more detailed illustrative manner. The key difference to the original ResNet implementation is that the shortcuts are added between every Convolution, rather than between large Convolution blocks.

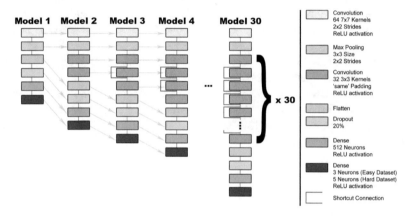

Fig. 2. Process of building and training the SiReN architectures. The layers correspond to the color legend to the right. Green arrows indicate how weights from shallower models were used to initialize the deeper ones, during training. Notably, from Model 3 onwards, only one layer is initialized with random weights (or, in this image, does not receive a green arrow).

ReLU activation was used in all cases, including the output layers given that the prediction corresponds to a regression task.

The loss function used was Mean Squared Error (MSE), the optimizer was Adam [4] with standard parameters, and each model was trained for 50 epochs. This number of epochs was arbitrarily set after preliminary tests showed that these models typically converge to a loss plateau within that time frame. Model checkpoints were used so that the network weights were saved in a separate file every time there was a validation loss improvement.

Alongside this MSE loss, the number of miscounts for each shape type was also computed as a performance metric. This value was then divided by the total numbers for each type of shape respectively, generating a miscount ratio that can easily be compared between datasets of different sizes with an arbitrary number of shapes of each type. As an illustrative example, test datasets with 100 and 1,000 blue circles, resulting in 20 and 200 blue circle miscounts, respectively, both have a 20% blue circle miscount ratio. Although the MSE is largely a smoother quantity, this miscount ratio is arguably more easily interpretable. As such, both are discussed within the Results section.

Additionally, same padding was used to allow for a large number of convolutions to be applied to the relatively small sized images. This procedure was also adopted in the original ResNet implementations.

Inter-Architecture Transfer Learning. It is worth noting that all SiReN models generated share similar layers in their structures. This motivates testing whether it is possible to initialize the weights of deeper models using those of shallower ones. This type of Inter-Architecture Transfer Learning (IATL) seems to be scarcely explored or discussed in literature, although transfer learning between identical architectures is a frequently adopted and well accepted strategy.

As a baseline, the 30 models were trained with random initialization, without using IATL, on the Easy dataset.

Furthermore after training, model30, the deepest one used for this manuscript, was initialized in the same fashion using the weights from other previous models besides model29 to see if it was still possible to make it converge and present a good performance.

3   Results

3.1   Easy Dataset

A consistent pattern emerged between the number of test set miscounts and certain key validation set loss plateaus. Taking the Easy dataset as an example, a validation MSE of around 6.5 was in all cases associated with a close to 100% miscount ratio for the test set. In that case the network fails to count all or almost all shapes in the test set, regardless of class. The next plateau is for a validation loss of about 2.5, which was in all cases associated with good performance for blue circles, but near 100% miscount ratios for green and red squares. Similarly, the third plateau is for validation losses of about 0.4, which occurred when there was a near 100% miscount ratio for red squares, with good performances for the other two shapes.

The preliminary results using commonplace CNN models (VGG, ResNet, Inception) were of an extremely high miscount number. In all preliminary tests the 6.5 loss plateau is quickly reached and maintained during the 50 training epochs. Indeed observing this behavior was one of the key aspects that motivated the generation of several SiReN models, as previously discussed in this manuscript. These plateaus seemed to correspond to local minima that the SiReN networks also had difficulty escaping in the case of random initialization. This is illustrated in top plot of Fig. 3.

Whenever the networks managed to break this 0.4 loss barrier, however, the validation losses converged quickly to values below 0.01, at which point the miscounts were in the order of about 0 to 10 for all shapes. This indeed happened for all models when IATL was applied, making all models from 1 to 30 be able to learn the task well. This is shown in the bottom plot of Fig. 3.

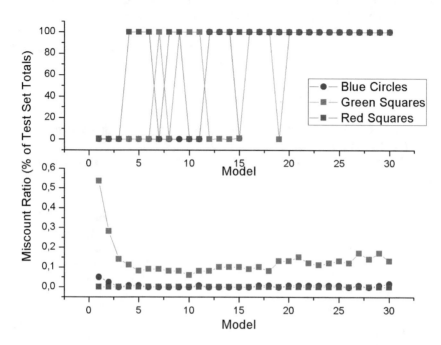

Fig. 3. Results for the Easy dataset, with random initialization of all models (top plot), and with the Inter-Architectural Transfer Learning approach (bottom plot). The shape of the plot labels correspond to the shapes used in the image datasets for easier comparison. The raw miscount numbers for a given shape class are divided by the total number of occurrences of that shape in the test set, to find a miscount ratio. Results for the Easy dataset, with random initialization of all models (top plot), and with the Inter-Architectural Transfer Learning approach (bottom plot). The shape of the plot labels correspond to the shapes used in the image datasets for easier comparison. The raw miscount numbers for a given shape class are divided by the total number of occurrences of that shape in the test set, to find a miscount ratio.

Notably the first few randomly initialized models do manage to converge and learn how to count with good performance (near 0% miscounts), although they still perform worse than the IATL trained models. After model3 the red square miscounts rise to about 100%. By model16 basically all following models have 100% miscounts for all class shapes.

With IATL the networks of all depths seem to quickly learn how to count red squares and blue circles. Indeed the red square miscount ratio is 0% for all models. The increase in depth at first massively reduces the miscount ratio for green squares, but adding convolutions beyond a certain point (around model10) seems to slowly increase the green square miscount.

Furthermore, initializing model30 with the weights from model7 yielded convergence below the 0.4 validation loss threshold. The number of test set miscounts was of 0 blue circles, 8 green squares and 0 red squares. This is striking compared to initialization with model29, which yielded worse test set miscounts of 2 blue circles, 13 green squares and 0 red squares. Models shallower than model7 could not initialize model30 successfully, with miscounts remaining at high values.

3.2   Hard Dataset

After obtaining these results, random initialization was foregone in favor of IATL. For the Hard dataset, this approach gave similar results to what was observed in the Easy case, with some key differences. Figure 4 summarizes these results.

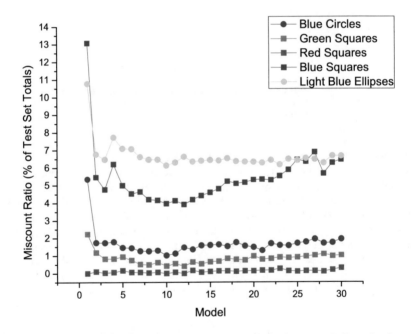

Fig. 4. Results for the Hard dataset with the Inter-Architectural Transfer Learning approach. The shape of the plot labels correspond to the shapes used in the image datasets for easier comparison. The raw miscount numbers for a given shape class are divided by the total number of occurrences of that shape in the test set, to find a miscount ratio.

Namely, a notable difference is that for the Easy Dataset the miscount ratios of green squares are higher than those of blue circles. This behavior is flipped for the Hard dataset.

Additionally, the miscount ratios for all 3 shapes present in the Easy Dataset are much higher on the Hard dataset (going from about 0, 0.1% to 1 or 2%).

After an initial drop in miscounts within the addition of a few layers, the deeper models seem to also have increasingly higher miscounts. This is however most noticeable in the blue squares, where the performance worsens beyond model10 in a way that is not seen for the other shapes.

4   Discussion

The behavior displayed by the SiReN models, seen in the results, raises some key points. The number of classes present in the images influences performances for individual classes to a large degree. This is most evident for the cases of blue circles, green squares and red squares, which are present in both the Easy and Hard datasets. The miscount ratios go from about 0.1% to about 2%.

Even with similar occurrence frequencies, the miscount ratios differ by an order of magnitude between the two datasets. This may justify extra preprocessing when preparing a dataset for regression counting tasks, to ensure that images don't contain too many classes at once.

Very irregular shapes also present a similar challenge at all model depths as seen with the ellipse miscounts in the Hard dataset (7% miscount ratio). However very large size variability may present as much of a challenge as depth increases. Although red squares were by far the least frequent shape in all datasets (about 12% and 6% of the total number of shapes in the Easy and Hard datasets, respectively), they were in all cases the shape with the least miscounts, possibly because they had constant size.

Experiments with the hard dataset further resulted in higher miscount ratios, ranging from 2% to 7%, for blue and light-blue shapes. This could indicate that populating the data with too many shapes on the same channel has some influence over performance. In other words, it is possible that the blue channel of the images is over-crowded compared to the other channels, leading to a more difficult generalization.

Additionally, increasing the depth of the models seems to improve performance up to a point, after which the performance either stagnates or worsens, depending on which class of shape is being counted. It seems that the additional layers can exacerbate the challenges inherent to some types of shapes more than others (e.g. blue squares in the Hard dataset). Trying to improve the results of deeper models in this case can be the aim of future works that might vary hyper parameters, number of training epochs, architectures, among other aspects.

Furthermore, Inter-Architectural Transfer Learning was shown to be a viable strategy for the task of regression counting. It allowed for deep models to converge to a good performance where it was impossible to do so with random initialization.

Since a model with more than 30 layers was initialized successfully by a much shallower model, this begs the question if shallow models could be constructed to successfully initialize the commonly very used deep models (such as ResNet or Inception). This could be a strategy to massively reduce training efforts while maintaining good performance, especially when there are no previously trained weights to draw from for a specific architecture.

5 Conclusion

An analysis of the task of regression counting using deep convolution based models was presented.

Both custom datasets and custom network models were generated. The datasets contained either 3 or 5 classes of shapes to be counted, and the networks consisted in simplified versions of ResNets.

The results showed a series of challenges posed by the characteristics of the several types of shapes generated. Namely, the influence of shape, size and occurrence frequency variability translated into different levels of miscount ratios in the results.

One of the interesting influences observed was that, for the hard dataset, the presence of a high number of blue shapes seems to correlate with a higher number of miscounts. Future experiments can focus on mitigation strategies that pre-classify shapes of similar colors before tackling the complete images.

The idea of using shallow models to initialize deeper similar ones was shown to work with very robust performance metrics and loss values. This strategy was termed Inter-Architectural Transfer Learning.

The previously cited related works often focus on different types of objects, and a universal benchmark for regression counting does not seem to exist as of yet. As such determining the state of the art regarding performance for this task is challenging. The datasets generated for this manuscript, or similarly customized ones, could potentially become initial benchmarks in this sense.

Furthermore the results found regarding IATL provide novel insights compared to the related approaches and open new possible questions. Future experiments can verify if the presented findings hold for tasks besides regression counting, and also repeat the tests here presented with more complex datasets and architectures.

The results presented may aid greatly in the process of training deep networks models when no weights are available for transfer learning, besides giving a direction towards building better datasets for training models in this type of counting task.

Acknowledgments. The authors acknowledge the National Council of Scientific Research and Development (CNPq) for partial funding of this project.

References

1. Aich, S., Stavness, I.: Object counting with small datasets of large images
2. Chollet, F., et al.: Keras (2015). https://keras.io
3. He, K., Zhang, X., Ren, S., Sun, J.: Deep residual learning for image recognition. In: Proceedings of the IEEE Conference on Computer Vision and Pattern Recognition, pp. 770–778 (2016)
4. Kingma, D.P., Ba, J.: Adam: a method for stochastic optimization. arXiv preprint arXiv:1412.6980 (2014)
5. Lehmussola, A., Ruusuvuori, P., Selinummi, J., Huttunen, H., Yli-Harja, O.: Computational framework for simulating fluorescence microscope images with cell populations. IEEE Trans. Med. Imaging 26(7), 1010–1016 (2007)
6. Lempitsky, V., Zisserman, A.: Learning to count objects in images. In: Advances in Neural Information Processing Systems (2010)
7. Rahnemoonfar, M., Sheppard, C.: Deep count: fruit counting based on deep simulated learning. Sensors 17(4), 905 (2017)
8. Rodriguez, A.C., Wegner, J.D.: Counting the uncountable: deep semantic density estimation from space. arXiv preprint arXiv:1809.07091 (2018)
9. Simonyan, K., Zisserman, A.: Very deep convolutional networks for large-scale image recognition. arXiv preprint arXiv:1409.1556 (2014)
10. Szegedy, C., Liu, W., Jia, Y., Sermanet, P., Reed, S., Anguelov, D., Erhan, D., Vanhoucke, V., Rabinovich, A.: Going deeper with convolutions. In: Proceedings of the IEEE Conference on Computer Vision and Pattern Recognition, pp. 1–9 (2015)
11. Veit, A., Wilber, M.J., Belongie, S.: Residual networks behave like ensembles of relatively shallow networks. In: Advances in Neural Information Processing Systems, pp. 550–558 (2016)
12. Venkatalakshmi, B., Thilagavathi, K.: Automatic red blood cell counting using hough transform. In: 2013 IEEE Conference on Information & Communication Technologies (ICT), pp. 267–271. IEEE (2013)

Improving Railway Maintenance Actions with Big Data and Distributed Ledger Technologies

Roberto Spigolon[1]($\boxtimes$), Luca Oneto[2], Dimitar Anastasovski[1], Nadia Fabrizio[1], Marie Swiatek[3], Renzo Canepa[4], and Davide Anguita[2]

[1] Cefriel - Politecnico di Milano, Milan, Italy
{roberto.spigolon,dimitar.anastasovski,nadia.fabrizio}@cefriel.com
[2] DIBRIS - University of Genoa, Genoa, Italy
{luca.oneto,davide.anguita}@unige.it
[3] Evolution Energie, Paris, France
marie.swiatek@evolutionenergie.com
[4] Rete Ferroviaria Italiana, Rome, Italy
r.canepa@rfi.it

Abstract. Big Data Technologies (BDTs) and Distributed Ledger Technologies (DLTs) can bring disruptive innovation in the way we handle, store, and process data to gain knowledge. In this paper, we describe the architecture of a system that leverages on both these technologies to better manage maintenance actions in the railways context. On one side we employ a permissioned DLT to ensure the complete transparency and auditability of the process, the integrity and availability of the inserted data and, most of all, the non-repudiation of the actions performed by each participant in the maintenance management process. On the other side, exploiting the availability of the data in a single repository (the ledger) and with a standardised format, thanks to the utilisation of a DLT, we adopt BDTs to leverage on the features of each maintenance job, together with external factors, to estimate the maintenance restoration time.

Keywords: Big data analytics · Distributed Ledger Technologies · Railway maintenance actions

1 Introduction

Railway Infrastructure Managers (IMs) are responsible of operating the existing rail infrastructures. Maintenance is, without any doubt, one of the main task of this job [3,5]: not properly maintained infrastructures are in fact more prone to failures, that in turn translate into disruptions of the normal execution of railway operations.

Maintenance operations are usually demanded to external contractors through framework contracts that guarantee the availability of specialized workers whenever there is a need, planned in advance or unexpected. Considering in

© Springer Nature Switzerland AG 2020
L. Oneto et al. (Eds.): INNSBDDL 2019, INNS 1, pp. 120–125, 2020.
https://doi.org/10.1007/978-3-030-16841-4_12

particular the planned maintenance operations, the actual work is scheduled on specific time slots, where the train circulations can be modified or suspended without causing major disruptions. An empirical estimation of the time needed to perform the jobs is used to plan the scheduling of all the operations on the available time slots.

Moreover, to ensure that each maintenance job is performed in the correct way, that all safety measures are put in place, and all responsibilities are clearly identified, IMs employ standardised procedures to guarantee that each action is executed in a proper order by the responsible actor, leaving a trace of that execution. To fulfill such requirements, each step in these procedures must be performed leaving a legally valid record of which actions were performed, by whom, and with which authorisations; thus leading to a lot of signed papers sent between the various actors via registered letters, and to recorded phone calls. A process involving paper documents can be inefficient, leading to increased waiting times between each step in the workflow. Also having a direct access to maintenance data, to assess the current status of a specific job or to perform data analysis [6,9], may not be straightforward.

In this context, BDTs and DLTs may bring a great benefit to the current management of maintenance jobs. From one side, the adoption of DLTs and smart contracts could enable the digitalisation of the process currently employed while maintaining all the required features, like a tamper-proof record for the tracking of all decisions and executed actions [2]; potentially allowing also the automatic enforcement of contractual clauses. From the other side, the analysis of historical data about previous maintenance operations could enable the development of a prediction algorithm able to accurately estimate the restoration time for each maintenance job, thus leading to a better planning of the operations. Moreover, the DLT enacts the gathering of all the data on a single repository (the ledger) and with a standardised format, allowing the periodic retraining of the prediction engine: such operation could hardly be done with data stored in isolated silos with a noninteroperable format.

For this reason in this paper we propose an architecture able to merge DLTs, to automate the railways maintenance workflows, and BDTs for improving the decision of IMs in executing railway operations.

2 DLT-Based Maintenance Management

The system described, currently under development, comprises two major components: a DLT peer-to-peer network and a prediction engine. The DLT network will be implemented using Hyperledger Fabric version 1.3[1]: a permissioned DLT, where only authorised peers may join the network [4]. The selection was conducted comparing the currently available solutions using a scoring model derived from the requirements of IMs; in particular we referred to the Italian Railway Network handled by Rete Ferroviaria Italiana (RFI) which defined a list of requirements.

---

[1] www.hyperledger.org/projects/fabric.

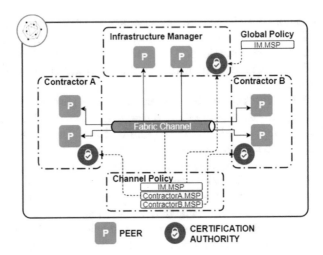

Fig. 1. A logical view of the Hyperledger Fabric network: each organization has its own peers authorized by the respective Certification Authority and connected to a single Fabric channel. Please note that the number of peers is not relevant in this view.

Figure 1 shows the logical architecture of the network, based on Hyperledger Fabric components [1]. Each organization participating in the maintenance operations management scenario has its own peers and a Certification Authority (CA). Each CA acts as a Membership Service Provider (MSP) for its own organization, and provides digital certificates to its related peers. The network is globally administrated by the IM in its role as ecosystem leader. All the organizations instead share the membership service of the dedicated logical channel where all the peers are connected, allowing each of them to add their own peers to the channel. Both the ledger and the chaincodes (smart contracts in Hyperledger Fabric) are replicated on every peer connected to the channel, providing redundancy of the data. The ordering service will be provided by a single orderer on a first testing phase, to be later developed to a crash tolerant Kafka cluster [1] on a later stage.

In our scenario, confidentiality of the data is not a critical requirement as the IM has a strong interest that all the process is transparent within all the authorized network participant. Nevertheless, Hyperledger Fabric v.1.3 enables the definition of Private Data within the same channel, to ensure that confidential data between two parties are not shared with the other participants: this is done using private databases separated from the main ledger, only broadcasting hashes of the private transactions on the main ledger. In addition, the network could easily support other scenarios and applications via the creation of different channels between the organizations, ensuring the complete separation of the respective ledgers.

The prediction engine is built as a separate component deployed outside the DLT network as it is not possible to implement it as a chaincode inside Fabric: the prediction engine needs to be able to automatically retrain (and therefore modify) the prediction algorithm, but chaincodes can only be updated manually. The interconnection between the two components will be developed through two REST APIs, as depicted in Fig. 2.

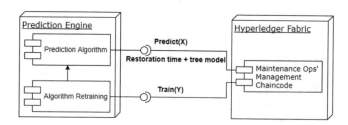

Fig. 2. Interconnection between the Hyperledger Fabric network and the Prediction Engine.

Once a maintenance operation is about to start, an operator has to officially state it committing a transaction to the DLT, where she inserts all the exogenous data, like the current weather condition, needed to the prediction engine. The chaincode therefore calls *Predict(X)*, where $\underline{X}$ stands for the required data for the prediction, getting in return the estimated restoration time and the tree model used to estimate it, that will be recorded as well on the ledger. It is important to note that the prediction algorithm does not need to retrieve additional data through external sources, since it gets everything from the chaincode. This is required to avoid non-determinism. Indeed, considering that the chaincode is executed by all the endorsing peers independently at potentially different times, retrieving data from external sources could change the results of each execution, since there is no control on external data, preventing the consensus from being achieved. *Train(Y)* is responsible of retraining the prediction algorithm using all the data stored on the ledger so far. In this case, there is no risk of having non-determinism as that API does not return anything; of course, retraining actions should be performed only when there are no maintenance operation starting, to avoid changing the prediction algorithm when it is predicting the restoration time of a job, leading to the non-deterministic condition explained before.

3   The Prediction Engine

The prediction engine is in charge of estimating the time to restoration for different assets and different failures and malfunctions. The predictive model needs to take into account the knowledge enclosed into maintenance reports, exogenous information such as the weather conditions and the experience of the operators in order to predict the time needed to complete a maintenance action over an asset and to restore its functional status. Moreover, the model should be interpretable enough to give insights to the operators on which are the main factors influencing the restoration time, to better plan the maintenance activities. This information will help IMs to assess the availability of the network, by estimating the time at which a section block including a malfunctioning asset will become available again, and properly reschedule the train circulation.

For this purpose we have built a rule-based model which is able to exploit real maintenance historical data provided by RFI, the historical data about weather conditions and forecasts, which is publicly available from the Italian weather services, and the experience-based model currently exploited by the train operators for predicting the restoration time of planned maintenance. Then we implemented the Decision Tree [7] using the

Table 1. Quality of the models.

Int.	MAE	MAPE	PCOR
RFI			
Maint.	30.5	31.5	0.75
Our proposal			
All	11.3 ± 1.1	10.7 ± 0.9	0.93 ± 0.03
Maint.	8.1 ± 1.0	7.8 ± 0.7	0.97 ± 0.03
Fail.	15.2 ± 1.3	14.3 ± 1.1	0.88 ± 0.04

MLlib [8] in Spark, because of the huge amount of data available (approximately 1 TB of data each year), and we deployed an infrastructure of four machines equipped with 32 GB of RAM, 500 GB of SSD disk space and 4 CPUs on Google Compute Engine[2]. We implemented the 10-fold cross validation for optimizing the number of points per leaf n_l by searching $n_l \in \{5, 10, 20, 50, 100, 200, 500\}$. All experiments have been performed 30 times to ensure the statistical robustness of the results. Table 1 reports:

- the error of the RFI model measured with the Mean Average Error (MAE), the Mean Average Percentage Error (MAPE), and the Pearson Correlation (PCOR) on the maintenance since no model for the failures is available to RFI;
- the error of the data-driven model measured with the MAE, MAPE, and PCOR on all the intervention, on the maintenance, and on the failures.

From Table 1 it is possible to note that the quality of our model is remarkably higher than the one of the RFI model.

4   Conclusions

The system described in this paper, built upon a permissioned DLT empowered with smart contracts and a prediction engine, permits the automated management of the highly regulated administrative workflow that each maintenance job has to deal with, while enriching it with the possibility to estimate the restoration time of each job, leading to a better planning of train disruptions. The main achievements of such system are twofold. The first one is to bring forward the

---

[2] https://cloud.google.com.

digitalisation of the workflow currently employed ensuring integrity and non-repudiation of every action performed inside the workflow thanks to the native features of DLTs; permitting, as a consequence, the instant retrieval of the status of each maintenance job. The second one is to allow to better plan the maintenance operations thanks to the availability of an estimated restoration time for each job. Additionally, the system could be extended to enable the enforcement of contractual clauses (i.e. penalties for delays) via automatic execution of disputation procedures backed by evidence stored in the audit-proof ledger.

Acknowledgments. This research has been supported by the European Union through the projects IN2DREAMS (European Union's Horizon 2020 research and innovation programme under grant agreement 777596).

References

1. Androulaki, E., Barger, A., Bortnikov, V., Cachin, C., et al.: Hyperledger fabric: a distributed operating system for permissioned blockchains. In: EuroSys (2018)
2. Benbunan-Fich, R., Castellanos, A.: Digitalization of land records: from paper to blockchain. In: International Conference on Information Systems (2018)
3. Budai, G., Huisman, D., Dekker, R.: Scheduling preventive railway maintenance activities. J. Oper. Res. Soc. **57**(9), 1035–1044 (2006)
4. De Kruijff, J., Weigand, H.: Understanding the blokchain using enterprise ontology. In: International Conference on Advanced Information Systems Engineering (2017)
5. Farrington-Darby, T., Pickup, L., Wilson, J.R.: Safety culture in railway maintenance. Saf. Sci. **43**(1), 39–60 (2005)
6. Fumeo, E., Oneto, L., Anguita, D.: Condition based maintenance in railway transportation systems based on big data streaming analysis. In: INNS International Conference on Big Data (2015)
7. James, G., Witten, D., Hastie, T., Tibshirani, R.: An Introduction to Statistical Learning. Springer, New York (2013)
8. Meng, X., Bradley, J., Yavuz, B., Sparks, E., Venkataraman, S., Liu, D., Freeman, J., Tsai, D.B., Amde, M., Owen, S.: MLlib: machine learning in apache spark. J. Mach. Learn. Res. **17**(1), 1235–1241 (2016)
9. Thaduri, A., Galar, D., Kumar, U.: Railway assets: a potential domain for big data analytics. In: INNS International Conference on Big Data (2015)

Presumable Applications of Deep Learning for Cellular Automata Identification

Anton Popov$^{(\boxtimes)}$ and Alexander Makarenko

Institute for Applied System Analysis, National Technical University of Ukraine
"Igor Sikorsky Kyiv Polytechnic Institute", Kiev, Ukraine
popovanton567@gmail.com, makalex51@gmail.com

Abstract. It is considered the prospects of deep learning applications for identification problems for cellular automata (CA): classical CA, CA with memory and with strong anticipatory property are examined. Some identification problems for neighbors, rules, coefficients of CA are discussed. One simplest example of deep learning for classical CA is described. Some possibilities of deep learning application for identification problems for CA with memory and anticipation are proposed. The case of of deep learning for the systems with multivalued behavior had been proposed.

Keywords: Deep learning · Cellular automata · Identification · Memory · Strong anticipation · Multivaluedness

1 Introduction

In modern computational neuroscience (and in other areas of science and technology), the use of models adequate to the behavior of a particular system. Note that, it seems, the most frequently used classes of models are the following: artificial neural networks; partial differential equations; models of related elements, differential equations systems or mappings; stochastic equations; network systems, etc. Remark that deep learning usage is one of common investigation instruments nowadays.

However, there is another important class of models, which is still much less common in computational neuroscientific problems, namely cellular automata (CA) (descriptions of classical CAs can be found in [9,10,15]). Regularity of space partitioning in CA simplifies computations but causes too rough description for some of real-world phenomenon.

Another problem is the choice of CA model rules that would allow a good representation of the behavior of real systems. The fact is that the "classical" fixed rules (for example, in the game "Life" [9,15]) do not always correspond well to the behavior of real systems. Approaches to identifying the rules of CA

© Springer Nature Switzerland AG 2020
L. Oneto et al. (Eds.): INNSBDDL 2019, INNS 1, pp. 126–135, 2020.
https://doi.org/10.1007/978-3-030-16841-4_13

have recently been used, based on real data on the temporal behavior of CA: Adamatsky [12], Bandini [3], Bolt [4] and others. In most cases, the rules of CA transitions from the current time layer to the next were used.

Relatively recently, CA have appeared that have a more complex structure of time dependence. The first group of such CAs are CA with memory, when the rules of transition from the time layer to the next layer take into account not only cell states at the current time, but also cell states from several previous time layers, Adamatsky, [1] and other works. Another group of new CAs are cellular automata with anticipation (taking into account the future unknown states, call them ACA). ACA's transition rules also include cell state values at some future points in time (Makarenko, Krushinskiy [5] and literature there). Note that the anticipation inclusion in an CA can lead to many-valued ACA solutions with interesting possibilities for interpretation [11].

All above leads to the need to use a much larger amount of information (data) in the CA transition, especially in the case of an ACA with multi-valued solutions. Fortunately, new approaches emerged to extract information and data structure (formerly primarily in models of artificial neural networks) - the so-called "deep learning" [8]. Multi-layered objects are used in deep learning. CA with memory and anticipation have a data structure that depends on the learning time and we can expect an application of deep learning for cell neighbourhood determination as well as CA rules.

Therefore, this work is devoted to description of the general problem of using deep learning to the above complex cellular automata, describing possible problem statements, as well as possible new approaches and interpreting solutions when applying deep learning.

The structure of the work is as follows. Section 2 is devoted to a discussion of the general problem of the identification of cellular automata and the possibility of using for this purpose methods and means of deep learning. Section 3 provides a brief description of cellular automata. Section 4 describes the new class of CA, namely cellular automata with strong antisipation. Draft of deep learning approach to CA identification problem is provided at Sect. 5. Section 6 contains considerations on the possible formulations of the problems of identification of the rules of the CA and the use of Deep Learning in the case of possible ambiguity (multy-valuedness) of solutions.

2   General Problem of Identification of Cellular Automatic Machines

In the introduction, we have provided (very short) description of the problem of identification of cellular automata. In this section, in according to existing papers, especially [4] we bring the general formulation of the task of identifying cellular automata, their types, and the possible role of Deep Learning for identifying cellular automata. As with most classical and non-classical methods of identification, it is assumed that there is an original process, a system, etc.

which generates certain data sets. It is assumed that for an approximate description of the behavior of such systems (the generation of approximations to exact,
in a certain sense, data), a model is used that provides such approximations.
Then the most common task of identification is to choose a class of models
for approximation, then choose a narrower subclass of models from a general
class and then select the parameters of such submodels, which ensures the least
divergence between real data sets and generated approximation models of solutions (data sets). We restrict ourselves to subclass of models with discrete time,
namely cellular automata, although some reasoning may be applicable to other
models, for example, to artificial neural networks. Then the general problem of
identification for such objects (CA) is as follows:

Identification task for cellular automata. Let there be some measure $E(U, V)$ of
the deviation of the approximated solution (or approximation data) V from the
exact solution (data set of the source system U). For cellular automata (CA) it
depends on many components: the geometry of the space of the cellular automaton, the neighborhoods of the elements, the rules for changing the state of the
cell, the state space of the cell and many others. Then the task of identification of
the cellular automaton is to determine the parameters of the cellular automaton
(or part of the parameters) at which the minimum deviation $E(U, V)$ is achieved
on a certain subset of the initial data.

$$(1)$$

The formulation of the identification problem in the form 1 allows for a very
large scope for detailing of specific tasks and used cellular automata. So, the
measure of deviation $E(U, V)$ can take a different form depending on the task
and purpose. In the simplest case $E(U, V)$ may coincide with L_2 or C. However,
even in this case $E(U, V)$ may depend on the values only at the final moment
of time, or on all the values of the cells on the trajectories at all points in time.
So the dimension of the space of adjustable (identifiable) parameters can be
relatively large even for relatively simple cellular automata. We also note that
for cellular automata with strong anticipation, we discovered the possibility of
many-valued solutions [5, 11], which in principle can significantly increase the
dimension of the parameter space. As already indicated in the introduction, in
the literature on the use and study of cellular automata, especially classical ones,
the idea of identification has already been encountered in various statements. So,
one of the methods was the use of genetic algorithms. Another obvious approach
is the use of gradient methods. It also seems reasonable to use artificial neural
networks for the identification goals. However, an increase in the dimension of
the parameter space and the number of cells in the CA with the rules connecting
the various time layers, leads to an even greater increase in the parameter space.
However, it is this relationship that provides an implicitly defined structure that
can be used to apply the approaches from the arsenal of deep learning. Of course,
there may be many such schemes. But, as far as authors of the article know, in

this formulation of the problem of cellular automata identification have not yet been considered. Therefore, to demonstrate the possibilities for using the ideas of deep learning, here we present only a few statements of such applications of deep learning to CA with memory. Due to the novelty of the subject, this article has so far described the possible most obvious statements that have analogies with the use of deep learning, primarily in pattern recognition.

Example 1. For the case of one-dimensional CA space, all trajectories of cells (their states at all intervals of discrete time ($t = 0, 1, 2, \ldots K$, $i = 1, 2, \ldots, N$ where K is the finite interval of discrete time and N is the number of cells of a one-dimensional cellular automaton) can be considered as image of pixel space of $K \times N$ dimensions and then apply method of convolutional networks, as is done in the theory of pattern recognition. Then, the outputs of such networks are the coefficients of the rules of the CA.

Example 2. One can consider the evolution of CA as a stream of CA configurations (a set of states of CA cells at fixed points in time). Then you can apply the methods of deep learning, adapted to work with dynamically changing scenes and their recognition.

Example 3. It is also possible to use approach of combination deep learning with reinforcement learning, as was used in video-games [2] or in Alpha Go [14]. Then the CA configurations at fixed times can be set in accordance with the scenes in video games or board configuration.

3   Cellular Automata Classes

3.1   Classical Cellular Automata Description

Cellular automata is a discrete model (see [9, 15]). It consists of a regular grid of *cells*, each in one of a finite number of states. The grid can be in any finite number of dimensions. For each cell, a set of cells called *neighborhood* is defined relative to the specified cell. An initial state (time $t = 0$) is selected by assigning a state for each cell. A new *generation* is created (advancing t by 1), according to some fixed *rule* (generally, a mathematical function) that determines the new state of each cell in terms of the current state of the cell and the states of the of the cells in its neighborhood.

3.2   Cellular Automata with Memory

Here we provide very brief description of CA with memory. First of all we follow the papers and books on CA (see for example [3, 6, 7, 9, 10, 12, 15]). We pose here the description of one-dimensional CA from [7] for example.

Also for further comparing and discussion we show the description of CA with memory from [7]: CA with *memory* extends standard framework of CA by allowing every cell x_i to remember some period of its previous evolution in the rules. Thus to implement a memory we design a memory function ϕ:

$$\phi(x_i^{t-\tau}, \ldots, x_i^{t-1}, x_i^t) \to s_i \tag{2}$$

Here x is a value of cell from finite alphabet Σ. Such that $\tau < t$ determines the degree of memory backwards and each cell $s_i \in \Sigma^n$ is a state function of the series of the states of the cell i with memory up to time-step. To execute the evolution we apply the original rule as follows:

$$\phi(\ldots, s_{i-1}^t, s_i^t, s_{i+1}^t, \ldots) \to x_i^{t+1} \tag{3}$$

In CA with memory, while the mapping ϕ remains unaltered, historic memory of all past iterations is retained by featuring each cell as a summary of its past states from ϕ.

4   Cellular Automata with Anticipation

4.1   Strong Anticipation Property

Here we pose shortly some materials from our previous investigations on CA with strong anticipation [5,11] for understanding the new prospects for deep learning applications. The term 'anticipation' had been introduced in active use in biology and applied math by Robert Rosen [13]. Since the beginning of 90th in the works by Dubois [6] the idea of strong anticipation had been introduced. Definition of an incursive discrete strong anticipatory system: an incursive discrete system is a system which computes its current state at time $t + 1$, as a function of its states at past times , $\ldots, t - 3, t - 2, t - 1$, present time t, and even its states at future times $t + 1, t + 2, t + 3, \ldots$

$$x(t + 1) = A(\ldots, x(t - 2), x(t - 1), x(t), x(t + 1), x(t + 2), \ldots p) \tag{4}$$

where the variable x at future times $t + 1, t + 2, t + 3, \ldots$ is computed in using the equation itself.

4.2   Cellular Automata with Strong Anticipation

The key idea is to introduce strong anticipation into CA construction. We will suppose that states of CA can depend on future (virtual) states of cells.

Here we for simplicity describe the system of such CA without memory and only with one-step anticipation. General forms of such equations ($k = 1$):

$$\phi(\bar{x}_i^t, \bar{x}_i^{t+1}) \to \bar{s}_i^{t+1} \tag{5}$$

$$\phi(\ldots, \bar{s}_{i-1}^{t+1}, \bar{s}_i^{t+1}, \bar{s}_{i+1}^{t+1}, \ldots) \to \bar{x}_i^{t+1} \tag{6}$$

The main peculiarity of solutions of 5, 6 is presumable multi-validness and existence of many branches of solutions. This implies also the existence of many configurations in CA at the same moment of time.

Here in the Fig. 1 we pose the results of computer simulation of Game of Life modification with accounting of strong anticipation (see details of description here [5]). Calculations corresponds to game with space dimension of 16 cells with periodic boundary conditions. The white cells corresponds to state 0 of cell and black cell corresponds to cell state 1. At the right side of the Fig. 1 we can see the existence of multiple states of cells (an configuration at a given moment of time). For simplicity initial state and only the next time step are presented.

Note that such multi-valuedness leads to the new problems of applications of Deep Learning.

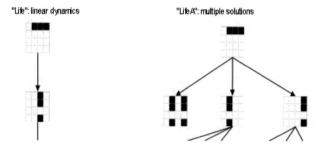

Fig. 1. Solutions of classical "Game of Life" (left side) and "Game of Life A" with strong anticipation (right side)

5   Structure of DL Approach for Simplest One Dimensional CA Identification

Here for illustration we provide an example for simples 1D CA. Problem of CA rules identification can be attributed to pattern recognition domain, where convolutional neural networks have established themselves as a solid and robust tool. One-dimensional CA at fixed time t is represented by 1d array. Sequence of such arrays form evolution of CA. So it's possible to represent evolution of CA with n dimensional neighborhood by $n + 1$ array - we add dimension of time. Then we can feed to CNN whole this evolution array or subset of slices as shown on Fig. 2.

Space of possible neighborhood combinations is defined by size of cells in the neighborhood n and number of each cell states m and equal to n^m. Then, total number of possible CA for fixed n and m is n^{n^m}. Therefore, even for small n and m we receive huge space of candidates CA.

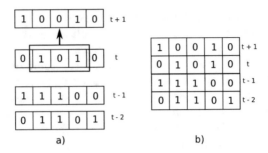

Fig. 2. Input data configuration for simplest one dimensional CA. Picture (a) treats CA evolution as sequence of CA states at each of moments of time $t-2$, $t-1$, t, $t+1$ etc. Picture (b) represents states of CA as a single batch.

6 Deep Learning for Cellular Automatas with Strong Anticipation

In the previous subsections, we discussed the application of deep learning methods to classical CA or CA with regard to memory effects. The usefulness of such approaches is due to the fact that in such systems a large number of time layers are used in the transition rules and, consequently, a large number of unknown coefficients in the transition rules (dynamic laws) of such CA should be found. However, described in Subsect. 4 of CA, taking into account strong anticipation, are completely new objects, and even in the absence of hyperincursion, i.e. with the uniqueness of solutions.

6.1 CA with Anticipation and with Unique Solution

Suppose a CA with strong anticipation is given, i.e. taking into account possible future values (4), but with a unique solution. For simplicity, we do not consider memory here. Assuming that the transition function ϕ in 2, 3, 5, 6 depends on a set of coefficients, local neighborhoods and possible states in $k, (k+1), (k+2), \ldots$ moment of time. For simplicity, we can further describe ideas using the simplest example of taking into account only one future layer, i.e. taking into account the states at a given moment k and only at the next $(k+1)$ moment of discrete time. For simplicity it is supposed that local neighborhood of each cell is known at all moments of the time and does not change over time. Although, in more complicated models, if necessary, it is possible to include neighborhood configuration in the list of defined values.

Then the formulation of the problem of finding the right CAs in this case (with unambiguous trajectories) coincides with the interpretation for the CA with memory. That is, some set of known trajectories of such CA and along these trajectories of the set of trajectories it is necessary to somehow determine the rules of CA, which minimize the deviations Δ on this set. In this case, there are certain differences from the case of classical CA or CA with memory, but they are not of a fundamental nature. So, if the behavior of the CA system is in the discrete time interval $[0, 1, 2, \ldots, T]$, where T is an integer (the measurement interval limit, then, due to the leading rule of the CA, you can calculate the CA trajectories only in the interval $[0, 1, 2, \ldots, T - m]$, since the CA state values are unknown for $t > T$. This may involve deep learning.

6.2   CA with Multi-valued Solutions

In case of multi-valued solutions, Eqs. 5, 6 are valid, but the meaning of all the found rules may vary. For simplicity, we will discuss here only the case of unambiguous and real coefficients in the considered CA rules (although in principle much more complicated variants are possible, for example, with the multiplicity of coefficients). Thus, for the time being we consider the case of a single-valued function ϕ. So, under certain conditions, with single-valued coefficients in the CA rules and single-valued functions ϕ, multi-valued solutions of Eqs. 5, 6 are possible, precisely because of its nonlinearity.

Consider what will change in the general scheme for determining the rules (coefficients) of the CA for a start on the example of the application of genetic algorithms. We note that here we give only sketches of ideas, and their complete formalization and use require a large sequence of labor-intensive processes. Since the CA trajectory in this case, is a multi-valued branching construct, the question arises how to estimate the deviations from the original trajectory, as well as the second question about the adequate measurement of these deviations. Here, first of all, the use of metrics for deviations of sets - for example, the Hausdorff metrics - is suggested. Or, perhaps, at some stage, not a scalar, but a vector measure of deviation will be required, and in the future, it may also be multi-valued.

The next task is to determine the transition rules of genetic algorithms: mutations, recombination of a crossover, taking into account the peculiarities of measuring deviations. Also in CA, probabilities may be required if the CA algorithm is rejected. In principle, they can implicitly depend on the structure of the distribution of multivalued branches of multivalued CA solutions. Note that if we also allow the polysemy of the coefficients of the rules of the CA and the function f, then there is the problem of constructing, substantiating and using multivalued genetic algorithms, which, as far as the authors know, have not yet been considered in the literature. At the same time, it should be noted that in the case of the ambiguity of solutions, completely new problems arise in the generalization of classical back propagation methods in the theory and practice of artificial neural networks. Fortunately, CAs generally do not use a differentiation procedure. Therefore, we can hope for a transfer of the concept of CA and a multi-valued case.

6.3   To the Perspectives of Applying Deep Learning to CA with Multi-valued Solutions

In the previous Subsect. 6.2 of this Sect. 6, some emerging features of CAs have already been described in the case of possible multivalued solutions. Let us now discuss the prospects for the use of DL for this case and possible necessity of generalizations of the deep-learning applications, which seem to lead to the need for generalizing DL to a multi-valued case. Here, we present only a few remarks that are being suggested, although the general multi-valued case requires a much larger amount of research on completely new aspects that may be new for deep-learning schemes.

The first remark concerns the scheme of applying DL to a new (as far as the authors know) class of objects with multi-valued (having many different states of system elements at the same moment in time (for examples, see [11])). We will give here only a few more or less obvious new features that we will try to illustrate here on the generally accepted example of DL - convolutional neural networks for pattern recognition. In this case, the base unit is not a single-valued image (for example, consisting of the states of screen pixels at a given moment (or scenes at successive moments of time)), but multi-valued images. Then, in the convolutional neural network scheme, the necessary step is the convolution of a multi-valued image with possible multi-valued output on the next layers. In this case, a lot of questions arise about the interpretation of results, measurements in real systems, averaging many-valued values. Note that in the multi-valued case, a completely new interpretation of previously solved problems (for example, theorems on the approximation of functions from C or L_2 by neural networks) is acquired. Returning to the problem of identifying CA, partially described in Sect. 4, we can point out the following possible modifications to the application of DL to the case of multi-valued CA. Thus, examples 1, 2, 3 at the end of Sect. 4 acquire a new quality taking into account possible multi-valued CA solutions. The case of many-valued functions in rules, many-valued coefficients, different local neighborhoods of a CA cell for different branches, etc., is much more complicated, but more interesting and completely unexplored. Note that in this case the computational complexity of the work significantly increases, taking into account the presence of many branches, states and the potential possibility of further branching during the evolution of the model (and, possibly, of the initial system itself). All this can lead to a significant increase in the amount of information that will need to be processed when solving the problems of identification of CA rules. On the other hand, the potential polysemy allows you to save fundamentally large volumes of information about the system while maintaining the same geometric configuration in the architecture of the model.

7   Conclusions

Thus in given paper we have proposed to consider some aspects of DL application. The main topics concerned presumable applications of DL to CA. The general structure of such applications are discussed as some concrete examples are considered. The general ideas had been illustrated on the example of simple one-dimensional cellular automata. Also the idea of deep-learning application to multi-valued behavior of the objects had been introduced. It should be remarkable that the same ideas on DL may be useful for other dynamical systems - artificial neural networks, differential equations etc.

References

1. Adamatsky, A.: Identification of cellular automata rules (1994)
2. Aggarwal, C.C.: Neural Networks and Deep Learning. A Textbook (2018). https://doi.org/10.1007/978-3-319-94463-0
3. Bandini, S., Mauri, G.: Multilayered cellular automata. Theor. Comput. Sci. **217**(1), 99–113 (1999). https://doi.org/10.1016/S0304-3975(98)00152-2
4. Bolt, W., Baetens, J., De Baets, B.: Idenmtifying CAs with evolution algorithms. In: Proceedings of the International Conference on AUTOMATA-2013. Giessen Univ Press, pp. 11–20 (2013)
5. Krushinsky, D., Makarenko, A: Cellular automata with anticipation. examples and presumable applications. In: Dubois, D.M. (ed.) AIP Conference Proceedings, USA, vol. 1303, pp. 246–254 (2010)
6. Dubois, D.M.: Generation of fractals from incursive automata, digital diffusion and wave equation systems. BioSystems **43**(2), 97–114 (1997). https://doi.org/10.1016/S0303-2647(97)01692-4
7. Dubois, D.M.: Computing anticipatory systems with incursion and hyperincursion. Int. J. Comput. Anticip. Syst. **2**, 3–14 (1998). https://doi.org/10.1063/1.56331
8. Goodfellow, I., Bengio, Y., Courville, A.: Deep Learning. MIT Press, Cambridge (2016). http://www.deeplearningbook.org
9. Ilachinski, A.: Cellular Automata A discrete Universe. World Scientific, Singapore (2001)
10. Kari, J.: Theory of cellular automata: a survey. Theor. Comput. Sci. **334**(1–3), 3–33 (2005). https://doi.org/10.1016/j.tcs.2004.11.021
11. Makarenko, A.: Multivaluedness aspects in self-organization, complexity and computatios investigations by strong anticipation. In: Stoop, R., Chedjou, J., Li, Z., Kyamakya, K., Mathis, W. (eds.) Recent Advances in Nonlinear Dynamics and Sinchronization, pp. 33–54. Springer, Cham (2018). https://doi.org/10.1007/978-3-319-58996-1_3
12. Martinez, G.J., Adamatzky, A., Alonso-Sanz, A.: Complex dynamics of elementary cellular automata emerging from chaotic rules. Int. J. Bifurc. Chaos **1274**(2) (2010). https://doi.org/10.1142/S021812741250023X
13. Rosen, R.: Anticipatory Systems. Philosophical, Mathematical, and Methodological Foundations, vol. 39. Springer, New York (2008)
14. Silver, D., Schrittwieser, J., Simonyan, K., Antonoglou, I., Huang, A., Guez, A., Hubert, T., Baker, L., Lai, M., Bolton, A., Chen, Y., Lillicrap, T., Hui, F., Sifre, L., Van Den Driessche, G., Graepel, T., Hassabis, D.: Mastering the game of Go without human knowledge. Nature **550**(7676), 354–359 (2017). https://doi.org/10.1038/nature2427
15. Wolfram, S.: A New Kind of Science. Wolfram Media, Champaign (2002). http://www.wolframscience.com

Restoration Time Prediction in Large Scale Railway Networks: Big Data and Interpretability

Luca Oneto[1(✉)], Irene Buselli[1], Paolo Sanetti[1], Renzo Canepa[2], Simone Petralli[2], and Davide Anguita[1]

[1] DIBRIS - University of Genoa, Via Opera Pia 13, 16145 Genova, Italy
{luca.oneto,irene.buselli,paolo.sanetti,davide.anguita}@unige.it
[2] Rete Ferroviaria Italiana S.p.A.,
Via Don Vincenzo Minetti 6/5, 16126 Genova, Italy
{r.canepa,s.petralli}@rfi.it

Abstract. Every time an asset of a large scale railway network is affected by a failure or maintained, it will impact not only the single asset functional behaviour but also the normal execution of the railway operations and trains circulation. In this framework, the restoration time, namely the time needed to restore the asset functionality, is a crucial information for handling and reducing this impact. In this work we deal with the problem of building an interpretable and reliable restoration time prediction system which leverages on the large amount of data generated by the network, on other freely available exogenous data such as the weather information, and the experience of the operators. Results on real world data coming from the Italian railway network will show the effectiveness and potentiality of our proposal.

Keywords: Railway network · Restoration time prediction · Big data · Data driven models · Interpretable models · Experience based models

1 Introduction

The functional behavior of railway infrastructure assets degrades for many different reasons [8]: age, extreme weather conditions, heavy loads, and the like. For example, the influence of snow on switches is critical, in particular when switch heating is not functioning properly. Even worse is the case of wind in combination with snowfall, when the assets belonging to a specific area can be significantly affected. Additionally, problems can be introduced unknowingly by performing maintenance actions, for example by a simple human error or as a reaction of the system to changes made on an object [1]. For instance, some maintenance activities (e.g. tamping or ballast dumping) performed close to a switch can change the track geometry, then other parts of this asset must be adjusted to the new situation. One of the most crucial pieces of information

© Springer Nature Switzerland AG 2020
L. Oneto et al. (Eds.): INNSBDDL 2019, INNS 1, pp. 136–141, 2020.
https://doi.org/10.1007/978-3-030-16841-4_14

needed to reduce the impact on train circulation of assets failures and maintenance is the restoration time, namely the time needed to restore the complete
functionality of the asset [2].

For this reason in this work we will investigate the problem of predicting the
time to restoration for different assets and different failures and malfunctions. In
other words, the objective of this analysis is to estimate the time to restoration
for planned (anticipating faults) and corrective maintenance (rectifying faults)
by looking at the past maintenance reports, correlated to the different assets and
different types of malfunctions. The predictive model needs to take into account
the knowledge enclosed into maintenance reports, exogenous information such
as the weather conditions (e.g. weather condition) and the experience of the
operators in order to predict the time needed to complete a maintenance action
over an asset and to restore its functional status. Moreover, the model should be
interpretable enough to give insight to the operators in what the main factors
influencing the restoration time are in order, for example, to better plan the
maintenance activities. This information will help the Traffic Management System to assess the availability of the network, for example by estimating the time
at which a section block including a malfunctioning asset will become available
again, and properly reschedule the train circulation.

For this purpose we will build a rule-based model which is able to exploit
real maintenance historical data provided by Rete Ferroviaria Italiana (RFI), the
Italian infrastructure manager that controls all the traffic of the Italian railway
network, the historical data about weather conditions and forecasts, which is
publicly available from the Italian weather services, and the experience-based
model currently exploited by the train operators for predicting the restoration
time of planned maintenance. Results on these real world data will show the
effectiveness and potentiality of our proposal.

2   Proposed Approach

The Restoration Time prediction problem can be easily mapped into a classical
regression [7] problem where we have an input space $\mathcal{X}$ and an output space
$\mathcal{Y} \subseteq \mathbb{R}$ and the purpose is to find the unknown relation μ between $\mathcal{X}$ and $\mathcal{Y}$. In
our case $\mathcal{X}$ is the description of the maintenance action or the failure, plus some
exogenous information such as the weather conditions, plus the experience of the
operator that, thanks to this last information, is sometimes able to provide, for
example in case of planned maintenance, an estimation of the restoration time. $\mathcal{Y}$,
instead, is the actual restoration time that we want to predict. In this framework,
we would like to find a model $h : \mathcal{X} \rightarrow \mathcal{Y}$ which approximates μ just based on
a finite set of observations of μ, called dataset $\mathcal{D}_n = \{(X_1, Y_1), \cdots, (X_n, Y_n)\}$,
composed of n tuples where $X \in \mathcal{X}$ and $Y \in \mathcal{Y}$. The model h should be a good
approximation of μ, but it should also be easy to understand or interpret by
a human operator who wants to get insights from the model h, and not just
a prediction of the restoration time. For this purpose, in order to measure the
quality of h in approximating μ we have to define one or more accuracy measures.

Since we will exploit $\mathcal{D}_n$ for building h, first we need to exploit an additional fresh dataset of cardinality m, called test set $\mathcal{T}_m = \{(X_1', Y_1'), \cdots, (X_m', Y_m')\}$ in order to be sure that the accuracy of h on $\mathcal{T}_m$ is an unbiased estimator of the true accuracy [6]. Then we will use as measures of accuracy the following quantities:

- the Mean Absolute Error: $\mathrm{MAE} = 1/m \sum_{i=1}^{m} |h(X_i') - Y_i'|$;
- the Mean Absolute Percentage Error $\mathrm{MAPE} = 100/m \sum_{i=1}^{m} |h(X_i') - Y_i'|/|Y_i'|$;
- the Pearson Product Moment Correlation Coefficient, (or bivariate correlation) $\mathrm{PPMCC} = \sum_{i=1}^{m}(h(X_i') - \bar{h})(Y_i' - \bar{Y}')/\sqrt{\sum_{i=1}^{m}(h(X_i') - \bar{h})^2}\sqrt{\sum_{i=1}^{m}(Y_i' - \bar{Y}')^2}$ where $\bar{h} = 1/m \sum_{i=1}^{m} h(X_i')$ and $\bar{Y} = 1/m \sum_{i=1}^{m} Y_i'$.

In order to ensure the interpretability of h, instead, we have to perform two different steps. First of all, we need a functional form of the model that is interpretable by construction, and hence the most reasonable choice is to use a rule based model, namely a decision tree. A decision tree can be efficiently and effectively learned from the data even when millions of samples are available [4,5]. The second step is to find how to map $\mathcal{X}$ in such a way that the model, learned on this mapping, still remains interpretable. More formally we have to find a function $\phi : \mathcal{X} \rightarrow \Phi$ where Φ must be, from one side, a rich representation of $\mathcal{X}$ and, from another side, it must contain features that are easy to understand, grounded, and physically connected with the nature of the problem. In this way the final functional form of the model $h \circ \phi : \mathcal{X} \rightarrow \mathbb{R}$ will be an interpretable model based on a rich and interpretable feature set. Note that Φ may contain both numerical and categorical features since decision trees can efficiently and effectively handle naively both types of feature spaces. In order to create this rich and interpretable feature space we operated a two step approach. First we have enriched the original space $\mathcal{X}$ of new features designed together with the RFI experts in order to include their knowledge inside the ϕ. Note that the experience of the operator is also included in $\mathcal{X}$ as a feature which estimates the restoration time for the planned maintenance based on a model developed by the RFI experts during the years[1]. As a second step we estimated the most important and easy to interpret statistical descriptors of the features designed by the experts (mean, variance, skewness, kurtosis, etc.) Finally, we learned over $\Phi \times \mathcal{Y}$ a decision tree, pruning the tree based on the 10-fold cross validation principle [6]. The pruning procedure has been performed optimizing the number of points per leaf [4]. In order to handle the size of the problem, as in our case n can count even millions of samples, we implemented everything in Scala using Spark with the Decision Tree library included in MLlib [5]. As a final remark we would like to underline that, given the nature of the problem which evolves in time, in $\mathcal{D}_n$ we have the data until December 2016 while in $\mathcal{T}_m$ we have the data from January 2017 on; in this way we are actually simulating to apply the model in the future based on data about the past.

---

[1] The details about this model cannot be disclosed because of confidentiality issues.

3   Experimental Evaluation

In order to test our proposal, we first have to describe the available data.

- RFI provided data from January 2015 until December 2018 about the Italian Railway Infrastructure[2]. They contain data about the location of the maintenance or fault, the stations and track involved, the estimated restoration time based on the RFI model for just the maintenance (RFI has not a model for the faults), and the actual duration of the maintenance. In the data there are also notes of the operators.
- From the Italian weather services [3] we retrieved data about the actual and predicted weather information regarding the same years of the RFI's data. From these services it is possible to retrieve the hourly information about wind, temperature, rain, snow, solar radiation, and fog.

From these data, together with the RFI experts we extracted the feature set Φ described in Table 1. Then we implemented the Decision Tree using the MLlib [5] in Spark and we deployed an infrastructure of four machines equipped with 32 GB of RAM 500 GB of SSD disk space and 4 CPU on the Google Compute Engine[3]. We implemented the 10-fold cross validation for optimizing the number of points per leaf n_l by searching $n_l \in \{5, 10, 20, 50, 100, 200, 500\}$. All experiments have been performed 30 times in order to ensure the statistical robustness of the results. Table 2 reports

Table 1. Feature set Φ extracted with the RFI experts.

Name	Meaning	Input/Output
Prov	Province of the intervention	Input
BegStat, EndStat	Geolocation of the beginning and end stations involved	Input
Track	What track are involved	Input
Type	Maintenance or Failure	Input
Intervention	Type of Intervention	Input
Day, Month, Hour	Information about the time of the beginning of the intervention	Input
Rain, Temp, Sun, Wind, Snow, Fog	Information about the weather at the beginning of the intervention	Input
PredictedTime	RFI estimated restoration time (just for the maintenance and not available for the failures)	Input
ActualTime	Actual restoration time (in minutes)	Output

---

[2] In order to give an idea of the number of maintenance and faults, just for a small region of the north of Italy we have more that 100.000 records; we cannot disclose more details because of confidentiality issues.

[3] https://cloud.google.com.

- the error of the RFI model measured with the MAE, MAPE, and PPMCC on the maintenance since no model for the failures is available to RFI on $\mathcal{T}_m$;
- the error of our model measured with the MAE, MAPE, and PPMCC on all the intervention, on the maintenance, and on the failures on $\mathcal{T}_m$.

Figure 1 reports the first three levels of the model derived from the data[4].

Table 2. First three levels of the model.

Int.	MAE	MAPE	PPMCC
RFI			
Maint.	30.5	31.5	0.75
Our proposal			
All	11.3 ± 1.1	10.7 ± 0.9	0.93 ± 0.03
Maint.	8.1 ± 1.0	7.8 ± 0.7	0.97 ± 0.03
Fail.	15.2 ± 1.3	14.3 ± 1.1	0.88 ± 0.04

Fig. 1. Quality of the models.

From Table 2 and Fig. 1 it is possible to note that the quality of our model is remarkably higher than the one of the RFI model. In particular for the maintenance we reach a MAPE lower than 10% which is more than 3× better than the accuracy of the RFI model. Even for the failures the accuracy of the restoration time model is remarkable if compared with the model of RFI. Note also that, by looking at Fig. 1, the model is very easy to interpret and reasonable (e.g. the type of intervention is on top, together with the location of the intervention and the weather information). Note also that the model of RFI is taken considerably into account in case of maintenance as expected since the RFI model has already a quite high predictive power.

4   Discussion

In this work we dealt with the problem of predicting the restoration time of a part of the railway network from an intervention on an asset based on data coming from the railway information system, exogenous variables like the weather information, and the experience of the operators. Moreover, given the particular application which is very human oriented, it was required to build a model as interpretable as possible in order to help the operators in taking decisions not just based on a prediction but also based on the functional form of the model. For these reasons, we proposed an approach which produces easy-to-interpret models, is computationally efficient and effective, and is able to handle a huge amount of historical interventions. Results on data provided by RFI coming from Italian Railway Network support our proposal both in terms of quality and interpretability of the derived models.

---

[4] The full model which has more levels is not reported for confidentiality issues.

Acknowledgments. This research has been supported by the European Union through the projects IN2DREAMS (European Union's Horizon 2020 research and innovation programme under grant agreement 777596).

References

1. Ahren, T., Parida, A.: Maintenance performance indicators (MPIs) for benchmarking the railway infrastructure: a case study. Benchmarking Int. J. **16**(2), 247–258 (2009)
2. Cacchiani, V., Huisman, D., Kidd, M., Kroon, L., Toth, P., Veelenturf, L., Wagenaar, J.: An overview of recovery models and algorithms for real-time railway rescheduling. Transp. Res. Part B Methodol. **63**, 15–37 (2014)
3. Italian Weather Services: Weather Data (2018). http://www.cartografiarl.regione. liguria.it/SiraQualMeteo/script/PubAccessoDatiMeteo.asp. Accessed 16 Oct 2018
4. James, G., Witten, D., Hastie, T., Tibshirani, R.: An Introduction to Statistical Learning. Springer, New York (2013)
5. Meng, X., Bradley, J., Yavuz, B., Sparks, E., Venkataraman, S., Liu, D., Freeman, J., Tsai, D.B., Amde, M., Owen, S.: MLlib: machine learning in apache spark. J. Mach. Learn. Res. **17**(1), 1235–1241 (2016)
6. Oneto, L.: Model selection and error estimation without the agonizing pain. WIREs Data Min. Knowl. Discov. **8**(4), e1252 (2018)
7. Shalev-Shwartz, S., Ben-David, S.: Understanding Machine Learning: From Theory to Algorithms. Cambridge University Press, Cambridge (2014)
8. Zoeteman, A.: Asset maintenance management: state of the art in the european railways. Int. J. Crit. Infrastruct. **2**(2–3), 171–186 (2006)

Train Overtaking Prediction in Railway Networks: A Big Data Perspective

Luca Oneto[1]([✉]), Irene Buselli[1], Alessandro Lulli[1], Renzo Canepa[2],
Simone Petralli[2], and Davide Anguita[1]

[1] DIBRIS, University of Genoa, Via Opera Pia 13, 16145 Genoa, Italy
{luca.oneto,irene.buselli,alessandro.lulli,davide.anguita}@unige.it
[2] Rete Ferroviaria Italiana S.p.A., Via Don Vincenzo Minetti 6/5, 16126 Genoa, Italy
{r.canepa,s.petralli}@rfi.it

Abstract. Every time two or more trains are in the wrong relative position on the railway network because of maintenance, delays or other causes, it is required to decide if, where, and when to make them overtake. This is a quite complex problem that is tackled every day by the train operators exploiting their knowledge and experience since no effective automatic tools are available for large scale railway networks. In this work we propose a train overtaking hybrid prediction system. Our model is hybrid in the sense that it is able to both encapsulate the experience of the operators and integrate this knowledge with information coming from the historical data about the railway network using state-of-the-art data-driven techniques. Results on real world data coming from the Italian railway network will show that the proposed solution outperforms the fully data-driven approach and could help the operators in timely identify and schedule the best train overtaking solution.

Keywords: Railway network · Train overtaking · Big data ·
Data-Driven Models · Hybrid models

1 Introduction

Railway Transportation Systems (RTSs) play a vital and crucial role in public mobility and goods delivery. In Europe the increasing volume of people and freight transported on railway is congesting the network [5]. The only fast and economically viable way to increase capacity is then to improve the efficiency of daily operations in order to be able to control a larger number of running trains without requiring massive public investments in new physical assets [23]. For this reason, in the last years, every actors of the RTSs has started extensive modernization programs that leverage on advanced information and communication solutions. The objectives are to improve system safety and service reliability, to enhance passenger experience, to provide higher transit capacity and to reduce operational costs.

© Springer Nature Switzerland AG 2020
L. Oneto et al. (Eds.): INNSBDDL 2019, INNS 1, pp. 142–151, 2020.
https://doi.org/10.1007/978-3-030-16841-4_15

In this work we focus on the problem of analyzing the train movements in Large-Scale RTSs for the purpose of understanding and predicting their behaviour. In particular, we will study the problem of the train overtaking prediction exploiting data-driven solutions leveraging on the huge amount of data produced and stored by the new RTSs information systems. Train overtaking prediction is the problem of predicting when it is required or preferable to make a train perform an overtaking in order to minimize the train delays and the penalty costs associated with them. The study of this problem allows to improve the quality of service, the train circulation, and the Infrastructure Managers and Train Operators management costs.

A large literature covering the prediction problems related to the train circulation already exists [10]. However, in general, the majority of the works focus on different problems: the running time prediction, the dwell time prediction, and the train delay prediction. These focus on predicting, respectively, the amount of time needed to traverse a section of railway between two checkpoints [1,6,11,13,15,17], the amount of time spent in a checkpoint and the difference between the actual arrival (or departing time) and the scheduled one in each of the stations composing the itinerary of a train [2,3,9,12,14,19,20,24].

The problem of train overtaking prediction, instead, has never been studied exploiting data-driven solutions. Current solutions model this problem as a complex optimization task [7,8,16] that is usually not easy to solve (or impossible to solve in large scale railway networks) and requires a lot of human effort during the modeling phase. For this reason, in practice, the current solution is to rely on the experience of the operators and on their knowledge of the network. We call this solution Experience-Based Model (EBM). A solution that we proposed here is to adopt a data-driven approach. In this framework, advanced analytic methods [10,13] can be exploited to analyze the historical data and to build Data-Driven Models (DDMs) which automatically predicts when it is better to perform the train overtaking. Unfortunately, also DDMs have their drawbacks since they do not handle easily the fact that prior knowledge about the problem may be available, apart from the historical data. For this reason, in this work, we propose an hybrid approach to the train overtaking prediction problem mixing together the EBMs and DDMs taking inspiration from our previous works, where we employed a similar idea to deal with the running time, the dwell time and the train delay prediction problems [18]. The combination of the two approaches allows us to create a model that shows the strengths of both EBMs and DDMs while limiting their weaknesses. On one hand, encapsulating the experience of the operators enables the creation of an interpretable and robust model which can be better exploited in a human-oriented environment like the one of the train operators. On the other hand, the exploitation of data-driven techniques allows to build more accurate predictive models. Results on real world data about the Italian railway network provided by Rete Ferroviaria Italiana (RFI - the Italian Infrastructure Manager) will show the effectiveness of our proposal.

2   The Train Overtaking Prediction Problem

In this section we introduce the notation needed to formally describe the problem of train overtaking. A railway network can be easily described with a graph. Figure 1 depicts a simplified railway network where two trains follow their itineraries. Let us consider the train at the station C_B, characterized by its itinerary (origin at station C_A, destination at station C_F, some stops and some transits). In the following, we will call checkpoint a station without differentiating where the train stops or transits and between actual stations and points of measure. The railway sections are the pieces of the network between two consecutive checkpoints and have also an orientation (e.g. transit from C_D to C_E is different from transit from C_E to C_D).

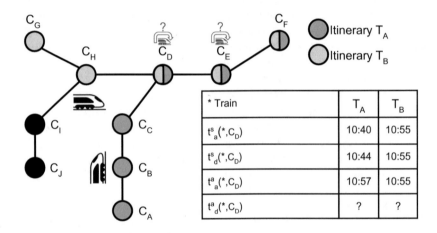

Fig. 1. Train *Overtaking*.

For any checkpoint C in the itinerary, the train T is scheduled to arrive and depart at different specified times, defined in the timetable, respectively, $t_a^s(T, C)$ and $t_d^s(T, C)$. The difference between the actual time (either for arrival $t_a^a(T, C)$ or for departure $t_d^a(T, C)$) and the scheduled time is defined as train delay. Each train and checkpoint have additional characteristics such as an unique identifier, the category of the train, and the category of the railway network. However, due to delays caused by many different reasons, it is common that two trains are in a wrong relative position along their itinerary.

Let us refer again to Fig. 1 for a graphical description of the problem. In our scenario, we say that two trains are in a wrong relative position at a checkpoint C_D when $t_a^s(T_A, C_D) < t_a^s(T_B, C_D)$ and $t_a^a(T_A, C_D) > t_a^a(T_B, C_D)$, i.e. T_A is expected to arrive before T_B in checkpoint C_D but, for some reason, train T_B arrives before train T_A in checkpoint C_D.

When an event like this occurs it is required to predict and enforce as soon as possible, with an overtake, the correct relative position of the trains for the purpose of minimizing delays and deviations from the timetable. In order to enforce the overtake it is necessary to predict what is the best subsequent station in the itinerary to perform the overtake minimizing the deviation from the timetable. Note that not all the checkpoints allow the overtake (e.g. no additional track is available to perform the overtake in all the checkpoints).

In the example of Fig. 1, the system detects the incorrect position in checkpoint C_D because train T_A is scheduled to arrive before train T_B on checkpoint C_D, but train T_A has a delay and, for this reason, train T_B arrives before train T_A in checkpoint C_D. After the detection of the incorrect relative position, the system starts evaluating if and when, on the current or subsequent checkpoints in common between the itineraries of train T_A and T_B, it is preferable to make train T_A overtake train T_B.

3   Our Proposal: An Hybrid Model

In this work, we propose an HM to tackle the train overtaking prediction problem. In particular, we mapped the problem of overtaking into a series of binary classification problems where the task is to predict if or not an overtake will be performed at a particular checkpoint. The idea is to leverage on both the experience of the operators (EBM) and historical data with the use of advanced data analytics methods (DDM). The goal is to build an accurate, dynamic, robust, and interpretable model able to support the decision of the operators.

For this purpose we propose a two level architecture. At the top level, we construct a tree following the suggestions of the operators, which captures the characteristics of the two trains under consideration for the overtake and additional information to better describe the scenario under examination. Such top level tree encapsulates the EBM developed by the operators during the years. At the bottom level, for each of the leaves composing the tree we have built a dataset with all the past occurrences of the overtake corresponding to that particular leaf. This dataset is richer, in terms of feature set, with respect to the top level tree and, leveraging on this we have built a DDM able to improve the accuracy of the top level tree.

More in detail, the top level decision tree encapsulates the experience of the operators in taking decisions when two trains are in the wrong relative position and one has to overtake the other (see Sect. 2). The proposed HM groups all the possible situations in subgroups based on a series of similarity variables (see Table 1), defined together with the RFI experts, which allow us to have, on one side, robust statistics, thanks to the possibility to learn from a reasonable group of similar overtake situations and, on the other side, a rich feature set, able to capture the variability of the phenomena. Then, in each leaf of this tree, we exploit a DDM able to learn from the historical data in that particular leaf, based on a super-set of features with respect to the one used in the tree (see Table 2). In particular, each leaf is a Random Forest (RF) classifier [4] (following the experience of the DDM developed in [21]), which predicts if the trains will perform the overtake in a particular checkpoint. The whole HM is built and updated incrementally as soon as new train movements are recorded. During the prediction phase, instead, we just visit the tree considering the particular overtake situation and we exploit the corresponding RF classifier to make the actual prediction.

Table 1. HM top level decision tree feature set (Cat. means Categorical)

Feature name	Cat.	Description
Railway Section	Yes	The considered railway section
Railway Checkpoint	Yes	The considered railway checkpoint
Train Type	Yes	The considered Train Type
Daytime	No	The time of the day with an hourly granularity
Weekday	Yes	The day of the week
Last Delay	No	The last known delay with the following granularity in minutes ($[0, 2]$, $(2, 5]$, $(5, 10]$, $(10, 20]$, $(20, 30]$, $(30, 60]$, $(60, 120]$, $(120, \infty)$));
Weather Conditions	Yes	The weather conditions (Sunny, Light Rain, Heavy Rain, Snow)

Table 2. HM bottom level RF feature set (Cat. means Categorical).

Feature name	Cat.	Description
Weather Information	Yes	Weather conditions (Sunny, Light Rain, etc.) in all the checkpoints of the train itinerary (for the already traveled checkpoints we use the actual weather while for the future checkpoints we use the predicted weather conditions)
Past *Train Delays*	No	Average value of the past *Train Delays* in seconds & Last known *Train Delay*
Past *Dwell Times*	No	Average value of the past differences between actual and scheduled *Dwell Times* in seconds & Last known difference between actual and scheduled *Dwell Time*
Past *Running Times*	No	Average value of the past differences between actual and scheduled *Running Times* in seconds & Last known difference between actual and scheduled *Running Time*
Network Congestion	No	Number of trains traversing the checkpoints of the train itinerary in a slot of 20 min around the actual and scheduled times respectively for the past and future checkpoints
Network Congestion Delays	No	Average *Train Delay* of the trains traversing the checkpoints of the train itinerary in a slot of 20 min around the actual and scheduled times respectively for the past and future checkpoints

4   Experimental Evaluation

In this section we will perform an extensive evaluation of the proposed HM to show its effectiveness based on real world data coming from the Italian railway network and provided by RFI. We will prove the effectiveness of our approach by using, as a baseline, a fully DDM based on the one derived in [18,21] for predicting delays, transit time, and dwell time.

4.1   The Available Data

The experiments have been conducted exploiting the real data provided by RFI about the Italian railway network[1]. In particular, RFI has provided

- data about train movements which contains the following information: Date, Train ID, Checkpoint ID, Actual Arrival Time, Arrival Delay, Actual Departure Time, Departure Delay and Event Type. The Event Type field can assume different values: Origin (O), Destination (D), Stop (F), Transit (T);
- timetables, including planning of exceptional trains, and cancellations.

For the purpose of this work, RFI provided the access to the data of 12 months (the whole 2016 solar year) of train movements of one critical (in the sense that many overtakes need to be planned every day) Italian region. The data are relative to more than 3.000 trains and 200 checkpoints. The dataset contains 5.000.000 train movements.

For improving the quality of the predictors, we also exploited as exogenous information, with a Big Data oriented approach, the weather data coming weather stations in the area, freely available from the Italian weather services [22]. For each checkpoint we consider the closest weather station. Then we collected the historical data relative to the solar radiation and precipitations for the same time span. From this data it is possible to extract both the actual and the forecasted weather conditions (Sunny, Rain, Heavy Rain, and Snow).

4.2   Our Baseline: A Data Driven Model

In order to better understand the potentiality and effectiveness of our HM we decided to use a purely DDM as a baseline. The DDM has been constructed removing the top level tree structure of HM described in Sect. 3, which models the experience of the operators, and by building a single DDM based on the whole set of available data and the features reported in Table 2. By removing the top level structure of the HM we basically remove from the HM the experience of the operators resulting in a fully DDM. Instead of creating sub-groups of overtake situations sharing similar characteristics like in the HM, we leave to the DDM the task to learn everything from the historical data and perform the predictions.

A comparison with a fully EBM could not be performed since the Italian RTS information system does not store the prediction made by the operators, who just rely on their intuition and experience.

---

[1] We cannot report all the details because of confidentiality issues.

4.3   Results

We start this analysis showing which checkpoints were involved in the most and the least number of overtakings in 2016, and the number of overtakings identified by the HM and the DDM. Table 3 depicts the 5 checkpoints in which the most and least number of overtakes happened.

Table 3. The checkpoints having the most and least number of overtakes in 2016.

Most Overtaking				Least Overtaking			
Checkpoint Name	Real	HM	DDM	Checkpoint Name	Real	HM	DDM
Checkpoint A	624	518	378	Checkpoint F	17	0	1
Checkpoint B	273	248	226	Checkpoint G	14	12	11
Checkpoint C	220	211	152	Checkpoint H	13	2	2
Checkpoint D	188	206	185	Checkpoint I	11	1	3
Checkpoint E	177	168	164	Checkpoint L	10	2	0

From Table 3 we can observe that:

- the 3 checkpoints which had the highest number of overtakes account for around the 50% of the total overtaking happened in 2016 in the area under examination;
- HM predictions are close to the real one;
- in general HM underestimates the number of overtakings, which is expected because the model requires a certain number of historical data before starting to work correctly;
- The DDM underestimates even more the overtakes in each checkpoint, reflecting that the amount of data required to learn how to correctly predict an overtake is larger.

It is possible to make use of the information of Table 3 in order to identify in which checkpoints it is possible an overtake. In fact the DDM need to learn this information directly from the data while, in the HM, we can exploit the experience of the operators and plug this information directly in the model. In the Italian RTS, overtakes are not possible in all the checkpoints, and, in the area under examination, only in 48 checkpoints it is possible to perform an overtaking.

At this point we are ready to show the precision of the HM in identifying the overtakings. Tables 4 and 5 depict the confusion matrix of, respectively, the DDM

Table 4. Confusion matrices for the overtakes predicted by the DDM (ALL: all trains. REG: regional trains. HS: high speed trains. FRE: freight trains).

ALL	Yes	No
Yes	2246	893
No	639	12657

REG	Yes	No
Yes	1082	476
No	347	7870

HS	Yes	No
Yes	1055	307
No	175	2599

FRE	Yes	No
Yes	109	110
No	117	2188

Table 5. Confusion matrices for the overtakes predicted by the HM (ALL: all trains. REG: regional trains. HS: high speed trains. FRE: freight trains).

ALL	Yes	No
Yes	2355	783
No	500	12827

REG	Yes	No
Yes	1121	436
No	318	7899

HS	Yes	No
Yes	1112	250
No	103	2671

FRE	Yes	No
Yes	122	97
No	79	2226

and the HM. We reported also the confusion matrices relative to the different typologies of trains. From Tables 4 and 5 we can observe that:

- the HM clearly outperforms the DDM;
- the HM is highly accurate in predicting when two trains must not swap their positions;
- freight trains are the ones which perform the least number of overtakes and also the ones affected by the largest error;
- high speed trains are the ones which perform the highest number of overtakes and the ones in which the HM performs the least number of false-positive and false-negative predictions.

Finally, Fig. 2 reports the accuracy of the HM and the DDM in detecting overtakings during the whole 2016.

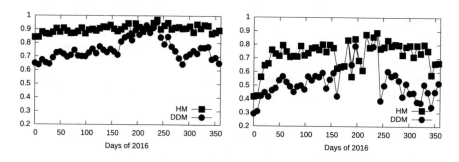

Fig. 2. Overtakes prediction accuracy in time. Overall accuracy (left). Recall (right).

We report both the overall accuracy and the recall (actual overtakes in a checkpoint correctly predicted). From Fig. 2 we can observe that:

- it is required 1 month of data to have a fully operational HM while the DDM requires more data.
- in general the HM shows higher accuracy with respect to the DDM and this advantage is costant over the whole year.

5   Conclusion

In this work we dealt with the problem of understanding and predicting the train overtakes. For this purpose, we exploited an hybrid approach which is able to encapsulate in one single model the knowledge about the network, the experience of the operators, the historical data, and other exogenous variables taking inspiration from the state-of-the-art approaches in this field of research. The result is a dynamic, interpretable and robust hybrid data analytics system able to handle non recurrent events, changes in the behaviour of the network, and to consider complex and exogenous information like weather information. Basically, the proposed approach preserves the strengths of the experience based methods and the data-driven methods and limits their weaknesses. Results on real world data coming from the Italian railway network show that the proposed solution provides remarkable results in addressing the train overtakes prediction problem.

Acknowledgments. This research has been supported by the European Union through the projects IN2DREAMS (European Union's Horizon 2020 research and innovation programme under grant agreement 777596).

References

1. Albrecht, T.: Reducing power peaks and energy consumption in rail transit systems by simultaneous train running time control. WIT Trans. State-of-Art Sci. Eng. **39** (2010)
2. Barta, J., Rizzoli, A.E., Salani, M., Gambardella, L.M.: Statistical modelling of delays in a rail freight transportation network. In: Proceedings of the Winter Simulation Conference (2012)
3. Berger, A., Gebhardt, A., Müller-Hannemann, M., Ostrowski, M.: Stochastic delay prediction in large train networks. In: OASIcs-OpenAccess Series in Informatics, vol. 20 (2011)
4. Breiman, L.: Random forests. Mach. Learn. **45**(1), 5–32 (2001)
5. Bryan, J., Weisbrod, G.E., Martland, C.D.: Rail freight solutions to roadway congestion: final report and guidebook. Transp. Res. Board (2007)
6. Daamen, W., Goverde, R.M.P., Hansen, I.A.: Non-discriminatory automatic registration of knock-on train delays. Netw. Spat. Econ. **9**(1), 47–61 (2009)
7. D'Ariano, A.: Improving real-time train dispatching: models, algorithms and applications. TRAIL Research School (2008)

8. D'Ariano, A., Pranzo, M.: An advanced real-time train dispatching system for minimizing the propagation of delays in a dispatching area under severe disturbances. Netw. Spat. Econ. **9**(1), 63–84 (2009)
9. Fang, W., Yang, S., Yao, X.: A survey on problem models and solution approaches to rescheduling in railway networks. IEEE Trans. Intell. Transp. Syst. **16**(6), 2997–3016 (2015)
10. Ghofrani, F., He, Q., Goverde, R.M., Liu, X.: Recent applications of big data analytics in railway transportation systems: a survey. Transp. Res. Part C Emerg. Technol. **90**, 226–246 (2018)
11. Goverde, R.M.P., Meng, L.: Advanced monitoring and management information of railway operations. J. Rail Transp. Plan. Manag. **1**(2), 69–79 (2011)
12. Hansen, I.A., Goverde, R.M.P., Van Der Meer, D.J.: Online train delay recognition and running time prediction. In: 2010 13th International IEEE Conference on Intelligent Transportation Systems (ITSC), pp. 1783–1788 (2010)
13. Kecman, P., Goverde, R.M.P.: Process mining of train describer event data and automatic conflict identification. In: Computers in Railways XIII: Computer System Design and Operation in the Railway and Other Transit Systems, vol. 127, p. 227 (2013)
14. Kecman, P., Goverde, R.M.P.: Online data-driven adaptive prediction of train event times. IEEE Trans. Intell. Transp. Syst. **16**(1), 465–474 (2015)
15. Ko, H., Koseki, T., Miyatake, M.: Application of dynamic programming to the optimization of the running profile of a train. WIT Trans. Built Environ. **74** (2004)
16. Lamorgese, L., Mannino, C.: An exact decomposition approach for the real-time train dispatching problem. Oper. Res. **63**(1), 48–64 (2015)
17. Lukaszewicz, P.: Energy consumption and running time for trains. Ph.D. thesis, Doctoral Thesis. Railway Technology, Department of Vehicle Engineering, Royal Institute of Technology, Stockholm (2001)
18. Lulli, A., Oneto, L., Canepa, R., Petralli, S., Anguita, D.: Large-scale railway networks train movements: a dynamic, interpretable, and robust hybrid data analytics system. In: IEEE International Conference on Data Science and Advanced Analytics (2018)
19. Marković, N., Milinković, S., Tikhonov, K.S., Schonfeld, P.: Analyzing passenger train arrival delays with support vector regression. Transp. Res. Part C Emerg. Technol. **56**, 251–262 (2015)
20. Milinković, S., Marković, M., Vesković, S., Ivić, M., Pavlović, N.: A fuzzy petri net model to estimate train delays. Simul. Model. Pract. Theory **33**, 144–157 (2013)
21. Oneto, L., Fumeo, E., Clerico, G., Canepa, R., Papa, F., Dambra, C., Mazzino, N., Anguita, D.: Advanced analytics for train delay prediction systems by including exogenous weather data. In: IEEE International Conference on Data Science and Advanced Analytics (2016)
22. Regione Liguria: Weather Data of Regione Liguria (2018). http://www.cartografiarl.regione.liguria.it/SiraQualMeteo/script/PubAccessoDatiMeteo.asp
23. Trabo, I., Landex, A., Nielsen, O.A., Schneider-Tilli, J.E.: Cost benchmarking of railway projects in europe-can it help to reduce costs? In: International Seminar on Railway Operations Modelling and Analysis-RailCopenhagen (2013)
24. Wang, R., Work, D.B.: Data driven approaches for passenger train delay estimation. In: 2015 IEEE 18th International Conference on Intelligent Transportation Systems (ITSC), pp. 535–540 (2015)

Cavitation Noise Spectra Prediction with Hybrid Models

Francesca Cipollini[1](✉), Fabiana Miglianti[2](✉), Luca Oneto[1](✉),
Giorgio Tani[2](✉), Michele Viviani[2](✉), and Davide Anguita[1](✉)

[1] DIBRIS - University of Genoa, Via Opera Pia 13, 16145 Genova, Italy
francesca.cipollini@edu.unige.it, {luca.oneto,davide.anguita}@unige.it
[2] DITEN - University of Genoa, Via Opera Pia 11A, 16145 Genova, Italy
fabiana.miglianti@edu.unige.it, {giorgio.tani,michele.viviani}@unige.it

Abstract. In many real world applications the physical knowledge of a
phenomenon and data science can be combined together in order to get
mutual benefits. As a result, it is possible to formulate a so-called hybrid
model from the combination of the two approaches. In this work, we
propose an hybrid approach for the prediction of the ship propeller cavitating vortex noise, adopting real data collected during extensive model
scale tests in a cavitation tunnel. Results will show the effectiveness of
the proposal.

Keywords: Cavitation noise prediction · Physical models ·
Data-Driven Models · Hybrid Models

1 Introduction

In the latest years, models combining knowledge of a physical phenomenon and
statistical inference are becoming of much interest in many practical applications [2]. In this context, ship propeller underwater radiated noise is an interesting field of application for these Hybrid Models (HMs), especially when the
propeller cavitates. Nowadays, Model Scale Tests (MSTs) are considered the
state-of-the-art technique to predict the cavitation noise spectra. Unfortunately,
they are negatively affected by scale effects which could alter the onset of some
interesting cavitating phenomena respect to the full scale propeller [6]; as a
consequence, for some ship operational conditions it is not trivial to correctly
reproduce the cavitation pattern in MSTs. Moreover, MSTs are quite expensive and time-consuming; it is not feasible to include them in the early stage
of the design. Nevertheless, data collected during these tests can be adopted in
order to tune Data-Driven Models (DDMs) [10] while the Physical Model (PM)
equations describing the occurring phenomenon can be used to refine the DDMs
prediction.

For this reason, in this paper we propose the adoption of HMs for the prediction of the cavitating vortex frequency and sound pressure level, in order to

© Springer Nature Switzerland AG 2020
L. Oneto et al. (Eds.): INNSBDDL 2019, INNS 1, pp. 152–157, 2020.
https://doi.org/10.1007/978-3-030-16841-4_16

take advantage of the best characteristics of both PMs and DDMs by combining them together [2]. In order to develop and test the models proposed in this paper, we have first collected a dataset by means of an extensive set of cavitation tunnel MSTs, and then tuned and validated the HMs on these real data showing satisfying accuracies.

2 Model Scale Tests

Experiments have been performed at the cavitation tunnel of the University of Genoa [9] for a total of 164 propeller loading conditions, subdivided almost equally between two propellers. The working conditions, for which noise samples are collected, have been chosen in order to provide an exhaustive characterization of cavitation noise. Quantities collected at the working points thus defined are used as input features for the developed HMs, and are summarized in Table 1.

The output data of the proposed HMs are represented by the resonance frequency and Radiated Noise Level (RNL) of the cavitating vortex, namely f_c and RNL_c shown in Fig. 1. It has to be remarked that both f_c and RNL_c can be determined adopting PMs, with an approach similar to the one presented in [1] for the resonance frequency and in [7] for its RNL. In Fig. 1, σ_{tip} is the cavitation index of the resultant velocity at the blade tip, Z the blades number, n the propeller rotation, K_t the thrust coefficient, σ_n the cavitation index based on rotational speed, and the parameter k has been chosen equal to 3. The values of τ and a_p are estimated by curve fitting to experimental data.

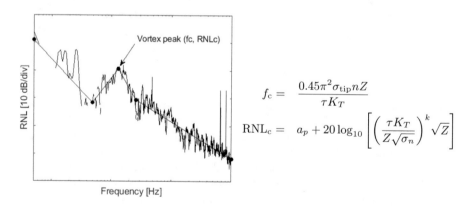

$$f_c = \frac{0.45\pi^2 \sigma_{\mathrm{tip}} n Z}{\tau K_T}$$

$$\mathrm{RNL}_c = a_p + 20\log_{10}\left[\left(\frac{\tau K_T}{Z\sqrt{\sigma_n}}\right)^k \sqrt{Z}\right]$$

Fig. 1. Cavitation noise spectra and f_c and RNL_c physical equations.

3 Hybrid Modelling

Let us consider an input space $\mathcal{X}$, an output space $\mathcal{Y}$, and a relation $\mu : \mathcal{X} \to \mathcal{Y}$ to be learned. In this context, we define an HM as $h : \mathcal{X} \to \mathcal{Y}$, which approximates μ, where $\mathcal{X}$ is the space of the features in Table 1 and $\mathcal{Y}$ is the space composed by

Table 1. Dataset input variables.

Propeller working parameters		Cavitation types	
Description	Unit	Description	Unit
Pitch ratio	[ ]	Suction side tip vortex	[ ]
Pitch setting	[°]	Detached tip vortex	[ ]
Advance coefficient	[ ]	Suction side tip vortex at 0°	[ ]
Thrust coefficient	[ ]	Suction side sheet	[ ]
Torque coefficient	[ ]	Suction side sheet at 0°	[ ]
Open water efficiency	[ ]	Suction side root bubbles	[ ]
Advance cavitation index	[ ]	Suction side bubbles	[ ]
Rotational cavitation index	[ ]	Vortex from sheet face	[ ]
Advance velocity	[m/s]	Pressure side tip vortex	[ ]
Propeller rotation	[rps]	Pressure side sheet	[ ]
Thrust	[kgf]	Pressure side root bubbles	[ ]
Torque	[kgf·cm]		
Relative pressure	[mBar]		
Wake parameters		Angle of attack	
Description	Unit	Description	Unit
Wake width at 0.7R	[°]	Max & Min angle of attack at 0.7R	[°]
Left wake slope at 0.7R	[°]	Average angle of attack at 0.7R	[°]
Right wake slope at 0.7R	[°]	Position of max angle of attack at 0.7R	[°]
Wake width at 0.9R	[°]	Max & Min angle of attack at 0.9R	[°]
Left wake slope at 0.9R	[°]	Average angle of attack at 0.9R	[°]
Right wake slope at 0.9R	[°]	Position of max angle of attack at 0.9R	[°]

f_c and RNL_c values. In order to adapt h to this problem, two sets of data $\mathcal{D}_n$ and $\mathcal{T}_m$ need to be exploited, to respectively tune h and evaluate its performances. $\mathcal{T}_m$ is needed since the error that h commits over $\mathcal{D}_n$ is optimistically biased since $\mathcal{D}_n$ has been used to tune h. For this work, the error that h commits on $\mathcal{T}_m$ in approximating μ is measured adopting the Mean Absolute Percentage Error (MAPE), by taking the absolute loss value of h over $\mathcal{T}_m$ in percentage.

Since h as an HM should be able to both take into account the physical knowledge about the problem and the information hidden in the data, it should learn from the data without being too different from the PM. This requirement can be mapped to a Multi Task Learning (MTL) problem [2,3], aiming at contemporary learning the PM and the available data, through a learning algorithm $\mathscr{A}_{\mathcal{H}}$ which exploits the data in $\mathcal{D}_n$ and the PM to learn h. Since a PM is needed to define h, we will first develop two PMs to estimate the relation between f_c and RNL_c and the input variables using the models of Fig. 1. In order to apply the MTL approach to this case, we will define triple of points

$\mathcal{D}_n = \{(\boldsymbol{x}_1, y_1, p_1), \cdots, (\boldsymbol{x}_n, y_n, p_n)\}$ where p_i is the output of the PM in the point $\boldsymbol{x}_i$ with $i \in \{1, \cdots, n\}$. Consequently, it is possible to modify a more general Kernel Regularized Least Squares (KRLS) formulation in order to contemporary learn a shared model $h(\boldsymbol{x}) = \boldsymbol{w}^T \boldsymbol{\varphi}(\boldsymbol{x})$ and some task specific models $h_i(\boldsymbol{x}) = \boldsymbol{w}_i^T \boldsymbol{\varphi}(\boldsymbol{x})$ with $i \in \{y, p\}$ which should be close to the shared one [8] and obtain the following cost function to be minimized

$$\boldsymbol{w}^*, \boldsymbol{w}_y^*, \boldsymbol{w}_p^* : \arg \min_{\boldsymbol{w}, \boldsymbol{w}_y, \boldsymbol{w}_p} \sum_{i=1}^{n} \left[\boldsymbol{w}^T \boldsymbol{\varphi}(\boldsymbol{x}) - y_i \right]^2 + \left[\boldsymbol{w}^T \boldsymbol{\varphi}(\boldsymbol{x}) - p_i \right]^2 \quad (1)$$

$$+ \left[\boldsymbol{w}_y^T \boldsymbol{\varphi}(\boldsymbol{x}) - y_i \right]^2 + \left[\boldsymbol{w}_p^T \boldsymbol{\varphi}(\boldsymbol{x}) - p_i \right]^2 + \lambda \|\boldsymbol{w}\|^2 + \theta(\|\boldsymbol{w} - \boldsymbol{w}_y\|^2 + \|\boldsymbol{w} - \boldsymbol{w}_p\|^2),$$

where λ is the regularization hyperparameter of KRLS and θ, is a hyperparameter forcing the shared model to be close to the task specific models. By exploiting the kernel trick it is possible to reformulate Problem (1), to obtain a formula for the resulting shared function $h(\boldsymbol{x}) = \boldsymbol{w}^T \boldsymbol{\varphi}(\boldsymbol{x}) = \sum_{i=1}^{n} (\alpha_i + \alpha_{i+n}) K(\boldsymbol{x}_i, \boldsymbol{x})$.

In order to set the $\boldsymbol{\varphi}$ and the kernel K, it is possible to apply to $\mathcal{D}_n$ a process of transformation, referred as Feature Mapping (FM) [8] where a function $\boldsymbol{\varphi} : \mathcal{X} \to \Psi$ is defined to map $\boldsymbol{x}$ to a new feature space where it is possible to learn a simple linear model. In this work, the FM was selected considering the acquired expertise of the cavitation noise phenomenon

$$\boldsymbol{\varphi}(\boldsymbol{x}) = \left[\prod_{i=1}^{5d} v_i^{k_i} : \sum_{i=1}^{5d} k_i = j, j \in \{0, 1, \cdots, p\} \right]^T \in \mathbb{R}^{\sum_{j=0}^{p} \binom{5d}{j}} \quad (2)$$

$$\boldsymbol{v} = \left[x_1, \cdots, x_d, \frac{1}{x_1}, \cdots, \frac{1}{x_d}, \ln(x_1), \cdots, \ln(x_d), e^{x_1}, \cdots, e^{x_d}, e^{-x_1}, \cdots, e^{-x_d} \right]^T \in \mathbb{R}^{5d}.$$

Note that p is the desired degree of the polynomial and c is a parameter trading off the influence of higher-order versus lower-order terms in the polynomial. The two p and c together with λ and θ are hyperparameters that need to be tuned in order to optimize the performance of the final model.

The many unnecessary features that have been generated during FM, are discarded thanks to a Feature Selection (FS) in order to increase the generalization performance of the model by selecting only the most informative of them [4]. For this purpose we will adopt the backward elimination techniques described in [4]. Since every MTL model is characterized by a set of hyperparameters $\mathcal{H}$, a proper Model Selection (MS) procedure needs to be adopted [5]. Several methods exist for MS purpose but resampling methods, like the nonparametric Bootstrap (BTS) [5] used in this work, represent the state-of-the-art. In BTS the original dataset $\mathcal{D}_n$ is resampled n_r times, with replacement, to build two independent datasets called training, and validation sets, respectively $\mathcal{L}_l^r$ and $\mathcal{V}_v^r$, with $r \in \{1, \cdots, n_r\}$ and $l = n$, where r is set to 500. Note that $\mathcal{L}_l^r \cap \mathcal{V}_v^r = \oslash$, $\mathcal{L}_l^r \cup \mathcal{V}_v^r = \mathcal{D}_n$. Then, the best combination the hyperparameters $\mathcal{H}^*$ in a set of possible ones $\mathfrak{H} = \{\mathcal{H}_1, \mathcal{H}_2, \cdots\}$ for the algorithm $\mathscr{A}_\mathcal{H}$ should be the one which allows to achieve a small error on the validation set.

4   Results and Discussion

In this section, the performances of the PMs and the HMs will be tested adopting the data described in Sect. 2, composed by 164 samples and 38 features. $\mathcal{D}_n$ and $\mathcal{T}_m$ have been created by splitting randomly the whole 164 samples keeping 90% of the data in $\mathcal{D}_n$ and the remaining 10% in $\mathcal{T}_m$. The set of hyperparameters of the HM tuned during the MS phase are $\mathcal{H} = \{p, c, \lambda, \theta\}$ chosen in $\mathfrak{H} = \{1, 2, \cdots, 10\} \times \{10^{-4}, 10^{-3}, \cdots, 10^{+4}\} \times \{10^{-4.0}, 10^{-3.8}, \cdots, 10^{+4.0}\} \times \{10^{-4.0}, 10^{-3.8}, \cdots, 10^{+4.0}\}$. All the experiments have been repeated 30 times and, Fig. 2 reports the scatter plots of the measured and predicted values of f_c and RNL_c for the PMs and HMs.

Looking at Fig. 2, it is clear that the HMs can predict with higher accuracy the f_c and RNL_c values with respect to PMs. Our results are surely preliminary and in future works we plan to make a deeper comparison in order to test the quality of our proposal. Nevertheless, in this work HMs have shown to be a promising tools for estimating the cavitating vortex, exploiting both physical knowledge and data science, on real world data collected in an extensive set of cavitation tunnel trials.

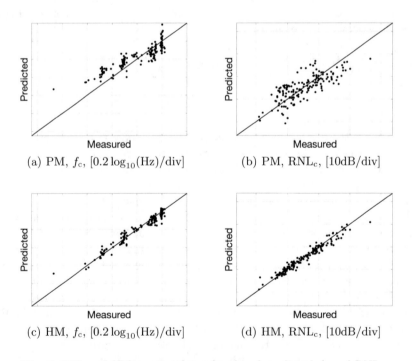

(a) PM, f_c, $[0.2 \log_{10}(\text{Hz})/\text{div}]$ (b) PM, RNL_c, $[10\text{dB}/\text{div}]$

(c) HM, f_c, $[0.2 \log_{10}(\text{Hz})/\text{div}]$ (d) HM, RNL_c, $[10\text{dB}/\text{div}]$

Fig. 2. PMs and HMs scatterplots of real and predicted f_c and RNL_c.

References

1. Bosschers, J.: Investigation of hull pressure fluctuations generated by cavitating vortices. In: Proceedings of the First Symposium on Marine Propulsors (2009)
2. Caruana, R.: Multitask learning. Mach. Learn. **28**(1), 41–75 (1997)
3. Evgeniou, T., Pontil, M.: Regularized multi–task learning. In: ACM SIGKDD International Conference on Knowledge Discovery and Data Mining (2004)
4. Guyon, I., Elisseeff, A.: An introduction to variable and feature selection. J. Mach. Learn. Res. **3**, 1157–1182 (2003)
5. Kohavi, R., et al.: A study of cross-validation and bootstrap for accuracy estimation and model selection. In: International Joint Conference on Artificial Intelligence (1995)
6. McCormick, B.W.: On cavitation produced by a vortex trailing from a lifting surface. J. Basic Eng. **84**(3), 369–378 (1962)
7. Raestad, A.: Tip vortex index-an engineering approach to propeller noise prediction. The Naval Architect, pp. 11–15 (1996)
8. Shalev-Shwartz, S., Ben-David, S.: Understanding Machine Learning: From Theory to Algorithms. Cambridge University Press, Cambridge (2014)
9. Tani, G., Aktas, B., Viviani, M., Atlar, M.: Two medium size cavitation tunnel hydro-acoustic benchmark experiment comparisons as part of a round robin test campaign. Ocean. Eng. **138**, 179–207 (2017)
10. Vapnik, V.N.: Statistical Learning Theory. Wiley, New York (1998)

Pseudoinverse Learners: New Trend and Applications to Big Data

Ping Guo[1(✉)], Dongbin Zhao[2], Min Han[3], and Shoubo Feng[3]

[1] School of Systems Science, Beijing Normal University, Beijing 100875, China
pguo@ieee.org
[2] Institute of Automation, Chinese Academy of Sciences, Beijing, China
dongbin.zhao@ia.ac.cn
[3] Faculty of Electronic Information and Electrical Engineering,
Dalian University of Technology, Dalian, China
minhan@dlut.edu.cn, fsb@mail.dlut.edu.cn

Abstract. Pseudoinverse learner (PIL), a kind of multilayer neural networks (MLP) trained with pseudoinverse learning algorithm, is a novel learning framework. It has drawn increasing attention in the areas of large-scale computing, high-speed signal processing, artificial intelligence and so on. In this paper, we briefly review the pseudoinverse learning algorithm and discuss the characteristics as well as its variants. Some new viewpoints to PIL algorithm are presented, and currently developments of PIL algorithm for autoencoder is presented under the framework of deep learning. Some new trends on PIL-based learning are also discussed. Moreover, we present several interesting PIL applications to demonstrate the practical advances on the big data analysis topic.

Keywords: Back propagation · Pseudoinverse learning ·
Multi-layer perception · Deep neural network · Big astronomical data

1 Introduction

Studied in mid-eighties of last century, the Multi-layer perceptron (MLP) is a kind of feedforward neural networks. MLP is called deep neural network (DNN) when it has more than three hidden layers. Now it has been successful used for lots of supervised learning applications. Both theoretical and empirical studies have shown the MLP has great power for pattern recognition [1]. The weight parameters of the network can be learnt by the error Back-propagation (BP) algorithm when there are few hidden layers, [2,3]. As we have known, BP algorithm has several disadvantages. It usually encounter a poor convergence rate and local minima problem [4]. In BP algorithm, the selection of hyper-parameters such as learning rate and momentum constant is often difficult.

This work is supported by the NSFC-CAS joint fund (No. U1531242).

L. Oneto et al. (Eds.): INNSBDDL 2019, INNS 1, pp. 158–168, 2020.
https://doi.org/10.1007/978-3-030-16841-4_17

In order to solve the above problems, Guo et al. [5] proposed a non-gradient descent algorithm, named pseudoinverse learning (PIL) algorithm [6]. Unlike BP, the PIL algorithm could exactly calculate the network weights directly, rather than optimize it iteratively. The PIL algorithm only adopts generalized linear algebraic methods, e.g., pseudoinverse operations and matrix inner products. Furthermore, it basically not need to set any hyper-parameters and it brings great convenience to the user.

An overview of PIL algorithm and application to big data analysis is given in this paper. The paper is organized as follows. In Sect. 2, we give the original pseudoinverse learning algorithm. In Sect. 3, we introduce several PIL variants. In Sect. 4, we discuss the new trend in PIL. In Sect. 5, we provide several applications to big data analysis by using PIL. Finally, we draw the conclusion in Sect. 6.

2   Original Pseudoinverse Learning Algorithm

BP algorithm is a great discovery in neural network learning, while it has several disadvantages, such as slow convergence performance or local minima. Non-gradient based algorithms have been considered as an alternative way, among which the PIL algorithm is a common used one [5,6]. In this paper, we would like to take the single hidden layer neural network (SHLN) as example to introduce the PIL algorithm.

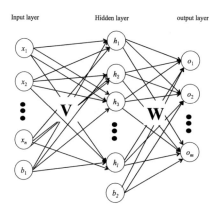

Fig. 1. A schematic diagram of single hidden layer neural networks.

The SHLN shown in Fig. 1 has three layers, including one input layer, one output layer and one hidden layer. There are $n + 1$ neurons in input layer, $l + 1$ neurons in hidden layer, and m neurons in output layer. We use $\mathbf{x} = (x_1, x_2, ..., x_n)$ to express n dimensional input vector, and $\mathbf{o} = (o_1, o_2, ..., o_m)$ stands for m dimensional output vector. b_1 in input layer is called bias neuron, and b_2 is the hidden layer bias neuron. $\mathbf{V}$ is the weight matrix connecting input and hidden neurons, and $\mathbf{W}$ is the weight matrix connecting the hidden and output neurons. The network mapping function is expressed as,

$$o_k = f_k(\mathbf{x}, \mathbf{\Theta})$$
$$= \sigma(\sum_{j=1}^{l} W_{k,j} g_j + b_2),$$
$$g_j = \sigma(\sum_{i=1}^{n} V_{j,i} x_i + b_1). \tag{1}$$

where $\mathbf{\Theta}$ stands for the network parameter group, including connecting weights $\mathbf{W}, \mathbf{V}$, and bias neurons b_1, b_2. While $\sigma(\cdot)$ is an activation function, such as sigmoid, hyperbolic, step, radial basis function, and so on.

From above equation, we can see that $\mathbf{g}(\mathbf{Vx} + \mathbf{b_1})$ is the hidden layer output, and $\mathbf{f}(\mathbf{Wg} + \mathbf{b_2})$ is the last layer output. If we let

$$\sum_{i=1}^{n} V_{j,i} x_i + b_1 = \sum_{i=0}^{n} V_{j,i} x_i$$
$$\text{with } V_{j,0} = b_1, x_0 = 1,$$

and

$$\sum_{j=1}^{l} W_{k,j} g_j + b_2 = \sum_{j=0}^{l} W_{k,j} g_j$$
$$\text{with } W_{k,0} = b_2, g_0 = 1,$$

now matrix $\mathbf{V}, \mathbf{W}$ become augmented matrix. In the literatures, the value of the bias is usually set to be $+1$, while some researchers also take it as a variable. For mathematical expression concise, hidden layer bias neuron b_2 is often omitted.

Given data set $D = \{\mathbf{x}^i, \mathbf{t}^i\}$, training the network is to find the weigh parameters $\mathbf{V}, \mathbf{W}$ with minimizing cost function.

$$E = \frac{1}{2N} \sum_{i=1}^{N} \sum_{j=1}^{m} \|f_j(\mathbf{x}^i, \mathbf{\Theta}) - t_j^i\|^2. \tag{2}$$

For simplifying, we can write this system error function in matrix form,

$$E = \frac{1}{2N} \|\mathbf{O} - \mathbf{T}\|_F^2. \tag{3}$$

where subscript F stands for Frobenius norm. Traditionally, it is learnt by BP algorithm. But the BP algorithm often encounter hyper-parameter problems as well as local minima, which is difficult for most beginners.

In order to overcome the drawbacks of the BP algorithm, Guo et al. [5] proposed the PIL algorithm to train a SHLN in 1995. In that work, the activation function is taken as the hyperbolic function $Tanh(\cdot)$. Minimizing following error function to find weight parameter matrix,

$$minimize\|\mathbf{YW} - \mathbf{B}\|^2, \tag{4}$$

where $\mathbf{Y} = Tanh(\mathbf{XV})$ is the output matrix of the hidden layer, $\mathbf{X}$ is the input matrix consisting of N input vectors as its rows and $d = n+1$ columns as input vector dimension n plus 1, $\mathbf{B} = ArcTanh(\mathbf{T})$, and $\mathbf{T}$ is the target label matrix which consists of N label vectors as its rows and m columns as target vector dimension.

Equation (4) is a least square problem, which could be solved by linear algebra. The formal solution to $\mathbf{W}$ in Eq. (4) is $\mathbf{W} = \mathbf{Y}^+\mathbf{B}$, where $\mathbf{Y}^+$ is the pseudoinverse of $\mathbf{Y}$. It also be written as:

$$\mathbf{YW} - \mathbf{B} = \mathbf{YY}^+\mathbf{B} - \mathbf{B} = \mathbf{0}. \tag{5}$$

If Eq. (5) holds, then $\mathbf{YY}^+ = \mathbf{I}$ will satisfy the requirement definitely. The number of hidden layer neurons is only one hyper-parameter for SHLN and perhaps it is the hardest problem for most beginners (see [9] for more details). In the work of Guo *et al.* [5], the l is set be N for the purpose of exact learning.

Therefore, we get a fast learning algorithm with non-iterative optimization. In most cases, it takes one step to get the optimal solution. The algorithm can be summarized as follows:

Algorithm PIL: Given a date set, we draw N pair samples $D = \{\mathbf{x}^i, \mathbf{t}^i\}_{i=1}^N$ as the training set, activation function $Tanh(\cdot)$, and set hidden neuron number be N,

> *Step 1*: Compute $\mathbf{V} = Pseudoinverse[\mathbf{X}]$ and hidden layer output matrix $\mathbf{Y} = Tanh(\mathbf{XV})$,
> *Step 2*: Compute $\mathbf{Y}^+ = Pseudoinverse[\mathbf{Y}]$ and $\mathbf{B} = ArcTanh[\mathbf{T}]$,
> *Step 3*: Compute the output weight matrix $\mathbf{W} = \mathbf{Y}^+\mathbf{B}$.

And the network output is $\mathbf{o} = Tanh(Tanh(\mathbf{xV})\mathbf{W})$.

As we know, the common practice is randomly initialize the weights, then update it with delta learning rules. While in PIL algorithm, the weights is computed with pseudoinverse solution and do not need to be adjusted further. In the work of Guo *et al.* [5], randomly set input weight V also has been investigated. Following sentence is copied from Ref. [5]:

> *A simple method is to set V as a random n by N matrix. In practice, this is not a proper method. As we mentioned above, we use Tanh[·] as the activate function. If the matrix $\mathbf{Z} = \mathbf{XV}$ contains elements of large values, it will result in complex numbers which was not desirable. So it is better to choose a proper matrix V so that Z has no elements of large value. One way is to set the values of the elements as small as possible.*

When writing above description in following form, we can regard it as a variant of the PIL algorithm:

Algorithm PIL0: Given a date set, we draw N pair samples $D = \{\mathbf{x}^i, \mathbf{t}^i\}_{i=1}^N$ as the training set, activation function $Tanh(\cdot)$, and set hidden neuron number as $l = N$,

> *Step 1*: Randomly assign input weight matrix $\mathbf{V}$ (the element values in $\mathbf{V}$ should in a small interval, say $[-1, +1]$),
> *Step 2*: Compute the hidden layer output matrix $\mathbf{Y} = Tanh(\mathbf{XV})$,
> *Step 3*: Compute the output weight matrix $\mathbf{W} = \mathbf{Y}^+\mathbf{B}$, where $\mathbf{B} = \text{ArcTanh}(\mathbf{T})$

Remarks

1. Set hidden neuron number $l = N$ is for exact learning, if the training error is allowed, we can set hidden neuron number $l < N$.
2. Activation function can be taken any nonlinear transformation function, such as sigmoid, Gaussian kernel, and so on.
3. When last layer activation function is taken as a linear function, we have $\mathbf{W} = \mathbf{Y}^+\mathbf{T}$ [6].

Recently, we find that Huang *et al.* created a name called extreme learning machine (ELM) for a SHLN learning algorithm. By analyzing weight parameter setting method, we can easily find that ELM is exactly the same with our PIL0 algorithm for SHLN in learning scheme [5]. So we believe that ELM algorithm is a **v**ariant cr**e**ated by **s**imple name al**t**ernation (VEST) of the PIL algorithm [7]. Also, please note that random vector functional link network (RVFL) [11] is different with SHLN, as Suganthan [12] points out: "*The major variation between RVFL and ELM is the presence and absence of direct connections between inputs and outputs, respectively.*"

3   PIL Variants

There are several variants for PIL in recent years. In 2001, Guo *et al.* [6] proposed a new paradigm that extended the architecture to multiple hidden layers. Later in 2003, Guo *et al.* extended their work by adding Gaussian noise to PIL [10].

3.1   MLP-PIL

The PIL algorithm for multilayer neural network is summarized as follows:

Algorithm MLP-PIL: Given a date set, we draw N pair samples $D = \{\mathbf{x}^i, \mathbf{t}^i\}_{i=1}^N$ as the training set, activation function $\sigma(\cdot)$, and set hidden neuron number be N,

> *Step 1*: Compute $(\mathbf{Y}^0)^+ = Pseudoinverse[\mathbf{X}]$,
> *Step 2*: Compute $\|\mathbf{Y}^l(\mathbf{Y}^l)^+ - I\|^2$. If it is less than the given error E, go to step 5. If not, go on to the next step.

Step 3: Let $W^l = (\mathbf{Y}^l)^+$. Feed forward the result to next layer, and compute $\mathbf{Y}^{l+1} = \sigma(\mathbf{Y}^l \mathbf{W}^l)$.

Step 4: Compute $(\mathbf{Y}^{l+1})^+ = Pseudoinverse(\mathbf{Y}^{l+1})$, set $l \leftarrow l + 1$, and go to step 2.

Step 5: Let final layer output weight matrix $\mathbf{W}^L = (\mathbf{Y}^L)^+ \mathbf{T}$.

And the network output is

$$o = \sigma(...\sigma(\sigma(\mathbf{x}\mathbf{W}^0)\mathbf{W}^1)...)\mathbf{W}^L. \tag{6}$$

There are some new viewpoints to PIL is as follow:

Remarks

1. Equation (6) showed a deep neural network architecture.
2. The depth of this DNN is dynamical growth and is data dependent.
3. If stopped at $\|\mathbf{Y}^l(\mathbf{Y}^l)^+ - \mathbf{I}\|^2 = 0$, we get an identity orthogonal projector $\mathbf{P} = \mathbf{Y}^L(\mathbf{Y}^L)^+$
4. When we let $\mathbf{T} = \mathbf{I}$, the PIL algorithm is an unsupervised learning algorithm, it realizes vector normalization in high dimensional space.

3.2   Gn-PIL

In the work of Guo *et al.* [10], another variant of PIL algorithm is discussed. It was stated in the discussion section of Ref. [10]:

"But if we intend to reduce the network complexity, we can add a same-dimension Gaussian noise matrix to perturb the transformed matrix in step 4 of the PIL algorithm. The inverse function of the perturbed matrix will exist with probability one because the noise is an identical and independent distribution. In such a strategy, we can constrain the hidden layers to at most two to reach the perfect learning".

Writing above description with mathematical algorithm form:

Algorithm Gn-PIL: Given a date set, we draw N pair samples $D = \{\mathbf{x}^i, \mathbf{t}^i\}_{i=1}^N$ as the training set, activation function $\sigma(\cdot)$, set hidden neuron number be N, and set Gaussian noise perturbation matrix be $\tilde{\mathbf{G}}$.

Step 1: Compute $(\mathbf{Y}^0)^+ = Pseudoinverse[\mathbf{X}]$,

Step 2: Compute $\|\mathbf{Y}^l(\mathbf{Y}^l)^+ - I\|^2$. If it is less than the given error E, go to step 5. If not, go on to the next step.

Step 3: Let $\mathbf{W}^l = (\mathbf{Y}^l)^+ + \tilde{\mathbf{G}}$. Feed forward the result to next layer, and compute $\mathbf{Y}^{l+1} = \sigma(\mathbf{Y}^l \mathbf{W}^l)$.

Step 4: Compute $(\mathbf{Y}^{l+1})^+ = Pseudoinverse(\mathbf{Y}^{l+1})$, set $l \leftarrow l + 1$, and go to [step 2].

Step 5: Let final layer output weight matrix $\mathbf{W}^L = (\mathbf{Y}^L)^+ \mathbf{T}$.

Remarks

1. Adding noise to input weight matrix is equivalent with adding noise to input data matrix, while training with noise is a kind of Tikhonov regularization [13].
2. Perturbation method also can be applied to replace the convolutional layer in the deep convolutional neural networks [8].

4   New Trends in PIL

4.1   AE-PIL

Autoencoder has been widely used in deep learning research and achieved satisfactory performance in various applications. Nowadays, the training of autoencoder usually adopts the variants of gradient descent. However, the methods based on gradient descent commonly encounter local minima or gradient vanishing. The PIL for autoencoder has been proposed and achieves comprehensively better performance in terms of training efficiency and accuracy.

Low Rank Constraint AE-PIL. In the work of Wang [14], a new training paradigm of autoencoder is presented to exploit the inherent information in input data with a fast and fully automated method. Different from the traditional training of autoencoder, the input weight matrix of each layer is obtained by the low rank approximation of the pseudoinverse of input feature matrix. The truncated singular decomposition is used for the low rank approximation to find the basis vectors of the sub-space. And the encoder weight and decoder weight are tied together. Furthermore, the rank of input data matrix is the guidance of the number of hidden units in autoencoder, which avoids several complicated hyper-parameter determination. This kind of low rank autoencoder pseudoinverse learning could automatically abandon the redundant information in the learning process. The performance on several applications shows the superiority on fast modeling process.

Sparse Constraint AE-PIL. Xu *et al.* [15] proposed a non-iterative learning algorithm based on pseudoinverse learning of sparse autoencoder. Sparse autoencoder are commonly used in the situation of large-scale data modeling when there are many irrelevant variables in input data. It constrains the activation rate of the units in each hidden layer. In the sparse constraint AE-PIL, the encoder weights are calculated by truncating the pseudoinverse matrix. Simultaneously, the biased ReLu activate function is used for mapping the input data in order to achieve sparsity of hidden units. In the sparse autoencoder, only the regularization parameter could influence the reconstruction error. Thus, it saves much time for hyper-parameter selection. The best optimization solution could be derived by the SC-PILAE without local minima.

Broad AE-PIL. Broad learning has been a hot research topic in recent years. The receptive function could be a powerful tool to map the original input data into high-dimension feature space. In the work of Xu *et al.* [16], they proposed the pseudoinverse learning paradigm for broad learning. Instead of optimizing the objective function by iterative gradient descent methods, they adopt the pseudoinverse learning for a faster calculation performance. The transformed feature could be derived by several receptive functions, such as kernel receptive function, function link receptive function, nonlinear transformation receptive function and random projection receptive function. The pseudoinverse broad learning provides an efficient and effective way to model big data in a both wide and deep structure.

4.2   Stacked Generalization

The stacked generalization means finding several primary trained networks and combining them together to reduce the bias of the generalizer with respect to the given data set. It has been a useful method to minimize the generalization error in network training, but it is a very computation intensive technique in implementation.

Stacking PILs. After Guo *et al.* successful realized stacked generalization with PIL in 2001 [17], Li *et al.* [18] recently propose a hierarchical model with pseudoinverse learning algorithm to deal with big data problem. Facing the imbalance of positive and negative samples problem, instead of using sampling methods, they utilize cross validation partition samples in stacked generalization to attain base learner classifiers. In order to achieve better generalization capacity by reducing over-fitting, gating network combined the output of base learners are used as meta learners. While training the feedforward neural network, which is also part of the base learner classifier, pseudoinverse learning algorithm and autoencoder trained by histogram of oriented gradient combined pseudoinverse learning are used to extract the feature of the data. Stacked generalization increase the accuracy of the model and pseudoinverse learning algorithm speed up the train process and avoid many problems such as gradient vanish, local minima and so on.

5   Applications to Big Data

After two decades of development, PIL algorithm has made a lot of progress. Its application area also extends from astronomical application to big data analysis.

With the development of the advanced astronomical instruments, spectral data have been widely used in modern spectroscopic surveys. However, the huge number of the data cause a difficulty while analyzing and calculating models. In the training process, BP algorithm performs a poorer efficiency. Considering the time-consuming problem and the complexity of the networks, Wang *et al.* [19] proposed a framework with local connection which can save time but with better accuracy. No need of iterative optimization algorithm, the network with "divide and conquer" strategy can use the basic structural units to extract the local features and unify these features. PIL algorithm is utilized to train the network and the performance of the application in astronomical spectrum recognition show the excellent effect. Compared with other baseline algorithms, PIL algorithm has obvious advantage in learning speed. For very high dimension big data set, both column number and row number of input data matrix are very large. In this case it is impossible to compute pseudoinverse of the input data matrix directly. Wang *et al.* [19] adopt "divide and conquer" strategy to partition huge input data matrix into hundreds of sub-matrix to implement pseudoinverse operation. The data dimension is divided with a locally connected network structure, while data number is divided with Bagging technique. Final the network structure is considered as an ensemble neural network, or stacked generalization [10] (or deep stacking network [21]).

In the big data modeling process of astronomical spectral data, in order to obtain a faster and more accurate classifier, Wang *et al.* [20] put forward a new method by PIL algorithm, which can automatically extract inherent features in spectral data sets and helps understanding the meanings behind the data. The multilayer architecture plays a crucial role in PIL algorithm to extract features. The result demonstrates an excellent performance with lower cost in real world data.

The pseudoinverse of the input data matrix can be computed with QR decomposition or singular value decomposition (SVD). However, the computational cost of SVD is several times higher than that of matrix multiplication. In practical implementation of PIL, a standard software library for numerical linear algebra LAPACK (Linear Algebra Package) is suggested to be utilized for solving systems of SVD. Furthermore, if the data dimension n is far less than the data number m, thin SVD should be considered because this is significantly quicker and more economical than that of the full SVD [22].

6   Conclusion

In the field of deep learning research, the weight parameters of a neural network are usually trained by gradient-based method or its variants. However, these methods often encounter several problems such as local minima or gradient vanishing. Fortunately, the PIL algorithm could be used to solve these problem. PIL is based on generalized linear algebra and pseudoinverse matrices, which does not have the above problems. After more than twenty-years development, PIL algorithm has achieved great advance and performs better. In the future, the non-gradient learning method based on PIL would become useful method for various machine learning task as well as big data analysis.

References

1. Bishop, C.M.: Neural Networks for Pattern Recognition. Oxford University Press, Oxford (1995)
2. Rumelhart, D., McClelland, J.: Learning internal representations by error propagation. In: Parallel Distributed Processing: Explorations in the Microstructure of Cognition: Foundations, pp. 318–362. MIT Press (1986)
3. Lecun, Y., Bottou, L., Bengio, Y., Haffner, P.: Gradient-based learning applied to document recognition. Proc. IEEE **86**(11), 2278–2324 (1998)
4. Wessels, L., Barnard, E.: Avoiding false local minima by proper initialization of connections. IEEE Trans. Neural Netw. **3**(6), 899–905 (1992)
5. Guo, P., Chen, CLP., Sun, Y.: An exact supervised learning for a three-layer supervised neural network. In: Proceedings of the International Conference on Neural Information Processing, pp. 1041–1044 (1995)
6. Guo, P., Lyu, M.: Pseudoinverse learning algorithm for feedforward neural networks. In: Advances in Neural Networks and Applications, pp. 321–326 (2001)
7. Guo, P.: A VEST of the pseudoinverse learning algorithm (2018). arXiv preprint: https://arxiv.org/abs/1805.07828
8. Xu, J., Boddeti, V.N., Savvides, M.: Perturbative neural networks. In: IEEE Conference on Computer Vision and Pattern Recognition (2018)
9. How many hidden units should I use? ftp://ftp.sas.com/pub/neural/FAQ3.html#A_hu, copyright (1997, 1998, 1999, 2000, 2001, 2002)
10. Guo, P., Lyu, M.: A pseudoinverse learning algorithm for feedforward neural networks with stacked generalization applications to software reliability growth data. Neurocomputing **56**(1), 101–121 (2004)
11. Pao, Y.H., Takefuji, Y.: Functional-link net computing: theory, system architecture, and functionalities. Computer **25**(5), 76–79 (1992)
12. Suganthan, P.N.: Letter: on non-iterative learning algorithms with closed-form solution. Appl. Soft Comput. **70**(1), 1078–1082 (2018)
13. Bishop, C.M.: Training with noise is equivalent to Tikhonov regularization. Neural Comput. **7**(1), 108–116 (1995)
14. Wang, K., Guo, P., Xin X., et al.: Autoencoder, low rank approximation and pseudoinverse learning algorithm. In: 2017 IEEE International Conference on Systems, Man, and Cybernetics, pp. 948–953 (2017)
15. Xu, B., Guo, P.: Pseudoinverse learning algorithm for fast sparse autoencoder training. In: 2018 IEEE Congress on Evolutionary Computation, pp. 1–6 (2018)
16. Xu, B., Guo, P.: Broad and pseudoinverse learning for autoencoder. In: IEEE International Conference on Systems, Man, and Cybernetics (2018)
17. Guo, P., Lyu, M.: A case study on stacked generalization with software reliability growth modeling data. In: Proceedings of ICONIP 2001, pp. 1321–1326 (2001)
18. Li, S., Feng, S., Guo, P., et al.: A hierarchical model with pseudoinverse learning algorithm optimization for pulsar candidate selection. In: 2018 IEEE Congress on Evolutionary Computation, pp. 1–6 (2018)
19. Wang, K., Guo, P., Luo, A. L., et al.: Deep neural networks with local connectivity and its application to astronomical spectral data. In: IEEE International Conference on Systems, Man, and Cybernetics, pp. 2687–2692 (2016)
20. Wang, K., Guo, P., Luo, A.L.: A new automated spectral feature extraction method and its application in spectral classification and defective spectra recover. Mon. Not. R. Astron. Soc. **465**(4), 4311–4324 (2016)

21. Deng, L., He, X., Gao, J.: Deep stacking networks for information retrieval. In: IEEE International Conference on Acoustics, Speech and Signal Processing (2013)
22. Guo, P.: Automatic determination of multi-layer perception neural net structure with pseudoinverse learning algorithm. In: Tutorial of ICONIP 2017 (2017). http:// sss.bnu.edu.cn/~pguo/pdf/2017/tutorial_ICONIP17.pdf

Innovation Capability of Firms: A Big Data Approach with Patents

Linda Ponta[1]([✉]), Gloria Puliga[1], Luca Oneto[2], and Raffaella Manzini[1]

[1] LIUC Cattaneo University, Castellanza, Italy
{lponta,gpuliga,rmanzini}@liuc.it
[2] DIBRIS - University of Genoa, Genova, Italy
luca.oneto@unige.it

Abstract. Capabilities and, in particular, Innovation Capability (IC), are fundamental strategic assets for companies in providing and sustaining their competitive advantage. IC is the firms' ability to mobilize and create new knowledge applying appropriate process technologies and it has been investigated by means of its main determinants, usually divided into internal and external factors. In this paper, starting from the patent data, the patent's forward citations are used as proxy of IC and the main patents' features are considered as proxy of the determinants. In details, the main purpose of the paper is to understand the patent's features that are relevant to predict IC. Three different algorithms of machine learning, i.e., Least Squares (RLS), Deep Neural Networks (DNN), and Decision Trees (DT), are employed for this investigation. Results show that the most important patent's features useful to predict IC refer to the specific technological areas, the backward citations, the technological domains and the family size. These findings are confirmed by all the three algorithms used.

Keywords: Innovation Capability · Patents' data · Least Squares ·
Deep Neural Networks · Decision Trees

1 Introduction

Capabilities are fundamental strategic assets for companies in providing and sustaining their competitive advantage. According to the seminal work of Barney et al. [3], an asset is a source of competitive advantage if it is valuable, rare, difficult to imitate and difficult to substitute. In these contexts, Innovation Capability (IC) is a special asset that has been defined in several ways and needs to account for wide and disperse scopes and levels in accordance with the firm strategy and competition environment [31]. It is tacit, non-modifiable and closely asset, correlated to the absorptive capacity of a firm to acquire and valorize interior experiences. Summarizing the most important definitions, IC is the firm ability to mobilize and create new knowledge [19] applying appropriate process technologies that will result in product and/or process innovation to satisfy

© Springer Nature Switzerland AG 2020
L. Oneto et al. (Eds.): INNSBDDL 2019, INNS 1, pp. 169–179, 2020.
https://doi.org/10.1007/978-3-030-16841-4_18

present and future market needs and to respond to unexpected opportunities created by competitors [1], through the use of different innovative assets [5]. However, it worth noting that the main idea behind the innovation is the successful bringing of new products or services to the market. According to R&D performance measurement literature, the importance of IC is clear and several studies are trying to analyse its determinants in deep [7,10]. However, this is still an open question.

Studies group the determinants into internal and external factors [28]. The former refers to the skills brought into the firm by the workforce, whereas the latter to the interactions of firms with other companies (suppliers, customers) or scientific actors (public assistance agencies, universities) that take place to complement internal knowledge.

Often, the number of patents has been used as measure of IC [23,30]. Nevertheless, nowadays, forward citations have been adopted as a more sophisticated indicator to measure novelty and quality of innovation [4,21]. It is worth remembering that the forward citations are the citations received by a patent in a given time range. In particular, the number of patent's forward citations represents the novelty, the quality and the market acceptance of the technology. First, forward citations have been widely recognized as a proxy of the importance of a technology and its evolution [21], permitting to assess the technological importance of a company's patent portfolio. In addition, empirical studies show how patents highly cited have stronger market value: an extra citation per patent boosting market value by 3% and that "unpredictable" citations have a stronger effect than the predictable portion [4]. According to the definition of IC previously provided, patents' forward citations represent a relevant and useful source for measuring IC. The main idea behind this paper is to use the patent data to understand the determinants of the companies' IC and the quality of their innovation process. Specifically, we aim to determine what drives the patents' forward citations, as proxy of IC, in terms of other patents' features that can be managed by companies and collected immediately after the patents are issued, as described in the economic and innovation literature [21]. The most scientific approaches to investigate the IC of firms use qualitative or linear regressions [28]. In order to overcome the limits of these classical methods, thanks to the large availability of patent data, a machine learning approach has been proposed in literature [21]. It worth remarking that, so far, machine learning approaches have been adopted in predicting the evolution of technologies [21], and very rarely as tool in supporting companies' decision making. The tenet of this research is the analysis of those patent features that can predict the forward citation and so the IC. For this, the main patent features, have been extracted from the Orbit Intelligence database provided by Questel[1]. Second, three different algorithms, Least Squares (RLS), Deep Neural Networks (DNN), and Decision Trees (DT) of machine learning are employed to capture the relationships between input and output features in a time period of interest [29].

---

[1] www.questel.com.

The paper is organized as follows: Sect. 2 presents the problem formalization, Sect. 3 the Data-Driven Approach, Sect. 4 describes the available data, Sect. 5 shows the computational experiments and the discussion of results. Finally, Sect. 6 provides the conclusion of the study.

2   The Problem Formalization

According to the aim of the paper, we seek to identify the features that determine the IC of companies by using, as proxy, the number of forward citations of each patent. We have grouped the patent's features, proxy of IC's determinants, into three categories: internal and external as suggested by Romijn et al. [28] and temporal as suggested by Jaffe [17]. The internal determinants represent the skills and knowledge owned inside the company [28]. According to that, the number of claims, the backwards citations, the technical concepts, other features strongly related to the patent structure (for example the number of words in the document or in the abstract), the technological classes based on the International Patent Classification (IPC) are proxy of the knowledge of the company [16,24,27]. It is worth remembering that the IPC is arranged in a hierarchical, tree-like structure, where each subdivision of digits (in our study, 4, 7 and 9 digits) provides a more detailed information about the technology. The external determinants are represented by the network created by the company and by its positioning [28]. For this reason, the number of assignees of the patent, the technological areas in which the firm operates, the countries where the patent has been extended are features representing the external environment of a firm. Moreover, with respect to Romijn et al. [28], we add temporal determinants because they provide relevant information about the citation evolution [17]. We selected features that characterize the patent in terms of time evolution, such as the date of first application, the date of publication and the number of months between the date of publication of youngest and the oldest patent of the patent's family.

The full set of patent features in the three categories, used in this analysis, is described in Table 1.

3   The Data-Driven Approach

The problem just described in Sect. 2 can be easily mapped into the conventional regression framework [29] where the goal is to identify the unknown relation $\mathfrak{R}$ between and input space $\mathcal{X}$, in our case the features described in Table 1, and an output space $\mathcal{Y} \subseteq \mathbb{R}$, in our case the number of forward patent's citations.

Note that, in general, the rule can be non-deterministic [29] and $\mathcal{X}$ can be composed by categorical features (the values of the features belong to a finite unsorted set) and numerical-valued features (the values of the features belong to a possibly infinite sorted set). In case of categorical features, we mapped it in a series of numerical feature thorough the one-hot encoding [15] and consequently the resulting feature space will be $\mathcal{X} \subseteq \mathbb{R}^d$.

Table 1. Features' description of patents used in the analysis

Symbol	Description	Determinants category	Use
F01	Date of the first application for the family	Temporal	Predictive
F02	Date of application of the patents	Temporal	Predictive
F03	Date of publication	Temporal	Predictive
F04	Date of grant	Temporal	Predictive
F05	Date of expire	Temporal	Predictive
F06	Technical concepts	Internal	Predictive
F07	Technological domain	Internal	Predictive
F08	IPC classification with 9 digits	Internal	Predictive
F09	First country of application	Internal	Predictive
F10	Backward citations	Internal	Predictive
F11	Independent claims	Internal	Predictive
F12	Dependent claims	Internal	Predictive
F13	Number of assignees	External	Predictive
F14	Months between the publication date of the youngest and the oldest patent of the family	Temporal	Predictive
F15	Number of Words in the document	Internal	Predictive
F16	Number of Technical terms in the abstract	Internal	Predictive
F17	Number of Words in the description	Internal	Predictive
F18	Number of Authorities	Internal	Predictive
F19	Number of Figures	Internal	Predictive
F20	Number of priorities	External	Predictive
F21	Size of the family	External	Predictive
F22	IPC classification with 4 digits	Internal	Predictive
F23	IPC classification with 7 digits	Internal	Predictive
F24	Technical concepts that occur at least 50 times	Internal	Predictive
	Number of citing patents		To be predicted

In the regression framework a set of data $\mathcal{D}_n = \{(\boldsymbol{x}_1, y_1), \ldots, (\boldsymbol{x}_n, y_n)\}$, with $\boldsymbol{x}_i \in \mathcal{X}$ and $y_i \in \mathcal{Y}$, are available.

A peculiar characteristic of our case is that n may be very large, even million of samples, wile $d \ll n$.

Moreover some values of $\boldsymbol{x}_i$ may be missing [9]. In this case, if the missing value is in a categorical feature, an additional category for missing values is introduced for that feature. If, instead, the missing value is associated with a numerical feature, as suggested in Donders [9], the missing value is replaced with the mean value of that feature and an additional logical feature is introduced to indicate if the value of that feature is missing or not for a particular sample.

The goal of the authors is to identify a model $\mathfrak{M} : \mathcal{X} \rightarrow \mathcal{Y}$, which best approximates $\mathfrak{R}$, through an algorithm $\mathscr{A}_{\mathcal{H}}$ characterized by its set of hyperparameters $\mathcal{H}$.

The accuracy of the model $\mathfrak{M}$ in representing the unknown relation $\mathfrak{R}$ is usually measured with reference to different indices of performance [6, 11]. In particular, authors adopted the Mean Absolute Percentage Error (MAPE).

Since the hyperparameters $\mathcal{H}$ influence the ability of $\mathscr{A}_{\mathcal{H}}$ to estimate $\mathfrak{R}$, a proper Model Selection (MS) procedure needs to be adopted [26]. Several methods exist for MS purpose but resampling methods, like the well-known k-Fold Cross Validation (KCV) [20] or the nonparametric Bootstrap (BTS) [2] approaches, represent the state-of-the-art of MS approaches when targeting real-world applications. Resampling methods rely on a simple idea: the original dataset $\mathcal{D}_n$ is resampled once or many (n_r) times, with or without replacement, to build two independent datasets called training, and validation sets, respectively $\mathcal{L}_l^r$ and $\mathcal{V}_v^r$, with $r \in \{1, \cdots, n_r\}$. Note that $\mathcal{L}_l^r \cap \mathcal{V}_v^r = \varnothing$, $\mathcal{L}_l^r \cup \mathcal{V}_v^r = \mathcal{D}_n$. Then, in order to select the best combination the hyperparameters $\mathcal{H}$ in a set of possible ones $\mathfrak{H} = \{\mathcal{H}_1, \mathcal{H}_2, \cdots\}$ for the algorithm $\mathscr{A}_{\mathcal{H}}$ or, in other words, to perform the MS phase, we have searched for the hyperparameters which minimize the MAPE of the model, trained on the training set, on the validation set

$$\mathcal{H}^* : \quad \min_{\mathcal{H} \in \mathfrak{H}} \frac{1}{n_r} \sum_{r=1}^{n_r} \frac{100}{v} \sum_{(\boldsymbol{x}_i, y_i) \in \mathcal{V}_v^r} \left| \frac{\mathscr{A}_{\mathcal{H}, \mathcal{L}_l^r}(\boldsymbol{x}_i) - y_i}{y_i} \right|, \tag{1}$$

where $\mathscr{A}_{\mathcal{H}, \mathcal{L}_l^r}$ is a model built with the algorithm $\mathscr{A}$ with its set of hyperparameters $\mathcal{H}$ and with the data $\mathcal{L}_l^r$. Since the data in $\mathcal{L}_l^r$ are independent from the ones in $\mathcal{V}_v^r$, the idea is that $\mathcal{H}^*$ should be the set of hyperparameters which allows to achieve a small error on a data set that is independent from the training set. In this work, authors will exploit the BTS procedure and consequently $r = 500$, if $l = n$ and the resampling must be done with replacement [26].

Finally, we need to estimate the error of the optimal model with a separate sets of data $\mathcal{T}_m = \{(\boldsymbol{x}_1^t, y_1^t), \cdots, (\boldsymbol{x}_m^t, y_m^t)\}$ since the error that our model commits over $\mathcal{D}_n$ would be optimistically biased since $\mathcal{D}_n$ has been used to find $\mathfrak{M}$. and for this reason we have to compute

$$\text{MAPE} = \frac{100}{m} \sum_{i=1}^{m} \left| \frac{\mathscr{A}_{\mathcal{H}^*, \mathcal{D}_n}(\boldsymbol{x}_i^t) - y_i^t}{y_i^t} \right|.$$

Another important property of $\mathfrak{M}$ is its interpretability, namely the possibility to understand how it behaves. In this context we have two options. The first one is to learn a $\mathfrak{M}$ such that its functional form is, by construction, interpretable [25] (e.g. Decision Trees and Rule based models). When, instead, the functional form of $\mathfrak{M}$ is not interpretable by construction [25] (e.g. Kernel Methods or Neural Networks) its interpretability must be derived a posteriori. A classical method for reaching this goal is to perform a feature ranking procedure [13] which gives an hint to the users of $\mathfrak{M}$ about the most important features which influences is results. A common approach for addressing this issue is to use the backward elimination techniques described in [13] where one feature at the time is removed from the input space, the performance of the different $\mathfrak{M}$ are tested against the one where that particular feature is present, and finally the feature are ranked based on their impact on the accuracy of $\mathfrak{M}$.

Given the above mentioned scenario, and the limitation that come from it, we will exploit three different learning algorithms: Regularized Least Squares (RLS) [29], Deep Neural Networks (DNNs) [12], and Decision Trees (DTs) [18]. All of them are able to easily handle million of samples, RLS are fast to train and easy exploit in practice, DNNs are quite powerful since they are able also to model non-linear relation between the features, and finally DTs are the most interpretable models available in literature.

3.1   Regularized Least Squares

The RLS [29] is a very simple algorithm consisting in finding the best weight of a linear model in $\mathcal{X}$ such that it minimizes the empirical mean square error over the available sample plus a regularization term, usually the norm of the weight. The trade-of between accuracy and complexity is regulated by a coefficient, the regularization coefficient, which balances the over- and under-fitting tendency of the learned model and must be tuned during the MS phase. Because of the linearity of the problem the weights can be effectively found, even when million of samples are available, with a distributed gradient descend easily implementable with the Spark MLlib library[2].

3.2   Deep Neural Networks

DNNs [12] are a learning algorithms which aim to emulate the components of the brain (the neurons) with a simple mathematical abstraction, the perceptron, and their interaction by connecting more perceptrons together. The neurons are organized in stacked layers connected together, their parameters are learned based on the available data via backpropagation. If the architecture of the DNNs consists of a single hidden layer it is called Shallow, while if it is composed by multiple layers stacked together, the network is defined as Deep. The number of layer and neurons regulates the complexity of the model and, also in this case, it is needed to find the best trade-of between accuracy and complexity of the network during the MS phase. Recently, many advances have been made in this field of research by developing new neurons, new activation functions, new optimization techniques, new regularization methods in order to reduce the overfit tendency of complex and deep networks. These advances allowed the researcher to successfully apply these methods on increasingly different and difficult real world problems. In particular the Keras library[3], running on top of Tensorflow[4], allows to easily deploy and train complex architecture, on both cluster of CPU or GPU, exploiting milion of samples.

---

[2] www.spark.apache.org/mllib.

[3] www.keras.io.

[4] www.tensorflow.org.

3.3   Decision Trees

The DTs [18] can be efficiently and effectively learned from the data even when millions of samples are available. In this case the trade-off between accuracy and complexity is regulated by the dept of the tree. The deeper is the three, more complex is the model. The depth of the tree must be tuned during the MS phase. As in the case of the RLS, given the linearity of the problem, learning a DT even with million of sample can be efficiently handled with the Spark MLlib library.

4   The Available Data

The data source used in this analysis is The Orbit Intelligence database, provided by Questel one of the world's leading intellectual property management companies. A total of 324,819 patents that are published by firms with registered office in Italy or by inventor with residence in Italy were collected over the period, 2005 to 2018. For each patent, the list of all considered features is shown in Table 1. In particular, we extracted from the database a total of 24 input features and 1 output feature.

5   Computational Experiments

In this section we will exploit the data described in Sect. 4 and the techniques described in Sect. 3 for the purpose of solving the problem described in Sect. 2. For this purpose we fist have to carefully describe the experimental setup:

- we first split the whole data in two parts $\mathcal{D}_n$ (70%) and $\mathcal{T}_m$ (30%) as described in Sect. 3;
- we learn the different models, using the MS procedure described in Sect. 3, where the following hyperparameters have been tuned
  - for what concerns RLS we search for the optimal value of the regularization hyperparameters in $\{10^{-6}, 10^{-5.8}, \cdots, 10^4\}$;
  - for what concerns the DNN we exploited Neurons with rectified linear units, we search for the number of hidden layers in $\{1, 2, 4, 8\}$, the number of neurons in each hidden layer in $\{10, 100, 1000\}$, the dropout for the hidden layer in $\{10^{-3}, 10^{-2}, 10^{-1}\}$. As optimizer we opt for the stochastic gradient descent;
  - for what concerns the DT we search for the depth of the tree in $\{3, 6, 12, 24\}$;
- then we compared the performance of the optimal models on $\mathcal{T}_m$, as described in Sect. 3, using the MAPE;
- finally we performed the feature ranking procedure described in Sect. 3 for RLS, DNN, and DT and reported the first four levels of the optimal DT for better understand what the models actually learned.

All the experiments have been repeated 30 times in order to ensure the statistical robustness of the results.

Table 2 reports the MAPE of the optimal models trained with RLS, DNN, and DT varying n. From Table 2 it is possible to observe that all the algorithms have a low error. More in detail, for small n, the best training algorithm is the RLS, whereas for high n, the DNN. Table 3 reports the top 10 features, sorted in descending order by importance, influencing the optimal models trained with RLS, DNN, and DT. From Table 3 it is possible to observe that all the three algorithms identify the same patent's features as important in predicting the forward citations and, so in estimating the IC. In particular, the most relevant features are the technological classes, based on the IPC, with 4 and 7 digits (F22 and F23, respectively), the technological domain (F07), the number of backward citations (F10) and the size of the family (F21). This result shows that the most important determinants to predict IC are mainly internal. This is an important result because it is the first step to investigate how firms should work for improving their IC. With respect to the IPC classes, firms must be aware that the classes may strongly determine the success of the technology in the market. In fact, it is worth noting that a large number of digits does not influence the capabilities of a company, meaning that small incremental changes give no information to predict IC, see for example Moaniba et al. [24]. Linked to that, also the kind and not the number of technical domains selected by the company is significant. Moreover, backward citations play a main role in predicting IC, so companies must be aware that their capacity to absorb previous knowledge and make use of it will strongly affect future innovation and the ability to be competitive [14, 16].

Moreover, the external determinants cannot be fully neglected. The size of the patents' family, is a valuable feature for understanding the market value of a patent, confirming the work of Harhoff et al. [16]. A slightly less important feature, even in the top 10 features, is represented by the temporal determinant of months between the date of publication of the youngest and the oldest patent of the family (F14). According to that, companies must pay attention to the routes that can be chosen to fill the patent, considering carefully national patent systems and international procedures. This is also underlined by Dechezlepr et al. [8]. In addition, this feature is more relevant than the internal determinant of the country of first application (F09). Thus, IC is more influenced by the strategy of internationalization of the company than from the country where it is first applied, that generally represents the origin of the knowledge [22]. Also the external determinants, such as the number of assignees (F13) helps in predicting IC, even if with less power compared to the previous described features. Conversely, some features have been not identified as predictive variables even often described in the literature as significant for the patent value [16]. Our study shows how claims are not an important feature in predicting the level of IC. Moreover, features that refer to the structure of the patents, such as the number of words in the abstract or in the description, are not relevant. Finally, Fig. 1 reports the first four levels (the complete tree is too complex for reporting it) of the optimal DT. From Fig. 1 it is possible to observe that the most relevant features have been confirmed. Specifically, technological classes at 4 digits represent the most relevant feature in predicting the number of citation, followed by the technological domains, the backward citations and the months between the of publication of the youngest and oldest patent of the family.

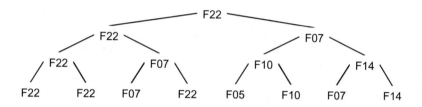

Fig. 1. First four levels of the optimal DT (splits are not reported for readability).

Table 3. Top 10 features, sorted in descending order by importance.

Rank	RLS	DNN	DT
1°	F22	F07	F22
2°	F07	F22	F23
3°	F21	F10	F05
4°	F10	F21	F07
5°	F23	F03	F14
6°	F03	F05	F10
7°	F05	F23	F21
8°	F09	F14	F03
9°	F13	F09	F09
10°	F14	F13	F13

Table 2. MAPE of the optimal models varying n.

n	RLS	DNN	DT
10^3	40.5 ± 5.2	45.3 ± 7.9	63.9 ± 5.3
10^4	25.2 ± 2.3	18.8 ± 3.4	35.3 ± 2.1
10^5	15.7 ± 0.9	10.1 ± 1.4	16.4 ± 1.1

6 Conclusion

This study has proposed a machine learning approach for understanding the IC's determinants using multiple patent features that can be identified immediately after the relevant patents are issued. To this end, three different algorithms, i.e. RLS, DNN and DT, have been employed to capture the complex nonlinear relationships between those patent features and the forward citations' count of a patent in a time period of interest. The specific case of patents issued by firms with registered office in Italy or by inventors with residence in Italy has verified that the proposed methods can identify the IC's determinants of firms.

The central tenet of the proposed approaches is that patent features, mainly internal determinants, can provide information about the firms' IC. The contributions of this research are two-fold. First, from a methodological perspective, this study shows the usefulness of machine learning approaches in reducing the number of features needed to explain IC, and so in reducing the complexity of the analyses by focusing only on the features with high predictive power. Second, from a managerial standpoint, the results suggest which few, but relevant, variables managers should look at in writing and issuing patents. Moreover, the study provides information about what variables companies should leverage to improve their IC, as it emerges from the patent data. Nevertheless, this study is not without limitations. Despite patents are widely used to measure IC, they represent a limited set of companies' innovation assets. Furthermore, in some sectors patents are not widely adopted as mechanisms to protect innovations. For this reason, other sources of information and data (for example expenditure in R&D) will be used to strengthen the analyses and to extend the scope of the study. The next step will be to investigate the dynamics of how and much the most important the patents' features influence the forward citations, proxy of the IC.

Acknowledgment. This work has been supported by LIUC - Cattaneo University under Grant "Data Analytics".

References

1. Adler, P.S., Shenbar, A.: Adapting your technological base: the organizational challenge. Sloan Manag. Rev. **32**(1), 25–37 (1990)
2. Anguita, D., Boni, A., Ridella, S.: Evaluating the generalization ability of support vector machines through the bootstrap. Neural Process. Lett. **11**(1), 51–58 (2000)
3. Barney, J., Wright, M., Ketchen Jr., D.J.: The resource-based view of the firm: ten years after 1991. J. Manag. **27**(6), 625–641 (2001)
4. Bessen, J.: The value of us patents by owner and patent characteristics. Res. Policy **37**(5), 932–945 (2008)
5. Christensen, J.F.: Asset profiles for technological innovation. Res. Policy **24**(5), 727–745 (1995)
6. Cincotti, S., Gallo, G., Ponta, L., Raberto, M.: Modeling and forecasting of electricity spot-prices: computational intelligence vs classical econometrics. AI Commun. **27**(3), 301–314 (2014)
7. Davila, T.: An empirical study on the drivers of management control systems' design in new product development. Acc. Organ. Soc. **25**(4–5), 383–409 (2000)
8. Dechezleprêtre, A., Ménière, Y., Mohnen, M.: International patent families: from application strategies to statistical indicators. Scientometrics **111**(2), 793–828 (2017)
9. Donders, A.R.T., van der Heijden, G.J.M.G., Stijnen, T., Moons, K.G.M.: Review: a gentle introduction to imputation of missing values. J. Clin. Epidemiol. **59**(10), 1087–1091 (2006)
10. Kerssens-van Drongelen, I., Nixon, B., Pearson, A.: Performance measurement in industrial R&D. Int. J. Manag. Rev. **2**(2), 111–143 (2000)

11. Ghelardoni, L., Ghio, A., Anguita, D.: Energy load forecasting using empirical mode decomposition and support vector regression. IEEE Trans. Smart Grid **4**(1), 549–556 (2013)

12. Goodfellow, I., Bengio, Y., Courville, A.: Deep Learning. MIT Press, Cambridge (2016)

13. Guyon, I., Elisseeff, A.: An introduction to variable and feature selection. J. Mach. Learn. Res. **3**, 1157–1182 (2003)

14. Hall, B.H., Thoma, G., Torrisi, S.: The marker value of patents and R&D: evidence from european firms. Acad. Manag. Proc. **2007**(1), 1–6 (2007)

15. Hardy, M.A.: Regression with Dummy Variables. Sage, Newbury Park (1993)

16. Harhoff, D., Scherer, F.M., Vopel, K.: Citations, family size, opposition and the value of patent rights. Res. Policy **32**(8), 1343–1363 (2003)

17. Jaffe, A.B., Trajtenberg, M.: Flows of knowledge from universities and federal laboratories: modeling the flow of patent citations over time and across institutional and geographic boundaries. Proc. Nat. Acad. Sci. **93**(23), 12671–12677 (1996)

18. James, G., Witten, D., Hastie, T., Tibshirani, R.: An Introduction to Statistical Learning. Springer, New York (2013)

19. Kogut, B., Zander, U.: Knowledge of the firm, combinative capabilities, and the replication of technology. Organ. Sci. **3**(3), 383–397 (1992)

20. Kohavi, R., et al.: A study of cross-validation and bootstrap for accuracy estimation and model selection. In: International Joint Conference on Artificial Intelligence (1995)

21. Lee, C., Kwon, O., Kim, M., Kwon, D.: Early identification of emerging technologies: a machine learning approach using multiple patent indicators. Technol. Forecast. Soc. Chang. **127**, 291–303 (2018)

22. Mahnken, T.A., Moehrle, M.G.: Multi-cross-industry innovation patents in the usa-a combination of patstat and orbis search. World Patent Inf. **55**, 52–60 (2018)

23. Manzini, R., Lazzarotti, V.: Intellectual property protection mechanisms in collaborative new product development. R&D Manag. **46**(S2), 579–595 (2016)

24. Moaniba, I.M., Su, H.N., Lee, P.C.: Knowledge recombination and technological innovation: the important role of cross-disciplinary knowledge. In: Innovation, pp. 1–27 (2018)

25. Molnar, C.: Interpretable machine learning (2018). https://christophm.github.io/book/

26. Oneto, L.: Model selection and error estimation without the agonizing pain. WIREs Data Min. Knowl. Disc. **8**(4), e1252 (2018)

27. Reitzig, M.: Improving patent valuations for management purposes-validating new indicators by analyzing application rationales. Res. Policy **33**(6–7), 939–957 (2004)

28. Romijn, H., Albaladejo, M.: Determinants of innovation capability in small electronics and software firms in southeast england. Res. Policy **31**(7), 1053–1067 (2002)

29. Shalev-Shwartz, S., Ben-David, S.: Understanding Machine Learning: From Theory to Algorithms. Cambridge University Press, Cambridge (2014)

30. Stern, S., Porter, M.E., Furman, J.L.: The determinants of national innovative capacity. Technical report National bureau of economic research (2000)

31. Ussahawanitchakit, P.: Innovation capability and export performance: and empirical study of the textile businesses in Thailand. J. Int. Bus. Strategy **7**(1) (2007)

Predicting Future Market Trends: Which Is the Optimal Window?

Simone Merello[1], Andrea Picasso Ratto[1], Luca Oneto[1(✉)], and Erik Cambria[2]

[1] DIBRIS - University of Genoa, Via Opera Pia 13, 16145 Genova, Italy
{simone.merello,andrea.picasso}@smartlab.ws, luca.oneto@unige.it
[2] Nanyang Technological University, Singapore, Singapore
cambria@ntu.edu.sg

Abstract. The problem of predicting future market trends has been attracting the interest of researches, mathematicians, and financial analysts for more then fifty years. Many different approaches have been proposed to solve the task. However only few of them have focused on the selection of the optimal trend window to be forecasted and most of the research focuses on the daily prediction without a proper explanation. In this work, we exploit finance-related numerical and textual data to predict different trend windows through several learning algorithms. We demonstrate the non optimality of the daily trend prediction with the aim to establish a new guideline for future research.

Keywords: Stock market prediction · Learning from data · Sentiment analysis · Optimal window

1 Introduction

Stock market prediction is a challenging problem to solve and its complexity is strictly related to multiple factors which could affect price changes. The sentiment information extracted from text has proven to be effective in predicting market movements [13] and technical analysis indicators can be considered as an attempt to extract the mood of market participants as well [8]. Contemporary, learning algorithms have been widely exploited for stock market prediction, starting from the regression of future prices to the more recent classification of market trends [7,14]. However, most of the works have focused on daily prediction without a proper explanation [7,14]. A minority of researchers has tried to forecast weekly trends [3] or intraday trends [11] but only few works in literature have focused on the time window of the predictions. Rao et al. [10] studied the correlation between tweets and trend windows of different size but the results were limited by a specific algorithm and a specific input space. Merello et al. [6] focused on the optimal time-lag between the publication of news articles and the prediction but the trend size was fixed a priori. We believe the selection of the

---

S. Merello and A. P. Ratto—Equal Contributors.

© Springer Nature Switzerland AG 2020
L. Oneto et al. (Eds.): INNSBDDL 2019, INNS 1, pp. 180–185, 2020.
https://doi.org/10.1007/978-3-030-16841-4_19

right prediction window a crucial point in the whole process. In this paper, we focus on understanding which is the optimal windows size to be predicted. The task of stock market prediction is mapped in a classification problem executed on different trend lengths. The evaluation of the models is performed using both data science metrics and financial metrics computed in a trading simulation. During the experimental phase, the best results were achieved on trends lasting more than one day, demonstrating the non optimality of the results achieved in the past research works.

2 Proposed Approach

The trend forecasts $\hat{y}_t$ were made at fixed and discretized time steps t relying on text published on dates $\bar{t}$. The labels y_t were defined according to the price movements during trends of length w: $y_t = \mathbb{1}(p_{t+w} - p_t)$ where $p_t \in \mathbb{R}^+$ is the close price at time step t and $\mathbb{1} : \mathbb{R} \to \{0,1\}$ represents the unit step function.

The information, available at time t relative to technical indicators and news articles, was considered as leading the trends. The single news article published in $\bar{t}$ was encoded in a feature vector defined as $\overline{n_{\bar{t}}} \in \mathbb{R}^d$ through the Loughran & McDonald dictionary [5] (L&Mc) or through the Affective Space embeddings [1] (AS). L&Mc is a tool specific for finance. It contains positive, negative and other listings of words sentiment-related thus, $\overline{n_{\bar{t}_i}}$ was defined as the count of words in the listing i found in the news article. AS was chosen to represent the concept-oriented sentiment contained in financial writings. The embeddings of the concepts contained in the article were averaged to represent the news. For each interval t, $I_t \in \mathbb{R}^f$ represented the value of technical indicators computed over the recent past prices p_t and volumes v_t. Table 1 reports the indicators considered in the experiments. The parameter s was chosen as $s \in \{30, 50, 100, 150\}$. The price fluctuation y_t was considered connected with news articles published in the recent past through $n_{t,\hat{w}} = \frac{1}{\hat{w}} \sum_{\bar{t} \in [t-\hat{w},t)} \overline{n_{\bar{t}}}$. The input of the models was defined as: $x_t = [N_t, I_t]$ where N_t was used to consider different input windows $N_t = [n_{t,1}, n_{t,5}, n_{t,10}, n_{t,15}, n_{t,20}, n_{t,30}, n_{t,50}]$.

Table 1. Technical analysis indicators used in the experiments. $\overline{p_{t,s}}$ represents the average value of the previous s prices, $\overline{v_t}$ is the average volume observed until t, u_t^{bb} and d_t^{bb} refer to the Bollinger Bands.

Name	Formula
Momentum	$\frac{p_t}{p_{t-s}} - 1$
SMA	$\frac{p_t}{\overline{p_{t,s}}} - 1$
Bollinger Bands	$\begin{cases} 1 & \text{if } p_t > u_t^{bb} \\ 0 & \text{if } p_t \in [u_t^{bb}, d_t^{bb}] \\ -1 & \text{if } p_t < d_t^{bb} \end{cases}$
Differentiated v_t	$v_t - \overline{v_t}$

The performances of four models were averaged for the prediction of trends lasting 1 h, 1 day but also 4, 5 and 7 days. Respectively Feed Forward Neural Network (FFNN) and Kernel Support Vector Machine (KSVM) algorithms were fed with two different input spaces where $\overline{n_t}$ was computed as the L&Mc or the AS representation of news articles in different experiments. The Gaussian kernel was chosen for the KSVM model. The FFNN architecture was defined by 3 hidden layers with the ReLU activation function, Batch normalization [4] and regularized through Max-norm and Dropout [12]. The output layer was defined by a sigmoid activation function and the whole model was trained with the Adam optimizer. Model selection was performed using cross validation on time dependent data according to the work of [6,9]. For each fold two subsequent intervals were used for training and evaluation respectively. The points used for evaluation were always chosen ahead the training set in time so that a possible look ahead bias was avoided. The last part of the dataset was considered as Test set. The data used for the experiments was affected from labels unbalance, caused by the trending tendency of the price movements. According to the results of [9], the SMOTE [2] balancing technique was adopted to avoid developing a biased classifier. Both the Training and the Validation set were balanced for the selection of the optimal parameters and hyperparameters respectively. Furthermore, model selection and the trend evaluation were held trough the Matthews correlation coefficient (MCC) for its property of skewness-independence. Sharpe ratio and Annualized gain were examined as results of the trading performance. The Sharpe ratio measures the average return earned in excess of the risk-free rate as a reward for the extra volatility endured in holding a riskier asset.

3 Experiments

During the experimental phase, the assets used to perform the market trend prediction were chosen as the most capitalized of the NASDAQ100 index to avoid liquidity problems during the trading simulation. Approximately eleven months of data were used for training, from 2017-04-03 to 2018-02-23 and four months were used for testing, from 2018-02-24 to 2018-06-21. It was made up of twenty tickers to make the results of this research statistically significant. The news articles were collected through the Intrinio API[1]. While prices and volumes were retrieved from the Google Finance API[2] with a frequency of 1 h. Thus, every hour a forecast for the future trend of length w was made.

In a first test, the MCC score achieved by the models was evaluated on average above all the considered stocks. Furthermore, the performances achieved by different models were averaged to avoid specific conclusions for a single classifier or a single input space. As second test, a real trading simulation was performed with a specific tool[3]. The evaluation of financial metrics as Annualized gain and

---

[1] https://intrinio.com/.
[2] https://pypi.org/project/googlefinance.client/.
[3] https://www.backtrader.com/.

Sharpe ration was considered necessary to measure the effectiveness of the predictions. The results achieved during the trading simulation depended on the selection of which stocks to trade (trading strategy) but also from the amount of capital to invest in each trade (sizing strategy). The trading strategy was defined according to the trading signal s_t computed as average of the last three predictions $s_t = \frac{1}{3} \sum_{i=0}^{2} \hat{y}_{t-i}$. Considering multiple predictions represents a careful behavior which avoids noisy decisions based only on one signal. The sizing strategy was designed so that, given the portfolio value P_t each time step t, an amount $\frac{P_t}{w \cdot n_{a,t}}$ was invested on the single asset. $n_{a,t}$ denotes the number of actions (*buy* or *sell*) to perform in t and w represents the trend length. A commission rate of 0.01% was considered and the risk-free rate required by the computation of the Sharpe ratio was set as the interest rate of ten-years U.S. Treasury bill in the period of evaluation.

According to Fig. 1, the MCC value relative to the trend length of seven trading days (49 h) achieved the best score. This result is considered significant since most of the state-of-the-art papers on return classification focus on daily trend prediction without a proper explanation. This experiment enlightened the sub-optimality of daily prediction with respect to other choices of the trend length. According to the average performance of the tested models, predicting trends lasting seven or four days allowed approximately to double the MCC score (respectively 0.10 and 0.12) in comparison to the daily prediction (MCC 0.05). During the trading simulation, the daily trend prediction has revealed to be a sub-optimal choice as well. The prediction of daily trends has lead to a 11% of annualized gain but the best result was achieved by the 4 days trend, reaching 23% of annualized gain. The Sharpe ratio highlighted similar conclusions, 1.5 score for the daily prediction, 2.3 for the 5 days trend. Overall the conclusions based on the MCC, a measure of the classification performances, have shown to differ slightly from the evaluation of financial metrics.

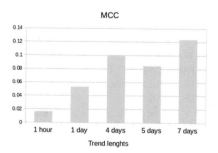

Fig. 1. MCC values averaged over the tested models. Trends lasting 1 h, 1, 4, 5 and 7 days are evaluated.

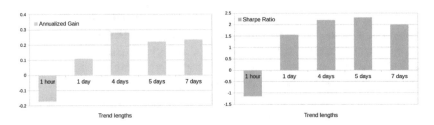

Fig. 2. Annualized Gain and Sharpe ratio averaged over the tested models. Trends lasting 1 h, 1, 4, 5 and 7 days are evaluated.

4   Conclusion and Discussion

This paper focus on the selection of the optimal trend to be predicted in financial forecasting. The results show that the performances change significantly according to this parameter thus, its value has to be carefully chosen. The prediction of trends lasting 7 days has led to the best classification performances.

References

1. Cambria, E., Fu, J., Bisio, F., Poria, S.: Affectivespace 2: enabling affective intuition for concept-level sentiment analysis. In: AAAI, pp. 508–514 (2015)
2. Chawla, N.V., Bowyer, K.W., Hall, L.O., Kegelmeyer, W.P.: Smote: synthetic minority over-sampling technique. J. Artif. Intell. Res. **16**, 321–357 (2002)
3. Huang, W., Nakamori, Y., Wang, S.Y.: Forecasting stock market movement direction with support vector machine. Comput. Oper. Res. **32**(10), 2513–2522 (2005)
4. Ioffe, S., Szegedy, C.: Batch normalization: accelerating deep network training by reducing internal covariate shift. arXiv preprint arXiv:1502.03167 (2015)
5. Loughran, T., McDonald, B.: When is a liability not a liability? Textual analysis, dictionaries, and 10-ks. J. Finance **66**(1), 35–65 (2011)
6. Merello, S., Picasso, A., Ma, Y., Oneto, L., Cambria, E.: Investigating timing and impact of news on the stock market. In: IEEE International Conference on Data Mining, International Workshop on Sentiment Elicitation from Natural Text for Information Retrieval and Extraction (ICDM) (2018)
7. Oancea, B., Ciucu, Ş.C.: Time series forecasting using neural networks. arXiv preprint arXiv:1401.1333 (2014)
8. Park, C.H., Irwin, S.H.: The profitability of technical analysis: a review. AgMAS Project Research Report 2004-04 (2004)
9. Picasso, A., Merello, S., Ma, Y., Oneto, L., Cambria, E.: Ensemble of technical analysis and machine learning for market trend prediction. In: IEEE Symposium Series on Computational Intelligence (SSCI) (2018)
10. Rao, T., Srivastava, S.: Analyzing stock market movements using twitter sentiment analysis. In: Proceedings of the 2012 International Conference on Advances in Social Networks Analysis and Mining (ASONAM 2012), pp. 119–123. IEEE Computer Society (2012)
11. Schumaker, R., Chen, H.: Textual analysis of stock market prediction using financial news articles. In: AMCIS 2006 Proceedings, p. 185 (2006)

12. Srivastava, N., Hinton, G., Krizhevsky, A., Sutskever, I., Salakhutdinov, R.: Dropout: a simple way to prevent neural networks from overfitting. J. Mach. Learn. Res. **15**(1), 1929–1958 (2014)
13. Xing, F.Z., Cambria, E., Malandri, L., Vercellis, C.: Discovering bayesian market views for intelligent asset allocation. arXiv preprint arXiv:1802.09911 (2018)
14. Xu, Y., Cohen, S.B.: Stock movement prediction from tweets and historical prices. In: ACL (2018)

F_0 Modeling Using DNN for Arabic Parametric Speech Synthesis

Imene Zangar[1(✉)], Zied Mnasri[1,3], Vincent Colotte[2], and Denis Jouvet[2]

[1] Electrical Engineering Department, University Tunis El Manar, Ecole Nationale
d'Ingénieurs de Tunis, Tunis, Tunisia
{imene.zangar,zied.mnasri}@enit.utm.tn
[2] Université de Lorraine, CNRS, Inria, LORIA, 54000 Nancy, France
{vincent.colotte,denis.jouvet}@loria.fr
[3] Università degli studi di Genova, DIBRIS, Genoa, Italy

Abstract. Deep neural networks (DNN) are gaining increasing interest in speech processing applications, especially in text-to-speech synthesis. Actually state-of-the-art speech generation tools, like MERLIN
and WAVENET are totally DNN-based. However, every language has
to be modeled on its own using DNN. One of the key components of
speech synthesis modules is the prosodic parameters generation module
from contextual input features, and more particularly the fundamental
frequency (F_0) generation module. Actually F_0 is responsible for intonation, that is why it should be accurately modeled to provide intelligible
and natural speech. However, F_0 modeling is highly dependent on the
language. Therefore, language specific characteristics have to be taken
into account. In this paper, we aim to model F_0 for Arabic speech synthesis with feedforward and recurrent DNN, and using specific characteristic features for Arabic like vowel quantity and gemination, in order
to improve the quality of Arabic parametric speech synthesis.

Keywords: Arabic parametric speech synthesis ·
Fundamental frequency F_0 · Deep neural networks ·
Recurrent neural networks

1 Introduction

Speech processing systems have gained increasing interest since a few years.
Actually, new technology advances have made possible to interact with machines,
through speech recognition and speech synthesis. Speech synthesis could be performed from text (Text-to-speech synthesis). In text-to-speech synthesis, the text
is processed to obtain a sequence of units (phonemes, diphones, syllables, etc).
The parameters of which are predicted by dedicated modules. These parameters
could be classified into prosodic parameters and acoustic parameters, mainly
mel-cepstrum coefficients like Mel-Generalized Cepstral coeffcients (MGC), Mel-
Frequency Cepstral Coefficients (MFCC), Linear Spectral Pair (LSP), etc. and

© Springer Nature Switzerland AG 2020
L. Oneto et al. (Eds.): INNSBDDL 2019, INNS 1, pp. 186–195, 2020.
https://doi.org/10.1007/978-3-030-16841-4_20

their temporal derivatives (Δ and $\Delta\Delta$). Prosodic parameters include segment duration and fundamental frequency (F_0). Both parameters are jointly responsible for rhythm and intonation.

F_0 modeling has always been the cornerstone of any speech synthesis system. In fact the evolution of F_0 is the manifestation of complex and interdependent phonological phenomena like intonation, accentuation and voicing. Therefore, an accurate F_0 model is necessary to produce intelligible and naturally sounding synthesis speech. So far, there has been a variety of F_0 models, highly accurate and successfully applied to speech synthesis systems. However, most of them are language dependent, since each F_0 model was established for a specific language. Furthermore, when a model is applied for another language, it should undergo many adjustments to fit the new target language. F_0 models could be classified into phonological vs. phonetic models. Phonological models can also be divided into tone-sequence and perceptual models. In tone-sequence models like ToBI (Tone and break indices) [1] the intonation of an utterance is described as a sequence of phonologically opposite tones, High (H) and Low (L). The different combinations of both tones give a finite state grammar. Then F_0 is regarded as an intonation event (high/low/rising/falling); whereas in perceptual models like IPO (Institute of perception research) model [2], the F_0 contour is described by a sequence of the most relevant movements like prominence. In this model, the intonation contour consists of a linear sequence of discrete intonational elements, i.e. the most relevant F_0 movements (but not tones). On the other side, i.e. the phonetic models, F_0 is regarded as a physical quantity to be measured/predicted. These models try to establish an analytic formulation or approximation of F_0 and/or its variations. Tilt model [3,4], and PaIntE, Parametric Intonation Event System model [5] are amongst the well known analytic F_0. The Tilt model attempts to label the utterance by one of four intonational events, i.e. pitch accents, boundary tones, connections and silence. The PaIntE model is similar to the Tilt model as it tries also to model the accents using a small set of parameters. Moreover, this model uses the sum of two sigmoid functions to represent F_0 contour locally.

With the development of machine learning, speech synthesis has been taking benefit from data-driven models to predict prosodic parameters, including F_0, directly from input features, using machine learning. Amongst data-driven models, HMM (Hidden Markov Models) have been used for parametric speech synthesis, since 1999 [7]. More recently, deep neural networks (DNN) have taken benefit of the development of GPU to become the leading technique in machine learning. With the recent advances of DNN, parametric speech synthesis and more particularly prosodic parameters prediction modules have been migrating from HMM-based model to DNN-based ones, in the aim to increase accuracy, and especially to avoid the over-averaging problem, noticed in HMM-predicted parameters [15]. Therefore, feedforward and recurrent DNN have been integrated to parametric speech synthesis for many languages, such Japanese and English [8]. Also for Arabic, DNN have been recently used to predict phone duration for Arabic parametric speech synthesis [10], through a dedicated DNN model for each class of Arabic phones, i.e. short vs. long vowels. and simple vs. geminated consonants. Initially, DNN were inserted in HMM-based parametric speech synthesis as a replacement of CART (classification and regression trees) which are used for HMM models clustering. In a second approach, DNN were used to predict raw prosodic parameters, that were injected to HTS models. For instance, DNN were proved to be more efficient than HMM to model segmental durations [9,10]. Also, state-of-the-art TTS tools like MERLIN [6] and WAVENET [11] are fully based on DNN. In this work, the studied problem is F_0 modeling using DNN and adding specific features for Arabic, i.e. vowel quantity and gemination, to the standard parametric speech synthesis feature set [10].

This paper is organized as follows: Sect. 2 reviews, F_0 modeling for parametric speech synthesis, with a special focus on DNN models; Sect. 3 details the proposed DNN models used in this work; and Sect. 4 presents, the experiments and the results.

2 F_0 Modeling for Parametric Speech Synthesis

Parametric speech synthesis was originally designed using HMM models in HTS (HMM-based text-to-speech synthesis) system [7], including many updates regarding the use of multi space probability distributions (MSD) [12], and the hidden semi-markov models [13]. However, since a few years, DNN models are used for parametric speech synthesis, and lead to an enhance the quality of the generated speech.

2.1 F_0 Modeling in HMM-Based Parametric Speech Synthesis

The parameters modeling in HMM-based parametric speech synthesis is based on a five-state left-to-right HMM modeling. Each state is modeled by a single Gaussian distribution with diagonal covariance. Decision trees are used to cluster HMM model according to the contextual input features. Finally, expectation-maximization (EM) algorithm is used to predict the HMM model output.

In HTS system, the prosodic parameters, including duration and F_0 were originally modeled using HMM. However for F_0, it is a special process, since F_0 is defined only in voiced regions of speech, where it is represented by one-dimensional continuous values. However, in unvoiced regions F_0 is replaced by a discrete label, so that it is impossible to apply the same HMM model for F_0 in voiced and unvoiced parts of the speech signal. Therefore, a multi-space probability distribution (MSD probability) was designed to model continuous and discrete F_0 respectively in voiced and unvoiced regions using HMM [14].

Nevertheless, F_0 modeling is still suffering from over-averaging, since the predicted Gaussian distributions tend to provide mean values of F_0. This phenomenon has been related in subjective tests of HTS vs. unit selection synthesis, where the latter technique was preferred [15], though it is older.

2.2 F_0 Modeling in DNN-Based Parametric Speech Synthesis

In [8], $log(F_0)$ was modeled by DNN. In this work, F_0 was first interpolated in the unvoiced regions to provide continuous F_0 values as set by [16]; then $log(F_0)$ is modeled using DNN. Besides, voicing decision labels were predicted.

In the preprocessing phase, 80% of silence data was removed from the training set. Input features were normalized to have a zero-mean and a unit-variance, whereas output targets were normalized to be in the interval $[0.01, 0.99]$. The DNN was trained using stochastic gradient descent (SGD), whereas the activation functions were all sigmoids, including the output layer. Different DNN architectures were trained on a speech corpus containing 33000 English utterances, covering nearly 13 h. The best performing model was a feedforward DNN containing 5 hidden layers, each with 2048 nodes. Results show that DNN gave more accurate voicing decision prediction than HMM, whereas HMM gave the lowest $log(F_0)$ $RMSE$ error [8].

In [17], F_0 was interpolated in unvoiced regions using an exponential decay function [18]. To predict the voicing decision label and $log(F_0)$ values many DNN architectures were investigated based on feedforward and recurrent layers, i.e. LSTM (longs short-term memory) and BLSTM (Bidirectional LSTM). For instance, a 6-feedforward-hidden-layer DNN with 512 nodes at each hidden layer, a 3-feedforward-hidden-layer DNN with 1024 nodes at each layer, a hybrid 3-feedforward-layer and 1-LSTM-layer and a hybrid 2-feedforward-layer and 2-BLSTM-layer with 512 nodes at each layer, for both hybrid architectures, and using sigmoid activation function were used. 5000 Chinese utterances covering 5 h of speech were used for training and development, and 200 utterances for test. Results show that in comparison to the baseline HMM model, used in HTS [16], the accuracy of voicing decision label prediction is nearly the same, whereas F_0 prediction $RMSE$ was lower for all DNN architectures, but without a significant difference between the different architectures, all varying between 15 to 16 Hz.

More recently, in [19], a hybrid 2-feedforward-layer and 2-LSTM-layer DNN with 512 nodes at each hidden layer, was trained on a 13-h Chinese speech corpus. The $RMSE$ calculated on the test set was 12 Hz.

It should be noted that in all these architectures, the standard features set of HTS system [24] was used. No specific language features were added to model the language characteristics.

3   Proposed DNN Models

F_0 modeling using DNN is a two-stage process, which requires first classifying speech segments into voiced and unvoiced regions, and then predicting F_0. Since both tasks are of different nature, i.e. classification and regression, we opted to train a different network for each.

3.1 Proposed Architectures for Discrete Voiced/Unvoiced Decision Classification and F_0 Values Prediction

Two different neural networks were used to predict voicing decision and F_0 values. For each DNN, many architectures were tried out, using dense layers only, dense layers with LSTM or BLSTM layers, and LSTM or BLSTM layers only (cf. Table 1). Actually, the intent from the choice of these architectures is to see whether recurrent neural network are more suitable than feedforward networks to model speech parameters, which are known to be highly recurrent.

Table 1. DNN architectures selected based on results on development set for voiced/unvoiced decision classification and for F_0 prediction.

DNN network	Number and types of hidden layers	Number of nodes in hidden layers
All-Dense	6 Dense layers	[512,512,512,512,512,512]
Dense-LSTM (1)	2 Dense layers + 4 LSTM layers	[256,256,512,512,1024,1024]
Dense-LSTM (2)	2 Dense layers + 4 LSTM layers	[512,512,512,512,512,512]
Dense-BLSTM (1)	2 Dense layers + 4 BLSTM layers	[256,256,512,512,1024,1024]
Dense-BLSTM (2)	2 Dense layers + 4 BLSTM layers	[512,512,512,512,512,512]
All-LSTM	6 LSTM layers	[512,512,512,512,512,512]
All-BLSTM	6 BLSTM layers	[512,512,512,512,512,512]

Also, it should be noted that after some experiments, all kept networks were implemented with *tanh* activation function for hidden layers, whereas linear output activation was used for F_0 and sigmoid output activation was used for voicing decision prediction. The *Adam* optimizer was used to train the voicing decision classification DNN whereas for the F_0 prediction DNN, the *Rmsprop* optimizer was used. The batch size was set to 100 in all the experiments.

3.2 Error Minimization Criteria

Since voicing decision prediction is a classification problem, cross-entropy loss function was used

$$L_{CE} = -\frac{1}{N} \sum_{i=1}^{N} v_i log(p(v_i)) \tag{1}$$

where N is the number of frames, v_i is the voicing decision value (0 or 1) of frame i, and $p(v_i)$ is its probability.

As far as continuous F_0 values prediction is concerned, MSE was selected as loss function, since it's a regression problem.

$$L_{MSE} = \frac{1}{N} \sum_{i=1}^{N} (log(F_{0i}) - log(\hat{F}_{0i}))^2 \tag{2}$$

where N is the number of frames, $log(F_{0i})$ and $log(\hat{F}_{0i})$ are the target and the predicted values of frame i, respectively.

To avoid overfitting, early-stopping option was used. Thus, if L_{CE} (in case of voicing decision network) or L_{MSE} (in case of F_0 prediction network) evaluated on the development set, don't improve after a certain number of epochs, set to 20 in our case, the training process is stopped.

4   Experiments and Results

4.1   Speech Corpus

To train the DNN-based F_0 model, an Arabic speech corpus of 1597 utterances was used [20]. The utterances correspond to news bulletin in Modern Standard Arabic (MSA), read by a native-Arabic male speaker. The signals were recorded at a sampling rate of 48 KHz and a precision of 16 bits. 70% of the utterances were used for training, 20% for development and 10% for test. To extract original voicing decision labels and F_0, SWIPE algorithm [23], included in SPTK toolkit [21] was used. Since the speaker is male, the F_0 tracking algorithm was bound between 80 Hz and 320 Hz. A single value of F_0 was extracted at each 5-ms frame. Voicing decision labels were deduced automatically using the fact that null F_0 values corresponds to unvoiced frames.

4.2   Features Selection and Preprocessing for F_0 Modeling

In addition to the standard HTS model input features set [24], classically used for parametric speech synthesis, two Arabic specific features, namely vowel quantity and gemination were added. Actually both features have been recently proved to enhance the quality of Arabic speech synthesis using DNN [22]. The input features were coded in different ways, with respect to their natures. Thus three types of feature encoding were used: yes/no features like stressed/non-stressed syllables were coded into binary values; class-wise features, such as phoneme identity, were coded with one-hot vectors and for unlimited-value features coarse coding was used. For example, with coarse coding, the relative position of phonemes in the syllable, i.e. beginning, middle or end, is encoded respectively, (1, 0, 0), (0, 1, 0) or (0, 0, 1) whatever the number of phonemes in the syllable. Thus, the input feature vector contains 439 coefficients. Moreover, the coefficients of the input features vector were normalized to have a zero-mean and a unit-variance.

On the other side, a linear interpolation was applied to the output $log(F_0)$ targets in the unvoiced parts of speech. Then, interpolated values were normalized to be within [0.01, 0.99] interval. However, output voicing decision labels were not normalized, since they are already binary.

4.3   Selected DNN Models

Various network parameters were experimented, i.e. various number and type of hidden layers, number of nodes in hidden layers and activation functions were empirically modified. The models were evaluated on the development set, using $RMSE(Hz)$ for predicted F_0 values and voicing decision error (VDE). The best models on the development set (cf. Table 2) were kept to be evaluated on the test set.

4.4   Objective Evaluation

To assess the quality of the predicted parameters, i.e. $log(F_0)$ and voicing decision label, four measures were performed first on the development set, to select the best performing model, and then on the test set:

- **Root mean square error** $(RMSE)$ between target and predicted F_0 values (cf.(3)) on voiced frames:

$$RMSE(Hz) = \sqrt{\frac{1}{N_v} \sum_{i=1}^{N_v} (F_{0i} - \hat{F}_{0i})^2} \tag{3}$$

  where N_v is the number of originally voiced frames and F_{0i} and $\hat{F}_{0i}$ are respectively the original and the predicted values of F_0 values at the originally voiced frame i.
- **Voicing Decision Error** (VDE), which represents the proportion of frames for which a wrong voiced/unvoiced prediction is made. Actually VDE is the percentage of bad-predicted labels, i.e. false positives and false negatives:

$$VDE(\%) = \frac{N_{V \to U} + N_{U \to V}}{N} \times 100 \tag{4}$$

  where $N_{V \to U}$, $N_{U \to V}$ and N are respectively the number of voiced frames predicted as unvoiced, the number of unvoiced frames predicted as voiced and the total number of frames.
- **Gross pitch error** (GPE), which is a measure which combines both F_0 prediction and voicing prediction classification. Actually GPE is the percentage of frames for which the F_0 relative error is higher than a certain threshold, set to 20% among the frames that are originally voiced and also predicted as voiced:

$$GPE(\%) = \frac{N_{GPE}}{N_{V \to V}} \times 100 \tag{5}$$

  where N_{GPE} and $N_{V \to V}$ are respectively the numbers of frames originally voiced and predicted as voiced for which $|F_0 - \hat{F}_0| > F_0 \times 0.2$ and $N_{V \to V}$ is the total number of frames originally voiced and predicted as voiced.
- **F0 Frame Error** (FFE), is a measure which combines Gross Pitch Error (GPE) and Voicing Decision Error (VDE). Actually FFE is the percentage of frames for which an error is made, either according to the VDE or GPE criteria:

$$FFE(\%) = \frac{N_{V \to U} + N_{U \to V} + N_{GPE}}{N} \times 100 \tag{6}$$

Table 2. Objective evaluation results for the proposed models

Proposed DNN model	Development set				Test set			
	RMSE (Hz)	VDE (%)	GPE (%)	FFE (%)	RMSE (Hz)	VDE (%)	GPE (%)	FFE (%)
All-Dense	33.41	3.51	22.14	14.14	37.86	3.35	26.98	17.48
Dense-LSTM (1)	17.65	3.08	5.05	5.50	19.23	2.78	6.82	6.38
Dense-LSTM (2)	**15.70**	2.92	**2.41**	**4.09**	18.14	2.71	4.43	5.06
Dense-BLSTM (1)	19.72	2.88	5.85	5.71	20.14	2.38	6.33	5.74
Dense-BLSTM (2)	19.93	**2.59**	4.98	5.04	21.39	2.51	6.39	5.93
All-LSTM	16.70	2.95	2.62	4.22	**17.71**	2.79	**3.38**	**4.59**
All-BLSTM	15.95	2.64	3.16	4.18	18.65	**2.32**	6.57	5.82

The objective evaluation consists in comparing the performance of DNN modeling to state-of-art models, i.e. HMM model as used in HTS [7], DNN model as used in MERLIN [6] (cf. Table 3). The DNN model as used in MERLIN is composed by 6 hidden layers with 1024 units each and *tanh* as activation transfer function. This model relies on the same set of features as HTS. It should be emphasized that the acoustic parameters are generated from HTS or MERLIN using the original duration to compare the original and predicted vector F_0.

Table 3. Comparison of objective evaluation results on test set for the best selected models on the development set and for the various modeling approaches

Models	RMSE(Hz)	VDE(%)	GPE (%)	FFE(%)
HMM model from HTS [7]	30.52	7.19	8.90	11.93
All-Dense DNN model from MERLIN [6]	36.93	4.73	9.11	9.46
Dense-LSTM (2)	**18.14**	2.71	**4.43**	**5.06**
Dense-BLSTM (2)	21.39	**2.51**	6.39	5.93

4.5 Discussion

In comparison to the state-of-the-art results (cf. Sect. 2.2), and taking into account the small size, only 3 h, of the used corpus, the $RMSE$ results are quite near of the 15-Hz-$RMSE$ obtained using only feedforward DNN trained on a 13-h English corpus [8], or the 12-Hz-$RMSE$ given by a 13-h Chinese corpus, trained with a DNN containing dense and LSTM layers [17].

In Table 2, looking to VDE values, it looks obvious that using recurrent networks gives better voicing decision prediction results. The Dense-BLSTM (2) model leads to the best performance on the development set, with a VDE of 2.59%. The corresponding 95% confidence interval is ±0.05%; which make this result significantly better than that of the other models. This model leads to a VDE of 2.51% (±0.07%) on the test set.

In Table 3, the results show that using Dense-LSTM and Dense-BLSTM to model respectively F_0 and voicing decision label, leads to better performance compared to state of the art models, i.e. HMM model as used in HTS [7], DNN model as used in [6]. Finally the combined F_0 and voicing decision measures like GPE and FFE show that using recurrent networks is more fitted to the problem of F_0 modeling. This could be explained by the recurrent nature of speech, where consonants, mostly unvoiced or partly voiced, are followed by voiced vowels.

5   Conclusion

In this paper, F_0 modeling was achieved using DNN for Arabic parametric speech synthesis. F_0 modeling is a two-fold process, which requires the prediction of the voicing nature (voiced or unvoiced) of speech segments, and then the prediction of F_0 values in voiced parts. An Arabic speech corpus was used to train two DNN models, one for voicing classification and one for F_0 prediction, using the standard features set for parametric speech synthesis, in addition to two Arabic-specific features, vowel quantity and gemination. Several architectures were tried out for both networks, using (i) feedforward (dense) layers only, (ii) a combination of feedforward and recurrent layers (LSTM or BLSTM) and (iii) recurrent layers only. Objective results using standard metrics for F_0 prediction quality, like $RMSE$, VDE, GPE and FFE show that the best F_0 prediction results are obtained with recurrent networks.

Acknowledgement. This research work was conducted in the framework of PHC-Utique Program, financed by CMCU (Comité mixte de coopération universitaire), grant No15G1405.

References

1. Pierrehumbert, J.: The phonology and phonetics of English intonation. Ph.D. Thesis, Massachusetts Institute of Technology (1980)
2. Hart, J., Collier, R., Cohen, A.: A Perceptual Study of Intonation. Cambridge University Press, Cambridge (1990)
3. Dusterhoff, K., Black, A.: Generating F_0 contour for speech synthesis using the tilt intonation theory. In: 3rd ESCA workshop on Intonation: Theory Models and Applications, pp. 107–110. Athens, Greece (1997)
4. Taylor, P.: Analysis and synthesis of intonation using the tilt model. J. Acoust. Soc. Am. **107**(3), 1697–1714 (2000)
5. Moehler, G., Conkie, A.: Parametric modeling of Intonation using vector quantization. In: 3rd ESCA Workshop on Speech Synthesis, pp. 311–316. Jenolan Caves, Australia (1998)
6. Wu, Z., Watts, O., King, S.: Merlin: An open source neural network speech synthesis system. In: 9th ISCA Workshop on Speech Synthesis, pp. 202–207. Sunnyvale, USA (2016)
7. Yoshimura, T., Tokuda, K., Masuko, T., Kobayashi, T. Kitamura, T.: Simultaneous modeling of spectrum, pitch and duration in HMM-based speech synthesis. In: 6th European Conference on Speech Communication and Technology, pp. 2347–2350. Budapest, Hungary (1999)

8. Zen, H., Senior, A., Schuster, M.: Statistical parametric speech synthesis using deep neural networks. In: 38th International Conference on Acoustics, Speech, and Signal Processing, pp. 7962–7966. IEEE, Vancouver, Canada (2013)
9. Chen, B., Bian, T., Yu, K.: Discrete duration model for speech synthesis. In: 18th Annual Conference of the International Speech Communication Association, pp. 789–793. Stockholm, Sweden (2017)
10. Zangar, I., Mnasri, Z., Colotte, V., Jouvet, D., Houidhek, A.: Duration modeling using DNN for Arabic speech synthesis. In: 9th International Conference on Speech Prosody, pp. 597–601. Poznan, Poland (2018)
11. Oord, A.V.D., Dieleman, S., Zen, H., Simonyan, K., Vinyals, O., Graves, A., Kalch-brenner, N., Senior, A., Kavukcuoglu, K.: Wavenet: a generative model for raw audio. arXiv preprint arXiv: 1609.03499 (2016)
12. Yoshimura, T.: Simultaneous modeling of phonetic and prosodic parameters, and characteristic conversion for HMM-based Text-to-Speech systems. Ph.D. Thesis, Department of Electrical and Computer Engineering, Nagoya Institute of Technology (2002)
13. Zen, H., Tokuda, K., Masuko, T., Kobayashi, T., Kitamura, T.: Hidden semi-Markov model based speech synthesis. In: 8th International Conference on Spoken Language Processing, pp. 1393–1396. Jeju Island, Korea (2004)
14. Tokuda, K., Masuko, T., Miyazaki, N., Kobayashi, T.: Multi-space probability distribution HMM. IEICE Trans. Inf. Syst. **85**(3), 455–464 (2002)
15. Zen, H., Tokuda, K. Black, A.W.: Statistical parametric speech synthesis. In: Speech Communication 2009, vol. 51, pp. 1093–1064. ELSEVIER (2009). https://doi.org/10.1016/j.specom.2009.04.004
16. Yu, K., Young, S.: Continuous F_0 modeling for HMM based statistical parametric speech synthesis. IEICE Trans. Inf. Syst. **19**(5), 1071–1079 (2011)
17. Fan, Y., Qian, Y., Xie, F. L., Soong, F. K.: TTS synthesis with bidirectional LSTM based recurrent neural networks. In: 15th Annual Conference of the International Speech Communication Association, pp. 1964–1968. Singapore (2014)
18. Chen, C.J., Gopinath, R.A., Monkowski, M.D., Picheny, M.A., Shen, K.: New methods in continuous Mandarin speech recognition. In: 5th European Conference on Speech Communication and Technology, pp. 1543–1546. Rhodes, Greece (1997)
19. Chen, B., Lai, J., Yu, K.: Comparison of modeling target in LSTM-RNN duration model. In: 18th Annual Conference of the International Speech Communication Association, pp. 794–798. Stockholm, Sweden (2017)
20. Halabi, N., Wald, M.: Phonetic inventory for an Arabic speech corpus. In: 10th International Conference on Language Resources and Evaluation, pp. 734–738. Slovenia (2016)
21. Speech Signal Processing Toolkit (SPTK). http://sp-tk.sourceforge.net/
22. Houidhek, A., Colotte, V., Mnasri, Z., Jouvet, D.: DNN-based speech synthesis for Arabic: modelling and evaluation. In: 6th International Conference on Statistical Language and Speech Processing, pp. 9–20. Mons, Belgium (2018)
23. Camacho, A., Harris, J.G.: A sawtooth waveform inspired pitch estimator for speech and music. J. Acoust. Soc. Am. **124**(3), 1638–1652 (2008)
24. Zen, H.: An example of context-dependent label format for HMM-based speech synthesis in English. The HTS CMUARCTIC demo (2006)

Regularizing Neural Networks
with Gradient Monitoring

Gavneet Singh Chadha$^{(\boxtimes)}$, Elnaz Meydani, and Andreas Schwung

Department of Automation Technology,
Southwestphalia University of Applied Sciences, Soest, Germany
{chadha.gavneetsingh,meydani.elnaz,schwung.andreas}@fh-swf.de

Abstract. Neural networks are the most evolving artificial intelligence method in recent times and have been used for the most complex cognitive tasks. The success of these models has re-scripted many of the benchmark tests in a wide array of fields such as image recognition, natural language processing and speech recognition. The state of the art models leverage on a large amount of labelled training data and a complex model with a huge number of parameters to achieve good results. In this paper, we present a regularization methodology for reducing the size of these complex models while still maintaining generalizability of shallow and deep neural networks. The regularization is based on the monitoring of partial gradients of the loss function with respect to weight parameters. Another way to look at it is the percentage learning evident in a mini-batch training epoch and thereafter removing the weight connections where a certain percentage of learning is not evident. Subsequently, the method is evaluated on several benchmark classification tasks with a drastically smaller size and better performance to models trained with other similar regularization technique of DropConnect.

Keywords: Gradient monitoring · Regularization · Neural Networks

1 Introduction

Since the seminal paper from Hinton *et al.* [5], Neural Networks (NN) and Deep NN have become the workhorse and the overwhelming favourite machine learning model for a wide variety of applications. The ground-breaking study conceptualized greedy layer-wise unsupervised training procedure (popularly termed as pre-training) to initialize an adequate set of parameters that help in steering the learning process towards a stable solution. The results from the study broke the myth surrounding the difficulty in training a NN with more than one hidden layer, also known as deep NN. Thereafter, Deep NN have gone on to perform at state-of-the art level on the majority of artificial intelligence applications such as image recognition, natural language processing and speech recognition [7].

Subsequently, a number of techniques have been developed to improve the generalization ability of deep NNs. Some of these techniques improve generalization by preventing over-fitting or counteracting the dreaded vanishing gradient

© Springer Nature Switzerland AG 2020
L. Oneto et al. (Eds.): INNSBDDL 2019, INNS 1, pp. 196–205, 2020.
https://doi.org/10.1007/978-3-030-16841-4_21

problem in Deep NNs. The two biggest rationales behind the immense success of Deep NNs can be attributed to the development of methods in the area of: (a) inducing regularization and (b) stable optimization. One of the methods for stable optimization is to use an activation function that does not saturate in its input domain [3] so that gradients can flow more easily through the network. The most prominent example in this case being the ReLU activation function [10] and its successors such as LReLU [9] and ELU [2]. As a consequence of these and many other recent techniques, Deep NN with multiple hidden layers can now be trained in a fast and efficient way, without the need of unsupervised pre-training.

In this study, we focus on inducing regularization in NN and Deep NN during training as in practice these models perform superiorly on the test set as compared to models without regularization. Regularization is beneficial in reducing the complexity of a NN so that it has enough expressivity to model the training set, without sampling the noise in the training set. Generally, regularization in NN can be achieved through numerous methods ranging from directly penalizing the weight parameters by adding a norm penalty to the loss function, to indirect penalizing with parameter sharing and sparse representations [4]. But the recent technique of Dropout [11] employs a new form of regularization wherein the training a Deep NN is accomplished by keeping only some randomly selected neurons activated in each layer per training case in a mini-batch. Since a portion of the units are deactivated during the training epoch, partial derivatives of the loss function are only calculated for the remaining activated units. A more generalized approach to Dropout is DropConnect [12] wherein rather than dropping a complete neuron, only some randomly selected weight connections to a neuron are deactivated. We propose a more principled approach of inducing regularization in a NN based on the amount of learning occurring at each of the weight connection. Based on this amount of learning, the insignificant weight connections between the input, hidden and output units are omitted. The idea is to keep the connections where there is high activity or high percentage of learning during training and drop the ones where there is relatively low percentage of learning. The proposed method speeds-up the training process and reduces the overall size of the network, keeping the accuracy constant or improving it with comparison to the original model. Furthermore, it is a deterministic regularization technique as compared to the stochastic approach of Dropout and DropConnect.

The rest of the paper is organised as follows. In Sect. 2 the related methods of Dropout and DropConnect are described shortly. Section 3 details the training methodology of NN together with gradient monitoring. Experimental results along with a detailed comparison on different classification datasets are illustrated in Sect. 4 with a short summary is Sect. 5.

2   Related Work

The basic goal of all machine learning models is to generalize and perform well on data which it did not see during training i.e. good test accuracy. The proposed methods of Dropout [11] and DropConnect [12] specifically target this aspect of model development wherein a NN model is pruned during training with the expectation that it does not over-fit the training set and has good performance at test time. In the following section the basic training procedure of these two methodologies is explained.

2.1 Dropout

The motivation of Dropout stems from the hypothesis that a combination of an exponentially large number of sub-models, the architecture of which is sampled stochastically from the original deep NN model are able to work well for a particular classification task as compared to the original model. Every one of this sub-network is created by randomly dropping out input and hidden units with a probability of 1-p (p being a hyper-parameter) for each pass of the mini-batch during training. As a consequence of a portion of the units being randomly dropped out during the forward pass, the error gradients are also passed through only the weight connections present for the backward pass. Accordingly, the magnitudes of the weight parameters of the dropped neurons are not changed for that mini-batch. At test time, averaging the predictions of such exponentially high number of models is however practically infeasible. Therefore, as an approximation to averaging the ensemble of NN models, all the neurons are present a test time but with the magnitude of the weight parameters being scaled to p times the magnitude after training.

2.2 DropConnect

DropConnect is seen as a generalization of the Dropout method. Compared to Dropout where a complete neuron is removed, DropConnect removes a neuron only partially. Specifically, Dropout masks the activation of a neuron with a binary mask drawn from a Bernoulli distribution with probability p whereas DropConnect masks directly the weight matrices. The approximation at test time is also present. Figure 1 shows explicitly the architectural differences between a standard fully connected NN, a Dropout based NN and a DropConnect NN.

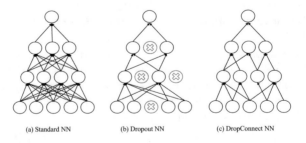

(a) Standard NN   (b) Dropout NN   (c) DropConnect NN

Fig. 1. Graphical representation of (a) standard, (b) Dropout and (c) DropConnect NN model

3   Training Description

Each weight connection of a NN is adjusted during the learning process based on the error gradient it receives from the back-pass of gradient descent algorithm. Since each weight connection is initialized randomly (normally from a uniform distribution), the influence of the gradient i.e the amount of change in the weight value is different on each of the connection. Consequently, the ratio of the error gradient to the original weight connection gives a better measure of the significance of a weight connection. The absolute value of this ratio is monitored during the learning process and the weight connections are dropped accordingly.

To illustrate the training procedure with gradient monitoring, consider a dataset D with $D = n$ where each data-point is of the form (x,y) consisting of input features x and the output label y. A NN with m (m $\geq$ 2) hidden layers maps a function

$$z_i = W_{m+1}\sigma(W_m\sigma(...\sigma(W_1x_i))), \tag{1}$$

where z_i is the predicted label for the input sample x_i, σ is a non-linear activation function and W_m is the weight matrix for the corresponding layer. The biases are implicitly included in the above formulation by augmenting each input sample x_i and the layer activations with ones. Furthermore, the above formulation is used for generating continuous values from the model. Since the output from a NN is a discrete variable in case of a classification problem, the output layer usually utilizes the softmax activation function which produces the probabilities associated with each class (lets say k be the number of classes). The softmax function thereby assigns the highest probability to the desired class with the size of the output vector from the softmax layer equivalent to the size of the output label y [4]. Formally, the softmax function is given by

$$softmax(z_i) = \frac{exp(z_i)}{\Sigma_k exp(z_k)}. \tag{2}$$

The softmax function exponentiates and normalizes z_i so as to obtain a value from the neural network which is as close as possible to the target label y_i. The output from the NN is usually compared with the target labels for classification tasks with the cross-entropy loss function which takes the probabilities generated by the softmax layer and the ground truth labels as input as

$$L(y, z) = -\Sigma_{i=1}^{k} y_i log(z_i). \tag{3}$$

The negative log-likelihood formulation of the loss-function is proposed to keep the gradient of the loss function with respect to the output units as large as possible which helps in steering the learning process towards a good solution. Output units with exponential functions saturate in their input domain where the input is very negative resulting in very small gradients. The *log* in the loss function negates this effect.

The goal of training is to minimize the loss function so that the predictions from the NN model are as close as possible to the ground truth labels. This is achieved by calculating the partial gradient of the loss functions with respect to each of the weight parameters recursively. The partial gradient for every weight parameter depends on the activation of the respective neuron it is connected to and the loss signal it receives from the loss function or from the neuron in the subsequent layer. Therefore, it is hypothesized that neurons with a smaller activation also have a smaller gradient causing slowness in learning of that neuron. Specifically, for the overall NN model, error gradients with respect to each of weight matrices namely L'_{W_1}, L'_{W_2}....L'_{W_m+1} over a mini-batch in the training set are calculated. Thereafter, these gradients are used to update the respective weight parameters via gradient descent optimization algorithms such as stochastic gradient descent (SGD), RMSprop, Adagrad or Adam [6]. The weight update equation for an arbitrary weight matrix W at mini-batch t via SGD is

$$W_{t+1} = W_t - \rho L'_{W_t}, \qquad (4)$$

where ρ is the learning rate. Since this gradient is the basis for all weight parameter update algorithms, in the next subsection we focus on how it evolves during the training process and propose the regularization based on these observations.

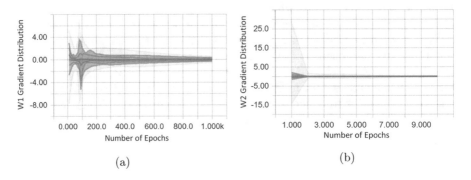

(a)   (b)

Fig. 2. Gradient distribution of (a) the weight matrix W_1 in Fisher Iris (b) the weight matrix at the output layer W_2 in MNIST

3.1   Distribution of Gradients

The results portrayed in Fig. 2 were obtained from a single hidden layer NN with 10 neurons for the Fisher Iris [1] dataset and two hidden layer NN with 256 neurons each for the MNIST [8] dataset. The distribution plots were similar even when deeper networks were chosen. The plots exemplify that the error gradient is distributed over a larger range around zero at the beginning of the training process epochs and as the training proceeds the distribution settles around zero. This points to the observation that most of the learning was accomplished in the initial stages and only minor updates to weight parameters were done after that. Therefore, we postulate to force those weights to zero which are not getting updated after certain training epochs.

3.2   Dropping Weights with Gradient Monitoring

The weight connections which have a low ratio of the error gradient to the original weight connection at a later stage in the training process are dropped as these weight connections do not contribute much to the overall output of the NN. The weight parameters are dropped with a masking matrix M. With this masking matrix, the new weight matrix is calculated as

$$W_{drop} = M * W, \tag{5}$$

where $*$ denotes the element-wise product. The masking matrix is sampled based on the absolute value of the ratio of the error gradient calculated on a mini-batch and the respective weight matrix. Specifically, M can be defined as

$$M = H\left(\left|\frac{L'_W}{W}\right| - \theta\right), \tag{6}$$

where H is the heaviside step function and θ is the threshold which ascertains the "amount" of learning that is expected from the training epoch. Consequently, all of the weights which fail to reach that amount are dropped with the help of this masking matrix. The biases are also masked out similarly at the same training epoch as the weights. Weights and biases once masked always remain zero, even at test time. This reduces the overall size of the model drastically and as we will see later on, without compromising the performance of the model. The dropped out weights and biases never participate in the training and inference of the model i.e. they always remain zero and gradient is passed only through the unmasked weights. Therefore, there is no separate inference rule and approximations like the ones used in Dropout and DropConnect. A summary of the learning procedure is provided in Algorithm 1.

Algorithm 1. SGD Training with Gradient Monitoring

1: **Input**: training data point (x_i, y_i), weight parameters W_{t-1} from step *t-1*, learning rate ρ, learning threshold θ, training iteration for masking η
2: **Output** Updated weight parameters W_t
3: Standard SGD training till iteration η
4: Sample masking matrix M : $M = H\left(\left|\frac{L'_W}{W}\right| - \theta\right)$
5: Compute new weights: $W_{drop} = M * W$
6: Standard SGD training with fewer weight parameters

---

4   Experiments

We keep the basis of our study to classification tasks for the MNIST hand written image classification [8] and the Fisher Iris multi-class classification task for classifying different types of flowers [1]. All the experiments are conducted by using SGD on batches of 100 for the MNIST dataset and full batch for Iris. The gradient monitored model is evaluated and compared with standard SGD and SGD with DropConnect. The number of learning epochs is the same for all methods with 25 epochs for the MNIST dataset and 1000 epochs for the IRIS dataset. The η is set at 1 and 200 epochs for MNIST and IRIS respectively. The number of hidden neurons in each hidden layer for MNIST and IRIS is 256 and 10 neurons respectively. The learning rate ρ and the learning threshold θ is kept contant at 0.001 and 1 respectively. The DropConnect probability is 0.5 with the Relu activation function being used for all the methods. The learning rate is kept constant throughout the training process and we report the classification accuracy and loss for the models trained.

4.1   MNIST

The MNIST dataset consists of 60,000 black and white training images each of size 28×28 containing a digit 0 to 9 (i.e. 10 classes). Additionally there are 10,000 images in the test set. The raw pixel values of the images after being scaled to zero mean and unit standard deviation, are taken as input to the NN model which has 784 input neurons. For the first experiment, we investigate the speed of convergence by observing the training loss of a single hidden layer neural network. Figure 3 exemplifies that the gradient monitored model not only reaches the lowest training loss but also reaches it much faster than in comparison to the other two models. To prove that the trained models are not over-fitting, the test accuracies for all the possible masking conditions of the weight connections are depicted in Fig. 4. All the models clearly show superior performance of the proposed method as compared to the standard SGD and DropConnect methods. Figure 5 depicts the testing accuracies of the different masked models including a deeper NN with two hidden layers. This figure exemplifies that masking the initial weight layer (W_1 in this case) of the model provides a better overall testing accuracy. This stems from the fact that the amount of error gradient being propagated to the initial layers is substantially lower than that in the deeper layers.

Furthermore, we analyse the effect of the learning threshold θ on the % of weights remaining and test accuracy. Figure 6 elucidates that increasing the learning threshold effects the % of weights remaining inversely, but the test accuracy increases first and then decreases after a certain value of threshold. the size of the original model is also drastically reduced by observing that the % of weights remaining for the best test accuracy is less than 5 and 10 % for W_1 and W_2 respectively. It must be noted here that the learning rate also plays a significant role in the % of weights remaining and Fig. 6 shows results with a learning rate of 0.01. Therefore, a key component for successfully masking weights with the proposed method is the selection of the threshold θ and the learning rate ρ.

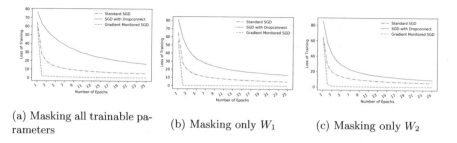

(a) Masking all trainable parameters

(b) Masking only W_1

(c) Masking only W_2

Fig. 3. Comparing training loss with different weight parameters masked on MNIST

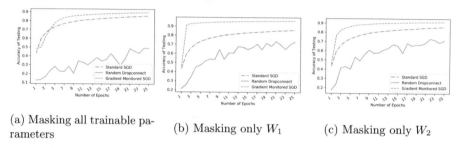

(a) Masking all trainable parameters

(b) Masking only W_1

(c) Masking only W_2

Fig. 4. Comparing testing accuracy with different weight parameters masked on MNIST

4.2 IRIS

The Fisher Iris is a well known benchmark test set consisting of a total 150 samples of three different classes of iris plants namely Iris Setosa, Iris Virginica and the Iris Versicolor. The four input features are sepal length, sepal width, petal length and petal width. The dataset is also scaled to zero mean and unit standard deviation and split in an 80–20 proportion for training and testing respectively. Since it is a small dataset we only investigate a single hidden layer NN. We investigate the training loss and testing accuracy for the three different models of standard SGD, SGD with DropConnect and gradient monitored SGD. Figure 7 shows the stochastic nature of the DropConnect which amplifies the

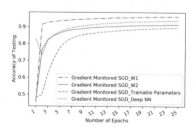

Fig. 5. Comparing different gradient monitored models

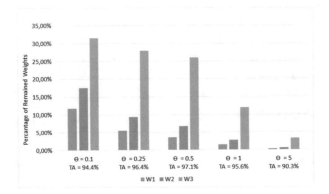

Fig. 6. Comparing percentage of weight remaining in W_1, W_2 and W_3 with learning threshold θ and test accuracy (TA) in MNIST

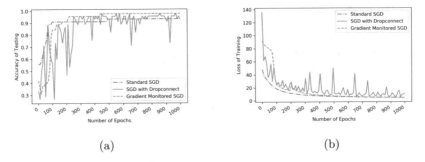

Fig. 7. Comparing (a) testing accuracy and (b) training loss with different models on the IRIS dataset

fluctuations in the learning process. The deterministic nature of the other two methods and the added advantage of faster and better convergence show the superior performance on the Iris dataset.

5   Discussion

In this study we propose a novel regularization techniques for NN and deep NN based on gradient monitoring. The deterministic approach does not comprise of any approximations such as the stochastic methods of Dropout and DropConnect. Gradient monitoring is justified empirically and the results proves that it helps in regularizing the NN significantly by reducing the number of trainable parameters drastically. Even with this reduced size, the proposed model outperforms the DropConnect technique on the MNIST and the Iris dataset. In the future, convolutional neural networks will be investigated with the same methodology for classification tasks.

References

1. Bezdek, J.C., Keller, J.M., Krishnapuram, R., Kuncheva, L.I., Pal, N.R.: Will the real iris data please stand up? IEEE Trans. Fuzzy Syst. **7**(3), 368–369 (1999). https://doi.org/10.1109/91.771092
2. Clevert, D.A., Unterthiner, T., Hochreiter, S.: Fast and accurate deep network learning by exponential linear units (elus). arXiv preprint arXiv:1511.07289 (2015)
3. Glorot, X., Bengio, Y.: Understanding the difficulty of training deep feedforward neural networks. In: Proceedings of the Thirteenth International Conference on Artificial Intelligence and Statistics, pp. 249–256 (2010)
4. Goodfellow, I., Bengio, Y., Courville, A.: Deep Learning, vol. 1. MIT Press Cambridge, Cambridge (2016)
5. Hinton, G.E., Osindero, S., Teh, Y.W.: A fast learning algorithm for deep belief nets. Neural comput. **18**(7), 1527–1554 (2006)
6. Kingma, D.P., Ba, J.L.: Adam: Amethod for stochastic optimization. In: Proceedings of the 3rd International Conference on Learning Representations (ICLR) (2014)
7. LeCun, Y., Bengio, Y., Hinton, G.: Deep learning. Nature **521**(7553), 436–444 (2015). https://doi.org/10.1038/nature14539
8. LeCun, Y., Cortes, C., Burges, C.J.: MNIST handwritten digit database. AT&T Labs**2** (2010). http://yann.lecun.com/exdb/mnist
9. Maas, A.L., Hannun, A.Y., Ng, A.Y.: Rectifier nonlinearities improve neural network acoustic models. In: Proceedings of the ICML, vol. 30, p. 3 (2013)
10. Nair, V., Hinton, G.E.: Rectified linear units improve restricted boltzmann machines. In: Proceedings of the 27th International Conference on Machine Learning (ICML-10), pp. 807–814 (2010)
11. Srivastava, N., Hinton, G., Krizhevsky, A., Sutskever, I., Salakhutdinov, R.: Dropout: a simple way to prevent neural networks from overfitting. J. Mach. Learn. Res. **15**(1), 1929–1958 (2014)
12. Wan, L., Zeiler, M., Zhang, S., Le Cun, Y., Fergus, R.: Regularization of neural networks using dropconnect. In: International Conference on Machine Learning, pp. 1058–1066 (2013)

Visual Analytics for Supporting Conflict Resolution in Large Railway Networks

Udo Schlegel[1]($^{(\boxtimes)}$), Wolfgang Jentner[1], Juri Buchmueller[1], Eren Cakmak[1],
Giuliano Castiglia[1], Renzo Canepa[2], Simone Petralli[2], Luca Oneto[3],
Daniel A. Keim[1], and Davide Anguita[3]

[1] DBVIS, University of Konstanz, Konstanz, Germany
{u.schlegel,wolfgang.jentner,juri.buchmueller,eren.cakmak,
giuliano.castiglia,daniel.keim}@uni.kn
[2] Rete Ferroviaria Italiana S.p.A., Rome, Italy
{r.canepa,s.petralli}@rfi.it
[3] DIBRIS, University of Genoa, Genoa, Italy
{luca.oneto,davide.anguita}@unige.it

Abstract. Train operators are responsible for maintaining and following the schedule of large-scale railway transport systems. Disruptions to this schedule imply conflicts that occur when two trains are bound to use the same railway segment. It is upon the train operator to decide which train must go first to resolve the conflict. As the railway transport system is a large and complex network, the decision may have a high impact on the future schedule, further train delay, costs, and other performance indicators. Due to this complexity and the enormous amount of underlying data, machine learning models have proven to be useful. However, the automated models are not accessible to the train operators which results in a low trust in following their predictions. We propose a Visual Analytics solution for a decision support system to support the train operators in making an informed decision while providing access to the complex machine learning models. Different integrated, interactive views allow the train operator to explore the various impacts that a decision may have. Additionally, the user can compare various data-driven models which are structured by an experience-based model. We demonstrate a decision-making process in a use case highlighting how the different views are made use of by the train operator.

Keywords: Visual and big data analytics · Decision support systems

1 Introduction

Railway Transport Systems (RTSs) play a crucial role in servicing the global society and the transport backbone of a sustainable economy. A well functioning RTS should meet the requirements defined in the form of the 7R formula [13,16]: *Right Product, Right Quantity, Right Quality, Right Place, Right Time, Right Customer,* and *Right Price.* Therefore, an RTS should provide: (i) availability of

L. Oneto et al. (Eds.): INNSBDDL 2019, INNS 1, pp. 206–215, 2020.
https://doi.org/10.1007/978-3-030-16841-4_22

appropriate products (the provisioning of different categories of train), (ii) proper number of executed transportation tasks (enough trains to fulfill the request), (iii) proper quality of execution of transportation tasks (safety, correct scheduling, and effective conflicts resolution), (iv) right place of destination according to a timetable (correct transportation routes), (v) appropriate lead time (reduced *train delays*), (vi) appropriate recipients (focused on different customer needs and requirements), and (vii) appropriate price (both from the point of view of the customers and the infrastructure managers).

The main responsibility of a train operator (TO) is to ensure the safety and smooth running of the trains and their scheduling. Although there exists a given train schedule, the TOs have to deal with derivations caused by delays, defects, or other unexpected impacting events. A conflict occurs when, due to the aforementioned reasons, regularly scheduled trains are delayed so that they would utilize the same railway-segment at the same time. Although this poses no immediate danger as the interlocking system will not allow both trains to drive, however, the operator must decide which train can go first. This decision is critical from the point of view that the criteria of the 7R formula should be respected. In the current state-of-the-art, a rule-based system proposes a solution on which train should go first. Yet, since the decision is not explained, most TOs don't use it. In fact, TOs mostly rely on their experience because most conflicts happen regularly, e.g., every day at the same time with the same two trains involved, as the schedule is mostly the same each day, and thus, vulnerable to the same challenges. This makes it difficult for TOs that lack experience to choose the optimal decision. This is also true for operators when they operate another region. Lastly, there is a risk that the selected decision of the TO is actually not optimal due to impacting cognitive biases such as the confirmation bias [1].

Dedicated Machine Learning (ML) techniques exist to predict the behavior of Large-Scale RTSs [15]. Such models are well suited to support the train operator in their decision-making process for conflict resolution. While their accuracy is convincing, the TOs have a rather hesitant stance towards these ML models as they typically do not reveal their inner workings nor explain why a specific decision on which train (A or B) should go first is recommended as superior to the second option. Visual Analytics (VA) has proven to be effective to overcome this boundary [18]. VA is a multi-disciplinary field which combines methods from machine learning, visualization, human perception, and human-computer interaction to support the useful understanding, reasoning, and decision making.

We propose a Visual Analytics Decision Support System (VADSS) for train operators that embeds these powerful ML models while providing access to the models through VA. This helps the TOs to make an informed decision whereas their experience is directly encoded into the system and not only resides in the operators' minds. We call this the *stateless operator*. By taking the operator's experience directly into the system, experience can be shared and pitfalls of knowledge only present in the operator's mind, such as misremembering, are mitigated. We identify four main aspects that are necessary for the TO to choose the optimal decision effectively. First, the complexity of the network and the manifold factors require an automatic model that optimizes on these various metrics. The TO needs to be able to inspect the models and the metrics on what these models optimize and must be able to compare this. Secondly, the TO needs to understand how the automatic models derive their decisions. Therefore, a model visualization is necessary. Thirdly, the impact of either decision (train A or B) must be visualized in the schedule visualization such that the impact on singular events or trains becomes visible which may be of higher priority than a global metric. Lastly, the historical decisions on the respective conflict must be available and accessible to the TO and must convey the historic information of the model and its predicted metrics, the actual metrics, as well as the impact of the decision to the schedule.

The proposed VADSS combines and integrates all these aspects. The next Section covers the first to aspects of the system and details which models are used to predict states and variables for large train networks. In Sect. 3, the combined system is described and also the second, third and fourth aspect of the systems are detailed. Furthermore, a use-case in Sect. 4 describes the decision making process and highlights the various aspects of the system. Section 5 draws some conclusions and outlines further research aspects.

2   Prediction Models for Large Train Networks

Train management systems can have a large number of different possible models to use to predict which decision in a critical case should be used or advised. First, there are Experience-Based Models (EBMs) [6,7], which are often generated out of rules from a train network and the experience of a train operator. Such models are often rule-based or based on decision trees (DT). Second, Data-Driven Models (DDMs) [14], which are created out of the historical data for the critical case. DDMs can be arbitrary machine learning models, such as random forests (RF) [3], Extreme Learning Machine (ELM) [8] or others. To be able to achieve the best of both worlds, Lulli et al. [12] propose a Hybrid Model (HM) incorporating the knowledge of the EBM to select an appropriate DDM. However, due to the nature of the task, it is necessary that a user can understand the decision basis of every model in a few seconds or minutes. Such complex decision models require the use of visual analytics to make them accessible to the train operators such that the visualizations highlight the important aspects of the model.

Through the nature of the proposed HM system by Lulli et al. [12], these visualization concepts focus on the decision making of a DT at first and more difficult algorithms later. By design, DTs are more understandable but can get quite complicated if the data which they are trained on is high-dimensional and complex. Highlighting a decision path through the first part of a HM to select a DDM is essential to show the user on which basis the first stage EBM decides for an underlying second stage DDM. This decision path can help a user to understand if the correct model was taken or if some other factors are weighted heavier than others. Features such as *railway section, railway checkpoint, train type, daytime, weekday, last delay and weather conditions* are just a few on which the decision for a DDM is made by the EBM. The feature-based reasoning of which DDM is chosen by the EBM significantly impacts the understanding of what priorities the DDM will have to base its prediction on.

A solution to this issue is to show a visualization of the DT that represents the EBM and to highlight the path the model took to choose a DDM. Through the visualization it is intuitive to understand what features are selected for the later prediction. For instance, the model could have chosen the *weekday* and the *daytime* to select a model and neglected the other features because in the case of the rush hour these are the essential features. Also, some outlier cases in which the DT takes a wrong turn could be identified. It could be that a DDM is chosen based on some path, which is not ideal, e.g. in a rush hour instead of *weekday* and *daytime, last delay* and *weather conditions*. A train operator can identify and fix it quickly with her experience and domain knowledge.

To further include the decisions of the underlying and possibly chosen DDM, the results of an explanation method are displayed. In the case of the HM, the chosen second stage DDM is a random forest and has its own feature set. The features for the RF are *weather information, past train delays, past dwell times, past running times, network congestion and network congestion delays*. Only visualizing all the DT of an RF is not sufficient as an RF in most cases consists of too many trees to show in limited space and due to their design as weak classifiers they are only useful in combination. There are, to some extend, methods to visualize RFs, for example, Forest Floor [19]. However, this method is difficult to interpret in a short time and requires a profound knowledge of the general technique to understand the decision the model suggests. A possible explanation method for an RF without a visualization would be to show the variable importance [3]. The variable importance of RFs show which features are most often used as a split criterion and thus more descriptive to separate and predict the data. For instance, the more often a feature such as *network congestion* is chosen as a split, the more descriptive this feature is to the dataset. To enable a fast understanding of the train operator, the system only shows the top three features in text ranked by importance, and some of the underlying information is hidden.

Other Machine Learning techniques, which are for instance in the proposed ELM by Oneto et al. [15], also have some possibilities to show such feature importance in a short, summarized way. In the case of an ELM, which is a type of a Neural Network, extracting an explanation is challenging due to its complexity but can be achieved through explanation methods such as LIME [17]. In general, showing the decision process of an algorithm, is a crucial part to build the train operators trust into the proposed decision and the model in general. Trust into the automated processes of a system is essential for the effective usage of the system [11]. While explaining and understanding the general approach of a model is crucial [9], visualization *is* required - especially for decision processes.

3   Visual Analytics Decision Support System

The interactive visual exploration of prediction models and the predictions themselves are essential for an operator to further learn about the impact of her
decision. Therefore, our primary goal is to present the operator interpretable
models and their results to gain insight for solving train conflicts. We propose
the Visual Analytics Decision Support prototype (see Fig. 1) which visualizes
the hybrid model as introduced in Sect. 2, the resulting costs, a visual preview
of the two predicted train schedules, and the experience-based data.

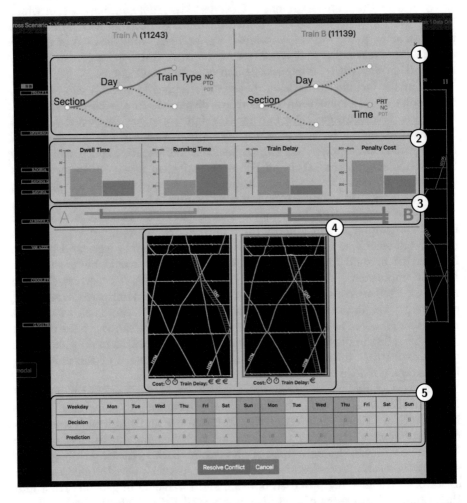

Fig. 1. Integrated view on the proposed conflict resolution interface containing the
reduced complexity model view (1), the metrics comparison view (2), the pipeline visualization (3), the side-by-side schedule prediction visualization (4) and the Historical
overview on previous decisions (5).

Our proposed Visual Analytics prototype supports proactive decision making, enabling the train management operator to investigate and assess decisions, to find optimal solutions, and to minimize risks. To achieve this goal, we extend existing visualizations in the control room setup following the VA principle introduced by Keim et al. [10].

In our work, we build upon Train Delay Prediction Systems (TDPS) as proposed by Oneto et al. [15]. TDPS can have different optimization strategies and rescheduling predictions, optimizing for different features and strategies such as experience driven [15], energy-efficient railway operations [2], passenger reliability perception [5], delay in the system [4] or others. The proposed hybrid model [15] combines EBMs and DDMs that are already in use by the RFI. In collaboration with the RFI experts, we have designed and implemented two views (see Fig. 1) to explore the HM and the resulting costs visually.

With our proposed solution, we intend to bridge the gap between model exploration and decision-making in a safety-critical environment. Bringing together Model and Solution Space Visualization at an operator's workstation entails careful consideration of the amount of information presented as well as the complexity of the provided interaction functionality. According to TMS experts, typically, operators consider only a narrow time frame of about 30 min after an occurring conflict to survey the results of the decisions they take. Any complications beyond this time frame is not considered due to the increasing complexity coming with each additional affected train.

Consequently, interfaces need to be minimalist in the amount of displayed information, yet able to always provide the right information, also dynamically, to an expert at the right time in the right granularity to rapidly make informed decisions and achieve the *stateless operator*. We determine four main components required for a rapid decision making interface for conflict resolution:

(1) Reduced-Complexity Model Representation
(2) Metrics comparison view
(3) Pipeline visualization for statistics exploration
(4) Side-by-side schedule prediction view
(5) Historical overview on previous decisions

Having an insight into the prediction model as discussed in Sect. 2 is essential for an operator to decide whether or not the model factors in all aspects the operator deems important. In other words, the visualization allows operators to decide, whether they should *trust* the model or not. The first view (see Fig. 1(1)) in our Visual Analytics prototype shows the EBMs that capture the knowledge of senior operators in the form of a DT. The two DT visualizations enable to identify, why a specific DDM was chosen by the EBM. Additionally, the operator can interactively explore alternative branches in both DTs by selecting decision nodes. The features of the EBM (e.g., railway section and train type) are included in the view to foster the interpretability of the model.

The visualization also enables to investigate the input features (hovering the root node) and shows only the highlighted decision path. We enable the exploration of the EBM to see why a specific DDM was chosen. The train operator can use a mouseover to investigate the decision path with all branching condition to understand the result of the EBM. Furthermore, next to the leaf nodes, we show the three specific features that are weighted important by the DDM. By displaying the feature importance, the operator is able to identify the most important features for the decision of the DDM. The operator is able to explore alternative branches in the DT by selecting nodes.

The second visualization (see Fig. 1(2)) displays the predicted resulting costs of both selected DDMs (e.g., running time and the dwell time). We linked the second view to the selection of the selected DDMs in the DT visualization. The results are shown in grouped bar charts enabling an intuitive comparison of the resulting cost of the two train predictions. By directly comparing the costs and the alternative path selections in the DT, we aim to help the operator to understand the HM, the predicted costs, and the trade-offs between them. Furthermore, the operator can look at alternative DDMs and include them in his decision. To improve the rapid interpretability of the confronted statistics, the pipeline visualization (see Fig. 1(3)) connects a line from the "winning side" for each feature to the respective train (either A or B).

Having selected the appropriate model, the third part of our proposed solution (see Fig. 1(4)) allows comparing the outcome of the two options the operator can choose from, namely which of the two trains A and B is allowed to move on first. For each of the two scenarios, the projected outcome is visualized in the same style as for the general schedule in the background. The red-shaded area indicates the consequence and deviation to the schedule if the operator decides for this solution. In addition, the main differences between the two models are expressed using an icon-based, simplified approximation, here at the example of cost and time. The predicted cost and delay time are binned to a simple representation of intuitive icons such as currency symbol and a clock. For each feature, one to five icons represent low to high cost or overall delay introduced through the decision.

At times it is possible that, despite the ability to explore and select the applied model, the expectation of the operator still differs from the proposed solution. As well, operators new on the job or the respective region or temporary replacements might have not yet enough experience to make informed decisions. To support our concept of the *stateless operator*, the last part of our suggested view provides a historical overview on previous similar situations and how they were decided. The user can see, what the model has predicted as the best solution versus which solution was finally picked by the operator. The overview on recorded decisions is suitable to either, ideally, reinforce the trust of the operator in the models or to hint at bad model quality depending on how many decisions have been met in accordance with the model in the past.

Our proposed visualization enables the train management operator to assess and evaluate the prediction model results to understand alternative solutions and impact on the train schedule. Furthermore, the visualization of the predicted costs enables the train management operator to gain knowledge about the uncertainty in the prediction. Supported by a clear global color-coding, the

operator is now able to make an informed decision for conflict resolution. We
do not expect operators to make use of all views for every decision, but instead
to adapt the applied models to current operating conditions. On a more general
level, the system is also intended to create more trust for the provided models, which, according to our experts, is improvable due to the currently applied
black-box model representations presented to the operators.

4   Use Cases

To show how an operator is able to interact with the system, this section provides
a use case scenario to show a step by step process of a TO. The task consists
of a critical conflict emerging if two trains need to leave or arrive at a station
at the same time, while there is no rail for both of them available. The TO
has to decide which train should have priority and solve the critical conflict. As
described before, a decision can have a rather big influence on the train network
and schedule overall. The TO wants to minimize the *penalty cost* at all times,
while still be able to have a low *train delay*.

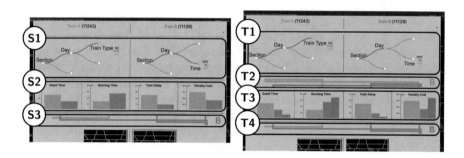

Fig. 2. (S1–3) shows a critical conflict with the model predictions and information.
(T1–4) shows another third model selected by the user to incorporate into the decision
of the TO.

When a conflict emerges, the system notifies the TO by showing her or him a
new modal with the relevant information described in previous sections. Either
train A or train B should have priority in a conflict to solve it. The HM in Fig. 2
(S1–3) shows a lower *penalty cost* for train B and thus advises the TO to choose
it. This lower *penalty cost* can either be seen in (S2) and the lower bar in the
metric comparison view or in (S3) in the pipeline visualization, which clearly
favors train B. However, as the TO wants to further investigate and understand
the decision, he inspects the visualizations of the available information. This
inspection means he takes a look at the DT visualization and the important
features for the RF prediction at (S1). With the decision process of the DT in
mind and the features of the RF, he decides that the *penalty cost* estimation

makes sense and that the model is correct. The TO decides to give train B priority to solve the conflict.

However, in another scenario Fig. 2 (T1–4), the TO could possibly be not satisfied with the model decisions. He thinks it is not advisable in the DT to use some of the features for a decision making. Other features are better for this specific case of solving the conflict, judging by his experience. In this case, the TO chooses another branch in the DT (T1) to take another DDM into account, which uses the data the operator thinks fits this situation and train better. This other DDM shows that after changing the model the other train has a higher *penalty cost* in the bar char and pipeline visualizations (T2–3). He can still compare how the different models decided in (T3) and by comparing the two pipelines in (T2) and (T4). Train A shows a better performance and the operator should pick it in this case.

5   Conclusion

Train operators constantly have to decide which train can go first in case of a conflict. Their decisions impact the schedule of large-scale railway transport systems and the associated performance metrics such as train delay and penalty cost. We present an integrated view for conflict resolution that makes use of the Hybrid Model proposed by Lulli et al. [12]. Our composed view provides train operators access to these models leveraging their power and increasing the user's trust into these models. The train operators are further enabled to explore a variety of models which are selected by an experience-based model. The train operators can explore what performance metrics are predicted by the different models. A Pipeline Visualization aids the train operators in how the various predicted metrics impact the current data-driven prediction for the decision. Furthermore, they can see how the future schedule will be impacted. A historical view shows past decisions of the operator in conjunction with the predicted decisions of the data-driven models. This supports the user to spot temporal and periodic patterns and shows if, for a particular conflict, train operators followed the proposition of the automated model. A use-case demonstrates how a train operator interacts with the integrated view to formulate her decision.

Acknowledgments. This research has been supported by the European Union through the project IN2DREAMS (European Union's Horizon 2020 research and innovation programme under grant agreement 777596).

References

1. Arnott, D.: Cognitive biases and decision support systems development: a design science approach. Inf. Syst. J. **16**(1), 55–78 (2006)
2. Bai, Y., Ho, T.K., Mao, B., Ding, Y., Chen, S.: Energy-efficient locomotive operation for chinese mainline railways by fuzzy predictive control. IEEE Trans. Intell. Transp. Syst. **15**(3), 938–948 (2014)

3. Breiman, L.: Random forests. Mach. Learn. **45**(1), 5–32 (2001)
4. Dollevoet, T., Corman, F., D'Ariano, A., Huisman, D.: An iterative optimization framework for delay management and train scheduling. Flex. Serv. Manuf. J. **26**(4), 490–515 (2014)
5. Dotoli, M., Epicoco, N., Falagario, M., Seatzu, C., Turchiano, B.: A decision support system for optimizing operations at intermodal railroad terminals. IEEE Trans. Syst. Man Cybern. Syst. **47**(3), 487–501 (2017)
6. Ghofrani, F., He, Q., Goverde, R.M., Liu, X.: Recent applications of big data analytics in railway transportation systems: a survey. Transp. Res. Part C Emerg. Technol. **90**, 226–246 (2018)
7. Hansen, I.A., Goverde, R.M., van der Meer, D.J.: Online train delay recognition and running time prediction. In: IEEE International Conference on Intelligent Transportation Systems (ITSC), pp. 1783–1788 (2010)
8. Huang, G.B., Zhu, Q.Y., Siew, C.K.: Extreme learning machine: a new learning scheme of feedforward neural networks. In: IEEE International Joint Conference on Neural Networks (2004)
9. Jentner, W., Sevastjanova, R., Stoffel, F., Keim, D.A., Bernard, J., El-Assady, M.: Minions, sheep, and fruits: metaphorical narratives to explain artificial intelligence and build trust. In: Workshop on Visualization for AI Explainability (2018)
10. Keim, D., Andrienko, G., Fekete, J.D., Görg, C., Kohlhammer, J., Melançon, G.: Visual analytics: definition, process, and challenges. In: Information Visualization, pp. 154–175 (2008)
11. Lee, J.D., See, K.A.: Trust in automation: designing for appropriate reliance. Hum. Factors **46**(1), 50–80 (2004)
12. Lulli, A., Oneto, L., Canepa, R., Petralli, S., Anguita, D.: Large-scale railway networks train movements: a dynamic, interpretable, and robust hybrid data analytics system. In: IEEE International Conference on Data Science and Advanced Analytics (DSAA) (2018)
13. Nowakowski, T.: Analysis of modern trends of logistics technology development. Arch. Civ. Mech. Eng. **11**(3), 699–706 (2011)
14. Oneto, L., Fumeo, E., Clerico, G., Canepa, R., Papa, F., Dambra, C., Mazzino, N., Anguita, D.: Advanced analytics for train delay prediction systems by including exogenous weather data. In: IEEE International Conference on Data Science and Advanced Analytics (DSAA), pp. 458–467 (2016)
15. Oneto, L., Fumeo, E., Clerico, G., Canepa, R., Papa, F., Dambra, C., Mazzino, N., Anguita, D.: Train delay prediction systems: a big data analytics perspective. Big Data Res. **11**, 54–64 (2018)
16. Restel, F.: The Markov reliability and safety model of the railway transportation system. In: Safety and Reliability: Methodology and Applications - Proceedings of the European Safety and Reliability Conference (2014)
17. Ribeiro, M.T., Singh, S., Guestrin, C.: "why should I trust you?": explaining the predictions of any classifier. In: ACM SIGKDD International Conference on Knowledge Discovery and Data Mining, pp. 1135–1144 (2016)
18. Sacha, D., Stoffel, A., Stoffel, F., Kwon, B.C., Ellis, G.P., Keim, D.A.: Knowledge generation model for visual analytics. IEEE Trans. Visual Comput. Graphics **20**(12), 1604–1613 (2014)
19. Welling, S.H., Refsgaard, H.H., Brockhoff, P.B., Clemmensen, L.H.: Forest floor visualizations of random forests. arXiv:1605.09196 (2016)

Modeling Urban Traffic Data Through Graph-Based Neural Networks

Viviana Pinto[1], Alan Perotti[1,2(✉)], and Tania Cerquitelli[3]

[1] aizoOn Technology Consulting, Turin, Italy
[2] Institute for Scientific Interchange, Turin, Italy
alan.perotti@isi.it
[3] Department of Control and Computer Engineering,
Politecnico di Torino, Turin, Italy

Abstract. The use of big data in transportation research is increasing and this leads to new approaches in modeling the traffic flow, especially for what concerns metropolitan areas. An open and interesting research issue is city-wide traffic representation, correlating both spatial and time patterns and using them to predict the traffic flow through the whole urban network. In this paper we present a machine learning based methodology to model traffic flow in metropolitan areas with the final aim to address short-term traffic forecasting at various time horizons. Specifically, we introduce an ad-hoc neural network model (GBNN, Graph Based Neural Network) that mirrors the topology of the urban graph: neurons corresponds to intersections, connections to roads, and signals to traffic flow. Furthermore, we enrich each neuron with a memory buffer and a recurrent self loop, to model congestion and allow each neuron to base its prediction on previous local data. We created a GBNN model for a major Italian city and fed it one year worth of fine-grained real data. Experimental results demonstrate the effectiveness of the proposed methodology in performing accurate lookahead predictions, obtaining 3% and 16% MAAPE error for 5 min and 1 h forecasting respectively.

Keywords: Traffic · Neural Networks · Transportation

1 Introduction

Around 50% of the global population lives in metropolitan areas, and this is likely to grow to 75% by 2050 [13]. Mobility within a wide metropolitan area is a complex system that needs to be modeled properly in order to predict the traffic flow and the consequent impact of congestions. Nowadays lots of data are available from detectors such as city cameras or induction loops: these devices can generally register the number of vehicles transitioned in an interval of time

The work presented in this paper is part of the SETA project, funded by the European Commission as part of the Horizon 2020 programme under contract 688082.

© Springer Nature Switzerland AG 2020
L. Oneto et al. (Eds.): INNSBDDL 2019, INNS 1, pp. 216–225, 2020.
https://doi.org/10.1007/978-3-030-16841-4_23

and their speeds. Furthermore these data can be integrated with information sourced from crowds, e.g. using mobile phone applications or sensors embedded into vehicles such as GPS. Data-driven approaches can be exploited to extract useful knowledge from those data and to effectively support short-term traffic forecasting. Traffic management systems can be trained on topological data and traffic measurements in order to estimate and predict the traffic flow. In this paper we use Neural Networks (NN) to model the traffic flow and to make short-term traffic forecasting. We present an original NN model based on the topology of the city (Graph Based Neural Network - GBNN), and we integrate topological and traffic data for a major Italian city (Turin) and we feed them to our GBNN model to perform traffic flow prediction.

The paper is outlined as follows. In Sect. 2 we present the state-of-the-art in traffic modeling. In Sect. 3 we discuss the GBNN model, including the general vision and the mathematical model, then we present experimental evaluation and results in Sect. 4. In Sect. 5, we end the paper with final remarks and directions for future work.

2   Related Work

In the last few years urban traffic data have received a great attention in different research areas such as transportation and machine learning. Transportation Research interest has been mainly focused on developing methodologies that can be used to model traffic characteristics such as volume, density and speed of traffic flows [7]. According to [15], most short-term traffic forecasting algorithms were built to function at a motorway, freeway, arterial or corridor level, not considering roads intersections at urban level. Predictions at a network level using data driven approaches remains a challenging task; the difficulty in covering a sufficient part of the road network by sensors, as well as the complex interactions in densely populated urban road networks, are among the most important obstacles faced in short-term traffic forecasting. In particular, NN have been widely applied to various transportation problems, due to their ability to work with massive amounts of multi-dimensional data, their modeling flexibility, and their learning and generalization ability: in [2] there is an early idea of application of NN to Transportation Engineering problems. A more structured idea of a possible application of NN in order to simulate and predict traffic flow can be found in [12], where a NN-based system identification approach is used to establish an auto-adaptive model for the dispersion of traffic flow on road segments or in [16], where a space-time delay NN is constructed to model autocorrelation of road traffic networks. Considering a section of freeway, a similar NN is built by [14] to model traffic flow and to obtain short-term prediction using real data; in particular, their work is focused on dealing with missing or corrupted data. Ma et al. [6] use traffic sensor data as input for a NN that predicts travel speeds, they consider a long period of time with high frequency data on an expressway in Beijing. Zhu et al. [17] study traffic volumes of three adjacent intersections, observing that traffic flows of single intersections are significantly affected by the traffic flows at the adjacent intersections. Polson et al. [11] focus on the use of road sensor data on a corridor in Chicago during a special event in order to study data representing a congestion; they develop a deep learning architecture that combines a linear model with a NN for the purpose of identify spatio-temporal relations among predictors and model nonlinear relations as well.

2.1 Neural Networks (NN)

Artificial neural networks [9] are computational paradigms composed by a series of interconnected processing elements that operate in parallel. The most used learning algorithm is the backpropagation algorithm, proposed and popularized by Hinton et. al in 1986 [3]. A recurrent neural network (RNN) allows for connections from the output layer to the input layer (compared with traditional feed-forward networks, where connects feed only to subsequent layers). Because RNNs include loops, they can store information while processing new inputs. This memory makes them ideal for processing tasks where prior inputs must be considered (such as time-series data). In particular, in Nonlinear Auto Regressive with eXogenous inputs (NARX) networks [8] each recurrent link includes a delay unit, so that the activation value of an output neuron O at time t is applied to an input neuron I at time $t + 1$.

3 Graph-Based Neural Network Model

In order to study the urban network, we model the city as a graph in which each junction corresponds to a node and each road to an edge. Turin's topological structure allows for this kind of model with small information loss due to the general straightness of roads. For the few curved roads, we choose to simplify the graph by representing them as straight edges. The resulting urban graph is strongly connected and, as we do not represent overpasses and underpasses, also planar. Ideally the graph should be direct with some multiple edges and loops; however, we decided to consider only the main roads and use an undirected graph.

We first create a neural network that matches the city graph, so that each graph node, representing a road intersection, corresponds to a neuron and each edge, representing a road, corresponds to a connection between neurons. The traffic flow is then modeled as signal propagation within the network - on ingoing/outgoing connections between neurons. We consider the possibility that not all the cars approaching a junction are able to leave it in a single time frame - we enable local congestion by means of a recurrent connection on each neuron. Note that the system we are considering is not a closed system: flow mass can be lost or generated, according to detected data. This can happen because of vehicles entering/leaving the city, or simply being parked or put back into circulation during the day. Starting from the gathering of real data from traffic detectors, we build the city graph with associated times series of traffic flow and we design a NN based on these information. Then we feed the traffic data to the NN and we analyze the results. In particular, we can divide the data processing in four steps, as follows:

1. *Data gathering and preprocessing* → collection of data referred to the topology of the city, collection of data referred to city detectors position and registered traffic flow, handling of missing data, propagation of data from city detectors to city nodes.
2. *Data exploration and visualization* → analysis of frequent patterns in the behavior of traffic flow in a day-interval and a week-interval.
3. *Neural Network model design* → creation of the GBNN model, definition of neuron-level and network-level parameters, learning procedure design.
4. *Model evaluation assessment* → performance analysis of the GBNN model.

3.1   Data Gathering and Preprocessing

Two different kinds of data are used in order to create the model. In the first phase we use topological data to shape the city graph; in the second phase we use data collected by road detectors to model traffic flow. We focus on the city of Turin, Italy, (Fig. 1 - left). We create the graph associated with the city using latitude-longitude data; we consider the main intersections and roads of the city center, thus obtaining a graph with 1074 nodes and 1874 edges (Fig. 1 - middle). The 201 considered detectors are positioned in the city as shown in Fig. 1 - right. Detectors data are given in time series with a granularity of five minutes in a period of one year (from June, 27th 2016 to June, 27th 2017) and the available data are the number of vehicles in a time interval and their average speed. We aggregate these data in a unique indicator, the flow. In order to use these data in the NN, the flow must be propagated from detectors to graph nodes (i.e. neurons of the NN), and we do it as follows:

$$f_i = \sum_{j=1}^{N} \frac{\hat{f}_j}{dist^2(i,j)}, \tag{1}$$

where f_i is the flow on vertex/neuron i, $\hat{f}_j$ is the flow on detector j, N is the number of detectors, and $dist$ is the euclidean distance between the node i and detector j. Turin city center area has 463 detectors, but we consider only detectors with almost complete data (>75%) with respect to time that are 201, in order to avoid as much as possible the artificial generation of data to fill the dataset. We filled each data gap with the average value of the last preceding and first following data-points.

3.2   Data Exploration and Visualization

According to the previously presented set-up, we studied the distribution of the flow during the day. We found three different emerging patterns: one for the week days, one for Saturdays and one for Sundays, as it is shown in Fig. 2, where the

Fig. 1. Turin city map, road network, and detectors distribution.

average of all data for each week day and for each time slice is represented. Week days present a peak in morning hours that completely disappear during the weekend. Sundays present a big flow decrease during lunch time. Note that data present a behavior out of the path in late-evening hours. This is probably due to a technical constraint of the considered detectors.

Some interesting patterns emerge from the distributions of Sundays and Mondays. In particular, the flow distribution of the Sundays presents some exceptions, that are days out of the general pattern as shown in Fig. 2 - top left: the green line represents January 1st and we can see an inverted behavior in early morning hours, while during the day the total amount of flow is smaller than other Sundays in the year; the red lines represent March 5th and April 2nd, ecological Sundays, in which the cars were not allowed to circulate from 10am to 6pm, thus the line perfectly follows the pattern until 10am, then drops until 6pm and then resumes matching the general pattern. For what concerns Mondays, Fig. 2 - bottom left shows some days with patterns similar to Sundays, with a smaller amount of traffic and with no morning peak, this lines represent holidays.

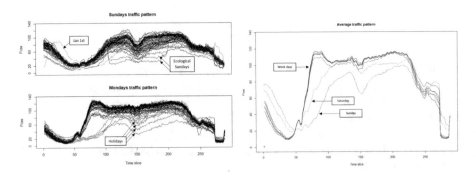

Fig. 2. Daily and weekly traffic patterns.

3.3   Neural Network Model Design

We create a recurring neural network able to perform short-term traffic prediction from available data. We associate each node of the urban graph, corresponding to a junction, to a neuron and each edge, corresponding to a road, to a neurons connection. Traffic flow is then modeled as propagating signal through the neural network. Since each neuron corresponds to a node of the urban graph, it is natural that neurons have multiple outgoing connections representing the outgoing roads of the junction. Each neuron has numerous ingoing connections, which are controlled by a delay layer representing the time window management; inputs are combined and an output is produced and then partitioned on the outgoing connections until the connections are saturated; the exceeding flow is inserted again in the neuron through the self-loop. The flow on this loop represents the traffic congestion on a junction. The starting urban graph is composed by nodes and edges, such that for each graph node i, $i = 1, 2, \ldots, N$, there exists a stack of previsions $f_i(t)$, $t \in [0, T]$, with flow data for each time interval. We build a NN with a neuron for each graph node and a connection for each graph edge. Furthermore, for each neuron i we create an artificial self loop (i, i), and for each connection (i, j) we set a maximum capacity $C_{i,j}$. In this work, we set this capacity to a flow value of 100, which is the third quartile of the flow distribution. Consider a time window of length $\tau + 1$ and let $F_i(t) = [f_i(t), f_i(t-1), \ldots, f_i(t-\tau)]^T$ be the vector of the time series associated with node i. Let M be the number of in-going edges of node i and K be the number of out-going edges from node i. For each node i a weight matrix $W^i(t)$ is defined, with K rows and τ columns. We initialized each weight matrix with exponential decay and Gaussian noise ($\sim \mathbf{N}(0, 1)$) as follows:

$$W^k(0) = \{w_{ij}^k(0)\}_{ij} = \left\{ \frac{1}{2^j} + noise \right\}_{ij}, \quad k = 1, 2, \ldots, N \tag{2}$$

so that each node initially routes the traffic equally to adjacent nodes (since $w_{i,j}$ does not depend on i), and weighing recent history more than previous one (as $w_{i,j}$ halves for every previous timestamp). We use the matrix $W^i(t)$ to estimate the K out-going flows from node i at time $t + 1$.

$$W^i(t) \cdot F_i(t) = \begin{bmatrix} w_{11}^i & w_{12}^i & \ldots & w_{1\tau}^i \\ w_{21}^i & w_{22}^i & \ldots & w_{2\tau}^i \\ \ldots & \ldots & \ldots & \ldots \\ w_{K1}^i & w_{K2}^i & \ldots & w_{K\tau}^i \end{bmatrix} \cdot \begin{bmatrix} f_i(t) \\ f_i(t-1) \\ \ldots \\ f_i(t-\tau-1) \end{bmatrix} = \begin{bmatrix} \tilde{f}_{i1}(t+1) \\ \tilde{f}_{i2}(t+1) \\ \ldots \\ \tilde{f}_{iK}(t+1) \end{bmatrix} = \tilde{\mathcal{F}}_i(t+1),$$

$$\tag{3}$$

where $\tilde{f}_{ij}(t+1)$ is the estimated flow at time $t + 1$ on edge (i, j) and $\tilde{\mathcal{F}}_i(t+1)$ is the vector of the estimated out-going flows from node i.

$$\begin{cases} \tilde{f}_i(t) = \sum\limits_{j=1}^{M} \tilde{f}_{ji}(t) + \tilde{f}_{ii}(t), \\ \tilde{f}_{ii}(t) = \max\left\{ \sum\limits_{j=1}^{K} \left(\tilde{f}_{ij}(t) - C_{ij} \right), 0 \right\}. \end{cases} \tag{4}$$

Formally, the estimated flow on node i at time $(t+1)$ is the sum of the estimated flows coming from its M in-going edges plus the already existing flow on its self-loop. The flow on the self-loop is the sum of the flows that cannot exit on edge (i, j) because the maximum capacity C_{ij} has already been reached.

For each node i we define an error $E_i(t)$ as follows

$$E_i(t) = f_i(t) - \tilde{f}_i(t). \tag{5}$$

Then we partition this error on the in-going edges on node i proportionally to the in-going flows as follows

$$E_{ji}(t) = \frac{\tilde{f}_{ji}(t-1) \cdot E_i(t)}{\sum_{j=1}^{M} \tilde{f}_{ji}(t-1)}. \tag{6}$$

We define the error $E^i(t)$ as the vector of the outgoing error from node i at time t.

$$E^i(t) = [E_{i1}(t), \ldots, E_{iK}(t)]. \tag{7}$$

Using the error vector $E^i(t)$, we can update backwards the matrix $W^i(t+1)$ for each node i as follows. Moreover, we add a parameter η, the learning rate.

$$W^k(t+1) = \{w_{ij}^K(t+1)\}_{ij} = \left\{ \frac{w_{ij}^K(t) + \eta \cdot E^K(t) \cdot f_k(t)}{\sum_{i,j} w_{ij}^K(t+1)} \right\}_{ij}. \tag{8}$$

3.4 Model Evaluation Assessment

In order to evaluate the predictive ability of the GBNN model, we adopted MAAPE, Mean Arctangent Absolute Percentage Error as proposed by [5]. MAPE (Mean Average Percentage Error) is the most widely used measure of forecast accuracy, but it is asymmetric and afflicted by the problem of the division by zero. To avoid this problem but to preserve the philosophy of MAPE, the MAAPE is introduced. It is defined as follows:

$$\text{MAAPE} = \frac{1}{N} \sum_{t=1}^{N} \arctan \left(\frac{|\text{ predicted} - \text{expected }|}{\text{expected}} \right). \tag{9}$$

The error belongs to $[0, \frac{\pi}{2}]$. In order to get a percentage value, we rescale the MAAPE by multiplying by $\frac{2}{\pi}$, such that now the error belongs to $[0, 1]$.

4 Experimental Results

We implemented our GBNN model in Java. We set the hyperparameters as follows: the dimension of the considered time windows (memory size) is 4, the maximum capacity of the edges is 100.00 for all links in the graph, the learning rate is 10^{-4}, the maximum time horizon for predictions is 12 (that is prediction 1 h ahead), the number of train iterations is 40. We proceeded to test the GBNN

Table 1. Table of MAAPE errors at different horizons

	Time horizons			
Day	5 m (%)	10 m (%)	30 m (%)	1 h (%)
Monday	2.89	4.05	9.43	15.85
Tuesday	2.95	3.99	9.24	15.62
Wednesday	2.82	3.90	9.09	15.33
Thursday	2.77	3.91	9.48	16.10
Friday	2.60	3.63	8.91	15.56
Saturday	2.39	2.96	6.78	12.13
Sunday	2.71	3.17	7.18	13.23
Average	2.73	3.66	8.59	14.83

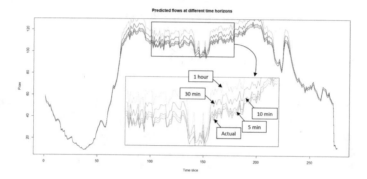

Fig. 3. Flow predictions analysis at different time horizons.

on each week day separately, using all days except one as training set and considering the remaining day as validation. We use the trained network to predict the traffic flow at different time horizons: that is, using the validation data at time t to predict the traffic flow at time $t + k$ minutes, with $k \in \{5, 10, 30, 60\}$.

We obtain very small errors for predictions 5 min ahead and the error is still under 10% for predictions 30 min ahead. GBNN is able to predict traffic flow 1 h ahead with an error around 15%. The results are almost constant for each day-based GBNN and for each time horizon, as shown in Table 1. We select a random neuron to analyze predicted flow values at different time horizons qualitatively. Figure 3 shows the actual flow and the predicted flows for the validation day on Mondays set. GBNN performs excellent prediction for the first and the last part of the day, while it is subjected to a bigger error during the middle hours. For our experiments, we used a Dell Precision M3800 (Intel Core i7-4712HQ, 16 GB RAM). Roughly, each training iteration requires 0.05 seconds for each day in the training set (288 time slices per day). We performed random search and grid search in order to find the optimal values for learning rate and memory size, that are the most incisive hyperparameters (see Fig. 4). In the first case, the lowest MAAPE was associated with a learning rate of 10^{-4}: concerning memory size, the value 4 (that is, to base the prediction upon the previous 20 min of data) was correlated with the lowest cross-horizon MAAPE. The shaded area in Fig. 4 represents the interval of values which minimise the average error. We decide to consider the trivial model, in which predictions are made by repeating the value of the known flow, to compare the results because classical statistical methods do not fit our urban traffic representation: incomplete data and instable distribution undermine the stability of statistical methods. We decide not to compare our model to an agent-based one because the approach completely differs: we consider the traffic mass as a whole, not as a set of individual agents. Moreover, there is no existing neural network model in literature which consider the entire urban network or the bidirectional traffic flow. The comparison of the GBNN model and the trivial one is shown in Fig. 5 for 5 min ahead predictions (left) and for 1 h ahead predictions (right), on the validation day for a randomly selected node. The top panels show the comparison of real flow, flow predicted by the trivial model (red line) and flow predicted by GBNN (blue line) for a time horizon of 5 min (left) and 1 h (right). The bottom panels display the error comparison for the models: red line for trivial model, black line for GBNN.

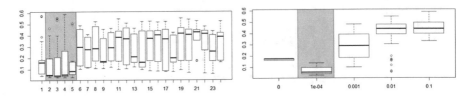

Fig. 4. MAAPE error analysis w.r.t. memory size (left) and learning rate (right).

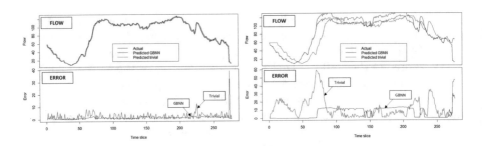

Fig. 5. Flows and errors comparison

5   Conclusions and Future Work

In this paper we have proposed an innovative neural network model, whose topology mirrors an urban graph and where each node is enriched with a recurrent connection in order to capture the temporal behavior. The resulting system is able to model the whole urban traffic flow, to take into account both space and time patterns in the data, and consequently to make network-wide traffic flow predictions. The prediction error is as low as 3% and 16% for 5 min and 1 h traffic forecasting respectively.

Future directions of work will focus on implementing peculiarities of urban road network that we omitted for the initial model, such as one-way street and roundabouts.

References

1. Chowdhuri, M., Sadek, A.W.: Advantages and limitations of artificial intelligence. Trans. Res. B., E-C168, 6–8 (2012)
2. Faghri, A., Hua, J.: Evaluation of artificial neural network applications in transportation engineering. Transport. Res. Rec. **1358**, 71–80 (1992)
3. Rumelhart, D.E., Hinton, G.E., Williams, R.J.: Learning Internal Representations by Error Propagation. MIT Press, Cambridge (1986)
4. Karlaftis, M.G., Vlahogianni, E.I.: Statistical methods versus neural networks in transportation research: differences, similarities and some insights. Transport. Res. C-Emer. **19**, 387–399 (2011)
5. Kim, S., Kim, H.: A new metric of absolute percentage error for intermittent demand forecasts. J. Forecast. **32**, 669–679 (2016)
6. Ma, X., Tao, Z., Wang, Y., Yu, H., Wang, Y.: Long short-term memory neural network for traffic speed prediction using remote microwave sensor data. Transport. Res. C-Emer. **54**, 187–197 (2015)
7. Moretti, F., Pizzuti, S., Panzieri, S., Annunziato, M.: Urban traffic flow forecasting through statistical and neural network bagging ensemble hybrid modeling. Neurocomputing **167**, 3–7 (2015)
8. Siegelmann, H.T., Horne, B.G., Giles, C.L.: Computational capabilities of recurrent NARX neural networks. IEEE Trans. Syst. Man Cybern. Part B **27**(2), 208–215 (1997)
9. Haykin, S.: Neural Networks: A Comprehensive Foundation. Prentice Hall, Upper Saddle River (1999)
10. Rosenblatt, F.: The perceptron: a probabilistic model for information storage and organization in the brain. Psychol. Rev. **65**, 386–408 (1958)
11. Polson, N.G., Sokolov, V.O.: Deep learning for short-term traffic flow prediction. Transport. Res. C-Emer. **79**, 1–17 (2017)
12. Qiao, F., Yang, H., Lam, W.H.K.: Intelligent simulation and prediction of traffic flow dispersion. Transport. Res. B-Meth. **35**, 843–863 (1999)
13. United Nations: Department of Economic and Social Affairs, Population Division. World urbanization prospects: the 2014 revision (2014)
14. van Lint, J.W.C., Hoogendoorn, S.P., van Zuylen, H.J.: Accurate freeway travel time prediction with state-space neural network under missing data. Transport. Res. C-Emer. **13**, 347–369 (2005)
15. Vlahogianni, E.I., Karlaftis, M.G., Golias, J.C.: Short-term traffic forecasting: where we are and where we're going. Transport. Res. C-Emer. **43**, 3–19 (2014)
16. Wang, J., Tsapakis, I., Zhong, C.: A space-time delay neural network model for travel time prediction. Eng. Appl. Artif. Intel. **52**, 145–160 (2016)
17. Zhu, J.Z., Cao, J.X., Zhu, Y.: Traffic volume forecasting based on radial basis fuction neural network with the consideration of traffic flows at the adjacent intersections. Transport. Res. C-Emer. **47**, 139–154 (2014)

Traffic Sign Detection Using R-CNN

Philipp Rehlaender[1]([✉]), Maik Schroeer[2], Gavneet Chadha[3],
and Andreas Schwung[3]

[1] Paderborn University, Paderborn, Germany
`rehlaender@lea.upb.de`
[2] Busch-Jaeger Elektro GmbH, Lüdenscheid, Germany
[3] South Westphalia University of Applied Sciences, Iserlohn, Germany

Abstract. Due to the increasing popularity of driving assistance systems, traffic sign detection has gained increased attention due to its role in traffic safety. This paper presents a traffic sign detection algorithm based on the regions with convolutional neural network (R-CNN) method. In this method, a segmentation algorithm proposes regions likely to contain a traffic sign. These regions are analyzed by a convolutional neural network providing a feature vector to a support vector machine. This method, however, suffers from a high computation time due to an excessive amount of regions proposed. To reduce the number of regions, this work proposes a perspective based logic that discards regions unlikely to contain traffic signs because of a size-position-mismatch. This results in an average 93% reduction of proposed regions. For faster training, the system consists of a two-layer support vector machine. The first layer is used to categorize the family of the picture, the second set of SVMs for classifying the traffic sign. The overall classification accuracy of this technique is 86%. Feeding the results to a tracking algorithm may result in an overall improvement because frames would compensate errors.

Keywords: Object detection · Convolutional neural network ·
Regions with convolutional neural networks (R-CNN) ·
Support vector machines (SVM)

1 Introduction

Advanced driving assistance systems have become more and more popular. Different sensors are used to assist the driver on the road such as cameras, radar sensors, etc. One particularly interesting driving assistance system is the detection of traffic signs. There are more than 100 different traffic signs on German roads. Not all of them are important for the safety of the driver. Usually, traffic sign detection consists of two algorithms: first, the picture is analyzed and potential traffic sign regions are proposed. Second, the proposed regions are evaluated to which group of signs they belong to. Literature has produced an excessive amount of detection methods. Convolutional neural network based methods

© Springer Nature Switzerland AG 2020
L. Oneto et al. (Eds.): INNSBDDL 2019, INNS 1, pp. 226–235, 2020.
https://doi.org/10.1007/978-3-030-16841-4_24

are presented in [5,6]. Mao et al. used in [5] a hierarchical convolutional neural network by stacking multiple convolutional layers on top of a classifier. The results proved to be very successful with a classification accuracy of 99.67%. Qian used in [6] a visual attribute learning approach through max pooling positions for attribute prediction. Support vector machine (SVM) based methods have been applied in [1,4,7]. In this paper, traffic sign detection is achieved using the R-CNN algorithm with a linear threshold logic (LTL) and a two-fold support vector machine (SVM) classification. The work is structured as follows: at first the applied R-CNN algorithm is explained. Section 3 provides a detailed explanation of the region reduction method. Section 4 explains the multi-class support vector machine algorithm. Finally, the paper is concluded and an outlook is given.

2   Regions with Convolutional Neural Networks

In this work, the regions with convolutional neural network (R-CNN) method introduced by Girshick in [2] is used for detecting and recognizing from 43 different traffic signs. For training and evaluation, the database provided by the German Traffic Sign Recognition Benchmark (GTSRB) [8] has been applied. The algorithm is three-fold. At first, a segmentation algorithm proposes regions likely to contain an object of interest. Secondly, a convolutional neural network samples the region and proposes a 4096 feature vector. Finally, a support vector machine classifies the proposed region to one of the 43 analyzed traffic signs or background. The R-CNN method has been extended by a perspective logic to discard regions unlikely to be a traffic sign due to their position-size ratio. Furthermore, the classification process has been modified to be two-fold: a first complete classification groups the sign into a subgroup for similar signs (white signs, speed signs, prohibition signs, warning signs, end signs and mandatory signs) or in case of signs that cannot be sub-grouped achieves a final classification. A second, sub-SVM for grouped signs, then achieves a detailed classification. This two-fold process resulted in a possible more detailed training since the training time of the sub-SVMs was substantially reduced, and thus to a better classification rate. The flow diagram of the complete implemented algorithm is depicted in Fig. 1.

If the first and second SVM provide the same class for the ROI, the class is selected as the final class. Otherwise, the ROI of the input image is resized up to 10% greater than the previous one and the process is repeated one time. In case of different results of both SVMs in the second cycle, the result of the sub-SVM is chosen. For classes which do not fit into one of the classes of the second SVMs, like the priority signs or stop signs, the result of the first SVM is chosen as the final one in the first cycle.

2.1   Segmentation Algorithm

In this work, the selective search algorithm proposed by [9] has been applied. It combines the strengths of exhaustive search and segmentation. By diversifying the search algorithm, this algorithm can deal with a variety of different image conditions.

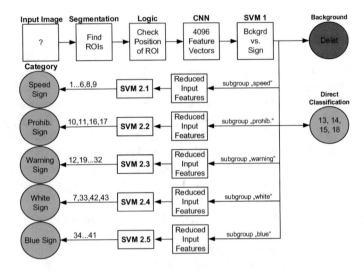

Fig. 1. Flow diagram of the implemented algorithm. A segmentation algorithm proposes regions, which are evaluated through a sequential threshold logic. The remaining regions are evaluated through a convolutional neural network that proposes a 4096 element feature vector to a first SVM. This SVM assigns the regions to one of five groups, background or provides a final classification.

2.2   Reduction of the Region Number

The size of a traffic sign in an image depends on its position in the image and on the position of the route inside the image. It is apparent that the segmentation algorithm returns redundant regions, which cannot be a traffic signs because their position in combination with their size mismatches. Through careful selection of boundaries for the width and height of a region, the number of regions submitted to the convolutional neural network can be substantially reduced. This is further explained in Sect. 3.

2.3   Convolutional Neural Network

The regions are processed by a convolutional neural network. Girshick recommends in [2] the usage of a pre-trained network and to fine-tune it for the specific task. This avoids huge training times and results in an improved performance [2]. For this purpose, the MatConvNet pre-trained CNN in [3,10] is used. Using this pre-trained network provides a state-of-the-art image classification basis. Since the CNN is applied for feature the extraction, the last fully connected layer of the CNN (layer no. 19) is used to extract the features of the images. This returns a 4096 feature vector. The resulting region proposals are a feature matrix with a size of $N \times 4096$ with N representing the number of regions proposed by the segmentation algorithm. Since the CNN can only process 227×227 RGB images, the extracted regions have been resized to fit this architecture. The features are computed by forward propagating to a mean-subtracted RGB image through five convolutional and two fully connected layers [2].

2.4 Classification Through a Two-Stage SVM

The 4096-feature vectors are transferred to the first SVM. The SVM is used as a predictor which classifies the ROIs according to its corresponding classes. Due to the fact that the misclassification rate within different groups of signs was higher than expected, a second SVM instance was added for groups with similarities like the group of speed signs, prohibition signs, warning signs, white signs and mandatory signs. For improved training, the features were reduced and modified by the lasso algorithm. The LASSO algorithm was able to reduce the number of features by more than 50%. The usage of this two-fold SVM resulted in a slight overall improvement of the classification rate. A detailed description is found in Sect. 4.

3 Reducing the Number of Regions Through a Linear Threshold Logic (LTL)

Most regions proposed by the segmentation algorithm do, on the one hand, not fit the side ratio of traffic signs and, on the other hand, they cannot be traffic signs because the position-size-mismatch does not fit the perspective of the analyzed frame. Figure 2 shows an illustration of the perspective theory. The vanishing point lies at the horizon. It can be very clearly seen that the size of the objects decreases the closer objects get to the vanishing point. Furthermore objects on the left and on the right side of the road have different sizes (Fig. 3).

Fig. 2. Visualization of the perspective. The size of all objects becomes linearly smaller the closer the objects get to the vanishing point. Regoins can be evaluated based on their position. Due to the size constraints, not all regions can be traffic signs and regions can be deleted. Figure adapted from [2].

Fig. 3. Visualization of the logic. The red border shows the minimum threshold, the yellow border, the maximum threshold. Region is in quadrant 1, the orientation point is the upper right. Figure adapted from [2].

The illustration shows that the picture can be splitted into four quadrants. For each quadrant the minimum and maximum threshold has to be defined individually. A proposed region is evaluated by its distinctive point which is in the first quadrant the point (x_2, y_2), in the second quadrant the point (x_1, y_2), in the third quadrant (x_1, y_1) and in the fourth quarter (x_2, y_1). This implies that a region can lie on all four quadrants at the same time. If at least one logic accepts the region, the region is kept valid and passed to the CNN.

Fig. 4. 6402 regions proposed by the segmentation. For the purpose of readability, only every tenth region is depicted. Figure adapted from [2].

The minimum and maximum threshold are defined by a linear function with respect to x and y. The origin of the coordinate system is moved to the center. Therefore the threshold function is defined for each quadrant q as follows:

$$\Delta x^q_{\min,x} = a^q_{\min,x} + b^q_{\min,x} \cdot |x| + c^q_{\min,x} \cdot |y| \tag{1}$$

$$\Delta y^q_{\min,y} = a^q_{\min,y} + b^q_{\min,y} \cdot |x| + c^q_{\min,y} \cdot |y| \tag{2}$$

$$\Delta x^q_{\text{max,x}} = a^q_{\text{max,x}} + b^q_{\text{max,x}} \cdot |x| + c^q_{\text{max,x}} \cdot |y| \tag{3}$$

$$\Delta x^q_{\text{max,y}} = a^q_{\text{max,y}} + b^q_{\text{max,y}} \cdot |x| + c^q_{\text{max,y}} \cdot |y| \tag{4}$$

For evaluating the region size, the 12 variables of Eqs. (1)–(4) have to be known. However, an evaluator can be tuned manually or through a machine learning tool. A carefully manually tuned logic was able to reduce the number of regions submitted to the CNN in average by 90%. The process of region reduction is depicted in Figs. 4 and 5. The initial number of 6402 proposed regions was reduced to a final number of 386. This accounts for a reduction of 94%.

Fig. 5. Suppression of the number of regions through the LTL to a total of 386. Figure adapted from [2].

4 Multiclass Support Vector Machine Classification

This chapter describes the design of the support vector machine. The superior SVM classifies the regions to sub-groups or background. For the sub-SVMs the LASSO feature regulation technique is used. Finally, sub-SVMS are trained for the inferior classification.

4.1 Primary SVM

The 4096 different feature vectors of each picture are given to this first SVM, which has the task to assign the most suitable class to every ROI which was given to the CNN. The SVM was trained on the training set of the GTSRB and the added background pictures of the CALTECH 101. The SVM was trained using Bayesian optimization producing the classification results depicted in Fig. 6.

4.2 Feature Regularization

Feature regularization using the LASSO algorithm was used to force features to zero. Through the use of this regularization technique features are forced to zero with the intention to avoid overfitting. The number of parameters forced to zero

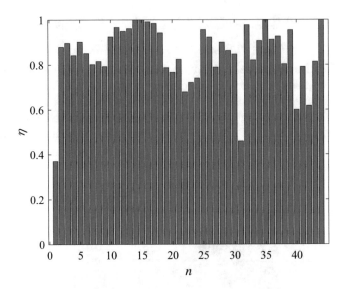

Fig. 6. Classification result η for the primary SVM with the class number n.

is depicted in Table 1. On a sub-trained SVM, this improved the classification accuracy in average slightly by 0.5%. Looking at the varying amount of features forced to zero, it is evident that different features are relevant for different sub-classes.

Table 1. Subset features forced to zero

Subset	Features forced to zero
Blue signs	2595
Prohibition signs	2601
Speed signs	2332
Warning signs	2469
White signs	3411

4.3 Secondary SVMs Design and Performance

The five sub-SVMs were trained on the reduced features space. Every sub-SVM was individually optimized to receive the best performance.

Blue Signs. The SVM has been trained using a royal basis function kernel, a dense random coding, Bayes-optimized regarding the kernel scale and the box constraint. The results of the classification are depicted in Fig. 7a. The overall classification accuracy of this sub-group is 86%. Further training with other kernel functions and coding matrices was not able to improve the classification rate.

Prohibition Signs. For this sub-SVM it was sufficient using a linear kernel with a one-vs-all coding trained with a 10-fold cross validation using random parameter search. The classification results are shown in Fig. 7b. The overall classification rate was 97%.

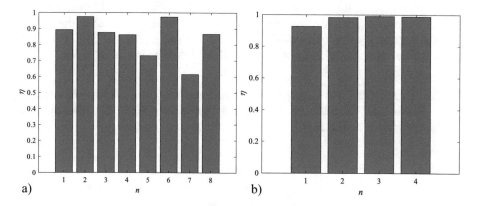

Fig. 7. Classification result η for the sub-SVM "blue signs" (a) and the sub-SVM "prohibition signs" (b) with the corresponding class number n.

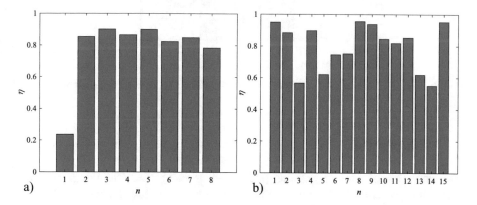

Fig. 8. Classification result η for the sub-SVM "speed signs" (a) and the sub-SVM "warning signs" (b) with the corresponding class number n.

Speed Limits. Extensive tests have been performed to achieve sufficient results on this class. The best results have been reached using a non-linear royal basis function kernel, Bayesian optimization and sparse-random coding. The results in Fig. 6 already showed that the speed limit of 20 km/h was very hard to classify. Analysis showed that this speed limit was often confused with 30 km/h and 50 km/h. The classification results are depicted in Fig. 8a. The overall classification accuracy of this group was 81%. The worst accuracy was reached for the 20 km/h speed limit with only 24%. This accuracy is not sufficient and needs further training. The accuracy of the detection for the other traffic signs is sufficient.

Warning Signs. The classification result of this predictor was in average 80%. The classification rates are shown in Fig. 8b. The results shows that the overall efficient accuracy some individual accuracies are not sufficient. This subgroup needs further training with i.e. the generation of synthetic sub-samples.

White Signs. The last classifier that has been programmed for white traffic signs. The overall classification rate of this group was 80%. The classification results are shown in Fig. 9.

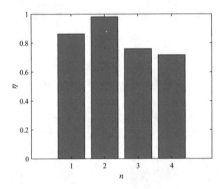

Fig. 9. Classification result η for the sub-SVM "white signs" with the class number n.

5   Conclusion

The detection of traffic signs was performed by using R-CNN. A segmentation algorithm proposed ROIs. The number of regions was reduced in average by 93% through a sequential linear logic to discard regions with a position-size-mismatch. A pre-trained CNN analyzed the remaining regions to extract a feature vector, which was used to classify the image by a two-fold SVM. A first SVM separated traffic signs from the background image and provided a first classification into sub-groups.

A set of inferior SVMs provided the final classification. This two-fold structure resulted in a slight improvement of the classification accuracy from 85 to 86%.

6 Outlook

The sequential linear logic proposed in this paper has been tuned manually to perform on a wide variety of driving situations. However, an automatic detection of the vanishing point and the position of the road inside the image could be used for automatic determination of the parameter set of (1) to (4). This would result in an improved suppression of redundant regions. The will be researched in future work.

References

1. Boi, F., Gagliardini, L.: A support vector machines network for traffic sign recognition. In: The 2011 International Joint Conference on Neural Networks, pp. 2210–2216. IEEE (2011). https://doi.org/10.1109/IJCNN.2011.6033503
2. Girshick, R., Donahue, J., Darrell, T., Malik, J.: Rich feature hierarchies for accurate object detection and semantic segmentation. In: 2014 IEEE Conference on Computer Vision and Pattern Recognition, pp. 580–587. IEEE (2014). https://doi.org/10.1109/CVPR.2014.81
3. Krizhevsky, A., Sutskever, I., Hinton, G.E.: ImageNet classification with deep convolutional neural networks. http://papers.nips.cc/paper/4824-imagenet-classification-with-deep-convolutional-neural-networks
4. Maldonado-Bascon, S., Lafuente-Arroyo, S., Gil-Jimenez, P., Gomez-Moreno, H., Lopez-Ferreras, F.: Road-sign detection and recognition based on support vector machines. IEEE Trans. Intell. Transp. Syst. **8**(2), 264–278 (2007). https://doi.org/10.1109/TITS.2007.895311
5. Mao, X., Hijazi, S., Casas, R., Kaul, P., Kumar, R., Rowen, C.: Hierarchical CNN for traffic sign recognition. In: 2016 IEEE Intelligent Vehicles Symposium (IV), pp. 130–135. IEEE (2016). https://doi.org/10.1109/IVS.2016.7535376
6. Qian, R.Q., Yue, Y., Coenen, F., Zhang, B.L.: Traffic sign recognition using visual attribute learning and convolutional neural network. In: 2016 International Conference on Machine Learning and Cybernetics (ICMLC), pp. 386–391. IEEE (2016). https://doi.org/10.1109/ICMLC.2016.7860932
7. Shi, M., Wu, H., Fleyeh, H.: Support vector machines for traffic signs recognition. In: 2008 IEEE International Joint Conference on Neural Networks (IEEE World Congress on Computational Intelligence), pp. 3820–3827. IEEE (2008). https://doi.org/10.1109/IJCNN.2008.4634347
8. Stallkamp, J., Schlipsing, M., Salmen, J., Igel, C.: Man vs. computer: benchmarking machine learning algorithms for traffic sign recognition. Neural Networks Off. J. Int. Neural Network Soc. **32**, 323–332 (2012). https://doi.org/10.1016/j.neunet.2012.02.016
9. Uijlings, J.R.R., van de Sande, K.E.A., Gevers, T., Smeulders, A.W.M.: Selective search for object recognition. Int. J. Comput. Vis. **104**(2), 154–171 (2013). https://doi.org/10.1007/s11263-013-0620-5
10. Vedaldi, A., Lenc, K.: MatConvNet: convolutional neural networks for MATLAB (2014). https://arxiv.org/pdf/1412.4564v3.pdf

Deep Tree Transductions - A Short Survey

Davide Bacciu$^{(\boxtimes)}$ and Antonio Bruno

Department of Computer Science, University of Pisa, Pisa, Italy
{bacciu,antonio.bruno}@di.unipi.it

Abstract. The paper surveys recent extensions of the Long-Short Term Memory networks to handle tree structures from the perspective of learning non-trivial forms of isomorph structured transductions. It provides a discussion of modern TreeLSTM models, showing the effect of the bias induced by the direction of tree processing. An empirical analysis is performed on real-world benchmarks, highlighting how there is no single model adequate to effectively approach all transduction problems.

Keywords: Structured-data processing · Tree transduction · TreeLSTM

1 Introduction

Structured transductions are a natural generalization of supervised learning to application scenarios where both input samples and predictions are structured pieces of information. Trees are an example of non-trivial structured data which allows to straightforwardly represent compound information characterized by the presence of hierarchical-like relationships between its elements. Within this context, learning a tree structured transduction amounts to inferring a function associating an input tree to a prediction that is, as well, a tree. Many challenging real world applications can be addressed as tree transduction problems.

The problem of learning generic tree transductions, where both input and output trees have different topologies, is still a challenging open research question, despite some works [2,21] have started dealing with learning to sample tree-structured information, which is a prerequisite functionality for realizing a predictor of tree structured outputs. Isomorphic transductions define a restricted form of tree transformations [7] which, nevertheless, allow to model and address several interesting learning tasks on structured data, including: (i) tree classification and regression, i.e. predicting a vectorial label for the whole tree; (ii) node relabeling, i.e. predicting a vectorial label for each node in a tree taking into account information from its surrounding context (e.g. its rooted subtree or its root-to-node path); and (iii) reduction to substructure, i.e. predicting a tree obtained by pruning pieces of the input structure.

Isomorph transductions have been realized by several learning models, starting from the seminal work on the recursive processing of structures in [13].

© Springer Nature Switzerland AG 2020
L. Oneto et al. (Eds.): INNSBDDL 2019, INNS 1, pp. 236–245, 2020.
https://doi.org/10.1007/978-3-030-16841-4_25

There it has been formalized the idea of extending recurrent models to perform a bottom-up processing of the tree, unfolding the network recursively over the tree structure so that the hidden state of the current tree node depends from that of its children. The same approach has been taken by a number of models from different paradigms, such as the probabilistic bottom-up Hidden Markov Tree model [4], the neural Tree Echo State Network [14] or the neural-probabilistic hybrid in Hidden Tree Markov Networks (HTNs) [3]. Another approach can be taken based on inverting the direction of parsing of the tree, by processing this top-down from the root to its leaves. This is diffused in particular in the probabilistic paradigm [12], where it represents a proper straightforward extension of hidden Markov models for sequences. More recently, within the deep learning community it has diffused a widespread use of Long Short Term Memory (LSTM) networks [15] for the processing of tree structured information. The so-called TreeLSTM model [20] extends the LSTM cell to handle tree-structures through a bottom-up approach implementing a specific instance of the recursive framework by [13]. Although defined for the general case of an n-ary tree, the TreeLSTM has been used on binary trees obtained by binarization of the original parse tree which, in practice, reduces significantly the contribution of the structured information to solving the task. Two recent works show example of a top-down TreeLSTM: one proposed in [21] to learn to sample trees and one in [6] showing the its use in learning simple non-isomorph transductions.

The aim of this paper is to present an orderly discussion of modern TreeLSTM models, assessing the effect of different stationarity assumptions (i.e. parameterization of the hidden state) on a full n-ary setting (i.e. without requiring a binarization of the structure). Also, since the choice of the tree parsing direction has a well-known effect on the representational capabilities of the model [4], we consider both bottom-up and top-down TreeLSTM models. In particular, we focus on benchmarking the TreeLSTM models on three different tasks associated with the three different types of isomorph tree structured transductions discussed above. We will show how each of such task has different assumptions and characteristics which cannot all be effectively addressed by a single approach.

2   Problem Formulation

Before delving into the details of the different LSTM-based approaches to deal with tree data, we formalize the problems addressed in the experimental assessment using the generic framework of structured transductions.

We consider the problem of learning tree transductions from pairs of input-output trees $(\mathbf{x}^n, \mathbf{y}^n)$, where the superscript identifies the n-th sample pair in the dataset (omitted when the context is clear). We consider labeled rooted trees defined by the triplet $(\mathcal{U}_n, \mathcal{E}_n, \mathcal{X}_n)$ consisting of a set of nodes $\mathcal{U}_n = \{1, \ldots, U_n\}$, a set of edges $\mathcal{E}_n \subseteq \mathcal{U}_n \times \mathcal{U}_n$ and a set of labels $\mathcal{X}_n$. The term $u \in \mathcal{U}_n$ denotes a generic tree node whose direct ancestor, called *parent*, is referred to as $pa(u)$. A node u can have a variable number of direct descendants (*children*), such that the l-th child of node u is denoted as $ch_l(u)$. The pair $(u,v) \in \mathcal{E}_n$ is used to denote an edge between a generic node and its child and we assume trees to have maximum finite out-degree L (i.e. the maximum number of children of a node). Each vertex u in the tree is associated with a label x_u (y_u, respectively) which can be of different nature depending on the application, e.g. a vector of continuous-valued features representing word embeddings or a symbol from a discrete and finite alphabet.

A tree transduction is defined as a mapping from an input sample to an output elements where both are tree structured pieces of information. Using $\mathcal{I}^{\#}, \mathcal{O}^{\#}$ to denote the input and output domains, respectively, then a structural transduction is a function $\mathcal{F} : \mathcal{I}^{\#} \rightarrow \mathcal{O}^{\#}$. We focus on transductions exploiting the following definition of tree isomorphism.

Definition 1. *Tree isomorphism Let* $\mathbf{x} = (\mathcal{U}, \mathcal{E}, \mathcal{X})$ *and* $\mathbf{x}' = (\mathcal{U}', \mathcal{E}', \mathcal{X}')$, *they are isomorphic if exists a bijection*
$f : \mathcal{U} \rightarrow \mathcal{U}'$ *such that* $\forall(u,u') \in \mathcal{E} \iff (f(u), f(u')) \in \mathcal{E}'$.

An equivalent definition can be given using the concept of *skeleton*.

Definition 2. *Skeleton tree Let* $\mathbf{x} = (\mathcal{U}, \mathcal{E}, \mathcal{X})$, *its skeleton is* $skel(\mathbf{x}) = (\mathcal{U}, \mathcal{E})$.

Following such definition, two trees are isomorphic if they have the same skeleton (labels are irrelevant, only structure matters).

A general *structured transduction* can be formalized by a learnable encoding-decoding process where $\mathcal{F} = \mathcal{F}_{out} \circ \mathcal{F}_{enc}$ with:

$$\mathcal{F}_{enc} : \mathcal{I}^{\#} \rightarrow \mathcal{H}^{\#} \qquad \mathcal{F}_{out} : \mathcal{H}^{\#} \rightarrow \mathcal{O}^{\#}.$$

The terms $\mathcal{F}_{enc}$ and $\mathcal{F}_{out}$ are the *encoding* and *output* transductions, while we assume the existence of a state space $\mathcal{H}^{\#}$ providing an intermediate and rich representation of the structured information, such as in the activations of the hidden neurons of a recursive neural model. Different types of transductions can be obtained depending on the isomorphism properties of the encoding and output mappings. In this work, we consider three types of tree transductions, each associated with a practical learning and prediction task. First, we consider a *tree-to-tree isomorphic transduction* where both encoding and output mappings are isomorphic. The second type is the *structure-to-element* or *supersource transduction* that map an input tree into a single vectorial element in the output domain, basically realizing a classical tree classification or regression task. The third type is the *structure-to-substructure* transduction which defines a restricted form of non-isomorphic transduction where the output tree $\mathbf{y}$ is obtained from the input tree $\mathbf{x}$ by pruning some of its proper subtrees. In practice, such a transformation can again be realized as an isomorphic transduction, where the encoding is isomorphic as in the previous cases. The output function, instead, isomorphically maps each element of the hidden encoding into an output node while using a specific *NULL* value as label of those nodes of the input structure that are non-existing in the output tree.

3   TreeLSTM for Constrained Tree Transductions

Several works have been dealing with extending Recurrent Neural Networks (RNN) to deal with tree structured data. Lately, most of these works focused on tree-structured extensions of LSTM cells and networks. Two sources of differentiation exist between the different models. One concerns the stationarity assumptions, that is how much tied are the network parameters with respect to topological aspects such as the position of a node with respect to its siblings. The second source of differentiation concerns the direction of tree processing (top-down or bottom-up for trees), which determines the context upon which a specific node is assessed (i.e. depending on the hidden state of the parent, for the top-down case, or depending on the states of its children, for the bottom-up case). In the following, we briefly review the TreeLSTM approaches in literature with respect to these two differentiating factors.

Top-Down (TD) TreeLSTM: in this model, tree processing flows from the root to the leaves. In literature, TD TreeLSTM models are mainly used in generative settings, where one wants to generate the children of a node based on the hidden state of the parent [2,21]. Their use as encoders of the full structure is not common, as this requires some form of mapping function summarizing the whole tree into a single encoding vector (e.g. the mean of the hidden states of all the nodes in the tree). Here, we consider the use of TD TreeLSTM in the context of learning isomorph transductions of the three types discussed in Sect. 2. In particular, we will assess the capabilities of this model in realizing non-generative tasks, highlighting limitations and advantages with respect to its more popular bottom-up counterpart. Formally, the activation of a TD TreeLSTM cell for a generic node u is regulated by the following equations

$$r_u = \tanh\left(W^{(r)}x_u + U^{(r)}h_{pa(u)} + b^{(r)} \right) \tag{1}$$

$$i_u = \sigma\left(W^{(i)}x_u + U^{(i)}h_{pa(u)} + b^{(i)} \right) \tag{2}$$

$$o_u = \sigma\left(W^{(o)}x_u + U^{(o)}h_{pa(u)} + b^{(o)} \right) \tag{3}$$

$$f_u = \sigma\left(W^{(f)}x_u + U^{(f)}h_{pa(u)} + b^{(f)} \right) \tag{4}$$

$$c_u = i_u \odot r_u + f_u \odot c_{pa(u)} \tag{5}$$

$$h_u = o_u \odot \tanh(c_u) \tag{6}$$

with the term x_u denoting unit input, $h_{pa(u)}$ and $c_{pa(u)}$ are, respectively, the hidden state and the memory cell state of the node's parent, σ is the sigmoid activation function and $\odot$ is elementwise multiplication. It can be seen that the formal model of this unit is that of a standard LSTM unit for sequences, but this will be unfolded over the tree in a TD fashion by following in parallel all the root to leaves paths.

The second type is Bottom-Up (BU) TreeLSTM, in which tree processing flows from the leaves to the root. In literature there are two types of BU TreeL-STM, which is mainly used as one-pass encoder for tree structured information [20]. The two BU TreeLSTM types differ in stationarity assumptions and the choice of which one to use depends on the specificity of the structured data at hand (e.g. finiteness of the outdegree, relevance of node positionality information). The choice of a bottom-up approach is motivated by consolidated results showing a superior expressiveness of bottom-up parsing with respect top-down approaches when dealing with trees [4]. When considered within the context neural processing of the structure this founds on the assumption that a node hidden state computed recursively from its children states is a "good" vectorial summary of the information in all the subtree rooted in the node. Another observation is that the bottom-up approach provides a natural means to obtain a state mapping function for the whole tree, by considering the hidden state of the tree as a good summary of the information contained in the whole structure.

Child-Sum TreeLSTM: this type of TreeLSTM is used to encode trees where the position of nodes (ordering) with respect to their siblings is not relevant for the task. Let $ch(u)$ be the set of children (of size K) of the generic node u, its state transition equation follows:

$$\tilde{h}_u = \sum_{k \in ch(u)} h_k \tag{7}$$

$$r_u = \tanh\left(W^{(r)}x_u + U^{(r)}\tilde{h}_u + b^{(r)} \right) \tag{8}$$

$$i_u = \sigma\left(W^{(i)}x_u + U^{(i)}\tilde{h}_u + b^{(i)} \right) \tag{9}$$

$$o_u = \sigma\left(W^{(o)}x_u + U^{(o)}\tilde{h}_u + b^{(o)} \right) \tag{10}$$

$$f_{uk} = \sigma\left(W^{(f)}x_u + U^{(f)}h_k + b^{(f)} \right), \forall k \in ch(u) \tag{11}$$

$$c_u = i_u \odot r_u + \sum_{k \in ch(u)} f_{uk} \odot c_k \tag{12}$$

$$h_u = o_u \odot \tanh(c_u) \tag{13}$$

with the term x_u denoting unit input, h_k and c_k are, respectively, the hidden state and the memory cell state of the k-th child, σ sigmoid function and $\odot$ the elmentwise product. As every LSTM it has the three gates, in particular there is a forget gate for every child but all of them share the same parameters.

N-ary TreeLSTM: this TreeLSTM variant allows to discriminate children by their position with respect to the siblings, while needing to fix a priori the maximum outdegree of the tree. Let u be the generic node, the associated TreeLSTM cell activations are as follows

$$r_u = \tanh\left(W^{(r)}x_u + \sum_{\ell=1}^{N} U_\ell^{(r)} h_{ch_\ell(u)} + b^{(r)}\right) \tag{14}$$

$$i_u = \sigma\left(W^{(i)}x_u + \sum_{\ell=1}^{N} U_\ell^{(i)} h_{ch_\ell(u)} + b^{(i)}\right) \tag{15}$$

$$o_u = \sigma\left(W^{(o)}x_u + \sum_{\ell=1}^{N} U_\ell^{(o)} h_{ch_\ell(u)} + b^{(o)}\right) \tag{16}$$

$$f_{uk} = \sigma\left(W^{(f)}x_u + \sum_{\ell=1}^{N} U_{k\ell}^{(f)} h_{ch_\ell(u)} + b^{(f)}\right),$$

$$\forall k = 1, 2, \ldots, N \tag{17}$$

$$c_u = i_u \odot r_u + \sum_{\ell=1}^{N} f_{u\ell} \odot c_{ch_\ell(u)} \tag{18}$$

$$h_u = o_u \odot \tanh(c_u) \tag{19}$$

with the term x_u denoting unit input, $h_{ch_\ell(u)}$ and $c_{ch_\ell(u)}$ are, respectively, the hidden state and the memory cell state of the ℓ-th child, σ is the sigmoid function and $\odot$ is the element-wise product.

The introduction of separate parameter matrices for each child allows the model to learn more fine-grained conditioning on the states of a unit's children. Equation (17) shows a parametrization of the k-th child's forget gate f_{uk} that allows more flexible control of information propagation from child to parent and it can be also used to control the influence between siblings.

4   Experiments and Results

In this section, we empirically evaluate the TreeLSTM types surveyed in Sect. 3 in three different classes of tree transduction tasks. All tasks are based on real-world data and have been chosen as they allow to further compare with other approaches in literature. Model selection choices have been performed on hold-out validation data and final performance is assessed on further hold-out test data. A L_2 penalization term has been added to the loss function for the sake of model regularization, using a penalization weight fixed to $\lambda = 10^{-4}$. The only hyperparameter in model selection is the number of LSTM units, chosen in $\{100, 150, 200, 250, 300, 350, 400\}$, and we have used a standard Adam optimizer [16]. For the sake of compactness, we only report test-set results, obtained by averaging on 10 independent runs with random weights initializations (variance is not reported, if less than millesimal).

The first experiment assesses TreeLSTM models on structure-to-element transductions by means of two tree classification tasks coming from the INEX 2005 and INEX 2006 competitions [11] (task INEX20xx in the following). Both dataset are multiclass classification tasks based on trees that represent XML documents from different thematic classes. The datasets are provided with standard splits in training and test sets: for model selection purposes we held out 10% of the former to define the validation set (by stratification, to preserve the dataset class ratios in the folds). All TreeLSTM models used a LogSoft-Max layer in output and a Negative Log-Likelihood loss. Performance on both INEX20xx datasets are evaluated in terms of classification accuracy, reported in Tables 1 and 2 for the model-selected configurations of each TreeLSTM model. These results highlight how the BU approaches outperform TD approaches in structure-to-element transductions. This is due to the fact that a BU encoding results in node hidden activations that summarize information concerning the whole subtree rooted on the node. When compared to other state-of-the-art approaches, the TreeLSTMs show very competitive results in INEX 2005 (where the best accuracy is 97.15%, attained by PAK-PT [1]) while the Child-Sum TreeLSTM has the best performance in INEX 2006 (the runner up being the Jaccard kernel in [5], with an accuracy of 45.06%).

Table 1. Inex05 test results

Model	Accuracy %
TD-TreeLSTM	62.05
ChildSum TreeLSTM	82.85
N-ary TreeLSTM	**96.89**

Table 2. Inex06 test results

Model	Accuracy %
TD-TreeLSTM	25.07
ChildSum TreeLSTM	**46.12**
N-ary TreeLSTM	38.57

The second experiment is focused on the assessment of TreeLSTM models on structure-to-substructure transductions based on the CLwritten corpus [9]. This is a benchmark dataset for sentence compression techniques which is based on sentences from written sources whose ground truth shortened versions were created manually. Here, the annotators were asked to produce the smallest possible target compression by deleting extraneous words from the source, without changing the word order and the meaning of the sentences.

The original corpus provides sentences in sequential form (see [9] for more details). The corresponding tree representation has been obtained using the (constituency) *Stanford Parser* [17] on the original sentences. Node labels in the resulting trees are of two kind: leaves are labeled with vocabulary words, represented through *word embeddings* obtained using the *word2vec* [18]. Internal nodes are labelled with semantic categories that are, instead, represented using a one-hot encoding. For model selection and validation purposes we have split the corpus in 903 training trees, 63 validation samples and 882 test trees, along the lines of the experimental setup defined for the baseline models in [10,19].

The TreeLSTM output is generated by a layer of sigmoid neurons, where an activation smaller than 0.5 represents the fact that the corresponding input node is not present in the output tree, while the opposite means that the node (and the corresponding word) is preserved. The associated loss is, indeed, the Binary

Cross Entropy. Performance on the corpus has been assessed using two metrics assessing different aspects of compression quality:

Accuracy or Importance Factor: this metric measures how much of the important information is retained. Accuracy is evaluated using *Simple String Accuracy* (SSA) [8], which is based on the string edit distance between the compressed output generated by the model and the reference ground-truth compression;

Compression Rate: this metric measures how much of the original sentence remains after compression. Compression rate is defined as the length of the compressed sentence divided by the original length, so lower values indicate more compression;

During training, early stopping decisions are taken by considering an hybrid metric, which trades off accuracy and compression according to the following definition

$$t = \frac{\text{accuracy}^2}{\text{compression rate}}.$$

Table 3 reports the performance values for the CLWritten task. Here it is evident that the TD approach outperforms both BU approaches. As in the first experiment, this is due to the characteristics of the transduction task. In a parse tree the words occur only at leaves and a disambiguation of their interpretation can be performed only by considering their context, which can only come from their ancestors because they have no children by definition. Hence, it follows that a parent-to-children information flow is more relevant than a children-to-parent one to determine if a word, represented necessarily by a leaf node, has to be included or not in a summary. The best result in literature for the task is obtained by a LSTM applied to the original sequential representation of the data, achieving less than 70% in accuracy and about 82% in compression. This marks the clear advantage of using a TD tree transduction approach, as the corresponding TreeLSTM outperforms the sequential model both in accuracy and compression.

Table 3. CLWritten summarization results: the reference compression value for the gold-standard compressions is 70.41.

Model	Accuracy %	Compression %
TD-TreeLSTM	**73.58**	**72.37**
ChildSum TreeLSTM	63.17	84.01
N-ary TreeLSTM	63.41	91.46

The last experiment assesses the TreeLSTM models on an isomorph structure-to-structure transduction. The task requires to relabel an input tree into an isomorphic structure with changed labels. This is applied to the problem of inferring the semantic categories of a parse tree (grammar induction), given the structure of the parse tree and knowledge on the labels of the leaves (i.e. the words in the sentences). To this end, we have used the Treebank2 dataset focusing, in particular, on the Wall Street Journal (WSJ) subset of trees. Due to the fact that output labels can be assigned taking values from a discrete set, the output layers used to label internal nodes is of LogSoftMax type (with Negative Log-Likelihood as training loss). For our experiments, we used the standard WSJ partition, where sections 2–21 are used for training, section 22 for validation, and section 23 for testing. The performance metric for the task is label accuracy (computed as the proportion of correctly inferred semantic categories for each parse tree). Table 4 reports the resulting accuracies for the TreeLSTM models: we were not able to obtain results for the N-ary TreeLSTM configuration within reasonable computing time (2 weeks) for this task, due to its computational complexity. Nevertheless, results highlight how the BU approach based on the ChildSum encoding outperforms the TD one. As in the first experiment, relevant information on this task flows from the leaves to the root, i.e. from the sentence words, which are observable, to the internal nodes, where we want to infer the missing semantic categories.

Table 4. Treebank2 test results

Model	Test accuracy %
TD-TreeLSTM	49.56
ChildSum TreeLSTM	**95.23**

Comparison with the start-of-the-art can't be done due to the fact we used a restricted set of the output labels, so comparisons would be unfair.

5   Concluding Remarks

We surveyed different TreeLSTM architectures and parameterizations, providing an empirical assessment of their performance on learning different types of constrained tree transductions. Our analysis, not unsurprisingly, concluded that there is no single best configuration for all transduction problems. In general, one must choose the model which computes the structural encoding following a direction of elaboration akin to the information flow in the structure. This said, in the majority of cases, the BU approach proved more effective than the TD, thanks to the fact that in BU approaches, the state of a node summarizes the information of its rooted subtree. Moreover, we showed that TreeLSTMs have competitive performances with respect to the state-of-the-art models in all types of structured transductions.

Acknowledgment. This work has been supported by the Italian Ministry of Education, University, and Research (MIUR) under project SIR 2014 LIST-IT (grant no. RBSI14STDE).

References

1. Aiolli, F., Da San Martino, G., Sperduti, A.: Extending tree kernels with topological information (2011)
2. Alvarez-Melis, D., Jaakkola, T.: Tree structured decoding with doubly recurrent neural networks. In: ICLR 2017 (2017)
3. Bacciu, D.: Hidden tree Markov networks: Deep and wide learning for structured data. In: IEEE-SSCI 2017, pp. 1–8 (2017)
4. Bacciu, D., Micheli, A., Sperduti, A.: Compositional generative mapping for tree-structured data; part I: bottom-up probabilistic modeling of trees. IEEE Trans. Neural Netw. Learn. Syst. **23**(12), 1987–2002 (2012)
5. Bacciu, D., Micheli, A., Sperduti, A.: Generative kernels for tree-structured data. IEEE Trans. Neural Netw. Learn. Syst. **29**(10), 4932–4946 (2018)
6. Bacciu, D., Bruno, A.: Text summarization as tree transduction by top-down TreeLSTM. In: Proceedings of IEEE SSCI 2018 (2018)
7. Bacciu, D., Micheli, A., Sperduti, A.: An input-output hidden Markov model for tree transductions. Neurocomputing **112**(suppl. C), 34–46 (2013)
8. Bangalore, S., Rambow, O., Whittaker, S.: Evaluation metrics for generation. In: Proceedings of the First INLG, vol. 14, pp. 1–8 (2000)
9. Clarke, J.: Global inference for sentence compression : an integer linear programming approach. J. Artif. Intell. Res. **31**, 399–429 (2008)
10. Cohn, T., Lapata, M.: Sentence compression as tree transduction. J. Artif. Intell. Res. **34**(1), 637–674 (2009)
11. Denoyer, L., Gallinari, P.: Report on the XML mining track at INEX 2005 and INEX 2006: categorization and clustering of XML documents. In: SIGIR 2007, vol. 41, pp. 79–90 (2007)
12. Diligenti, M., Frasconi, P., Gori, M.: Hidden tree Markov models for document image classification. IEEE TPAMI **25**(4), 519–523 (2003)
13. Frasconi, P., Gori, M., Sperduti, A.: A general framework for adaptive processing of data structures. IEEE Trans. Neural Networks **9**(5), 768–786 (1998)
14. Gallicchio, C., Micheli, A.: Tree echo state networks. Neurocomputing **101**, 319–337 (2013)
15. Hochreiter, S., Schmidhuber, J.: Long short-term memory. Neural Comput. **9**(8), 1735–1780 (1997)
16. Kingma, D.P., Ba, J.: Adam: A method for stochastic optimization. In: ICLR 2015 (2015)
17. Klein, D., Manning, C.D.: Fast exact inference with a factored model for natural language parsing. In: Advances in Neural Information Processing Systems, vol. 15, pp. 3–10 (2003)
18. Mikolov, T., Chen, K., Corrado, G., Dean, J.: Efficient estimation of word representations in vector space. In: Proceedings of Workshop at ICLR (2013)
19. Sakti, S., Ilham, F., Neubig, G., Toda, T., Purwarianti, A., Nakamura, S.: Incremental sentence compression using LSTM recurrent networks. In: 2015 IEEE Workshop on ASRU, pp. 252–258 (2015)
20. Tai, K.S., Socher, R., Manning, C.D.: Improved semantic representations from tree-structured long short-term memory networks. In: Proceedings of the 53rd Annual Metting on ACL and the 7th IJCNLP, pp. 1556–1566 (2015)
21. Zhang, X., Lu, L., Lapata, M.: Tree recurrent neural networks with application to language modeling. In: Proceedings of NAACL-HLT 2016, pp. 310–320 (2016)

Approximating the Solution of Surface Wave Propagation Using Deep Neural Networks

Wilhelm E. Sorteberg, Stef Garasto$^{(\boxtimes)}$ (ID), Chris C. Cantwell (ID),
and Anil A. Bharath (ID)

Department of Bioengineering, Imperial College London, South Kensington Campus,
London SW7 2AZ, UK
stef.grs@gmail.com

Abstract. Partial differential equations formalise the understanding of
the behaviour of the physical world that humans acquire through experience and observation. Through their numerical solution, such equations are used to model and predict the evolution of dynamical systems.
However, such techniques require extensive computational resources and
assume the physics are prescribed *a priori*. Here, we propose a neural network capable of predicting the evolution of a specific physical
phenomenon: propagation of surface waves enclosed in a tank, which,
mathematically, can be described by the Saint-Venant equations. The
existence of reflections and interference makes this problem non-trivial.
Forecasting of future states (i.e. spatial patterns of rendered wave amplitude) is achieved from a relatively small set of initial observations. Using
a network to make approximate but rapid predictions would enable the
active, real-time control of physical systems, often required for engineering design. We used a deep neural network comprising of three main
blocks: an encoder, a propagator with three parallel Long Short-Term
Memory layers, and a decoder. Results on a novel, custom dataset of
simulated sequences produced by a numerical solver show reasonable
predictions for as long as 80 time steps into the future on a hold-out
dataset. Furthermore, we show that the network is capable of generali-
sing to two other initial conditions that are qualitatively different from
those seen at training time.

Keywords: Deep learning for physical systems ·
Recurrent neural networks · Representation learning

Supported by the Rosetrees trust.

W. E. Sorteberg and S. Garasto—These authors are joint first authors.

L. Oneto et al. (Eds.): INNSBDDL 2019, INNS 1, pp. 246–256, 2020.
https://doi.org/10.1007/978-3-030-16841-4_26

1   Introduction

Partial differential equations (PDEs) are used to model physical systems in many different fields, including aeronautical, mechanical, electrical and electromagnetic systems design. The algorithms traditionally used for their numerical solution can achieve high accuracy, but can be computationally very expensive. Solvers may also require extensive parameter tuning and the dynamics are governed by the PDEs and therefore prescribed *a priori*. There has been recent interest in using deep networks and machine learning to learn the behaviour of physical systems directly from data and to complement the role of traditional solvers [2,3, 9–11]. For example, using approximate, but computationally low-cost, solutions could reduce the time needed for design iteration or provide real-time active control in complex systems. To achieve this, it is critical for the deep neural network to have the ability to generalise to many different initial conditions, even those that differ substantially from the data used during training, and to sustain long-term predictions without resorting back to numerical simulation software.

The main contribution of the present paper is the investigation of the use of deep learning to perform state-evolution prediction on the rendered amplitudes of surface waves enclosed by solid wall boundary conditions, using a novel modification of an existing architecture. This physical system is described by the Saint-Venant equations and used in many important applications, such as avalanche [7] and urban flood modelling [8]. It is non-trivial to solve, given the complexity of the equations, and the presence of reflections and interference phenomena. To the best of our knowledge, state evolution of surface wave phenomena using deep networks has not been previously attempted. The network acts solely on visual input, without knowing the parameters of the operating environment. From snapshots of rendered wave-amplitude observed at 5 instants in time (at the constant but arbitrary sampling rate of 100 Hz), we forecast the wave patterns over the next 10 time steps (i.e. 100 ms) using an encoder-propagator-decoder architecture [2]. Then, the last 5 predictions are repeatedly used as new inputs to enable long term predictions. The test of prediction quality is ambitious: after training on predictions of the next 20 frames, we propagate for 80 time steps (i.e. 800 ms) into the future at test time. Finally, we test the generalisation ability of the network on qualitatively new initial conditions.

2   Related Work

Previous uses of deep neural networks to predict the evolution of states of a system can be found within a variety of fields. Examples include predictions of multi-body dynamics [10], trajectories [2], a robot end-effector's interaction with its surroundings [3,4], and the evolution of chaotic systems [9]. In fluid dynamics, two pivotal studies that showed how neural networks can help to speed-up numerical simulations for divergence-free fluids governed by the incompressible Navier-Stokes equations are those by Yang et al. [15] and Tompson et al. [12]. In particular, the latter focused on accelerating the pressure step of the numerical solver [12]. More recently, Wiewel et al. used an LSTM-based architecture (Long Short-Term Memory), alternated with numerical simulations for stability, to predict the evolution of the 3-dimensional pressure field of a fluid [14]. Moreover, Kim et al. presented a generative neural network that synthesises parameterisable, but not fully arbitrary, velocity fields for fluids from a low-dimensional parameters set [5]. It is worth noting that most of the neural networks applied to the Navier-Stokes equations are trained to predict either the pressure or the velocity fields, or both, rather than the motion of particles within the fluid. However, it is harder to measure these fields experimentally as opposed to particle tracking. Thus, other works have applied deep learning to model particle motion [11]. Finally, Long et al. [6] use a hybrid algorithm combining a numerical solver based on Cellular Neural Networks with a LSTM that forecasts the evolution of external forces applied to the fluid. To the best of our knowledge, we are the first to predict the evolution of a system described by the Saint-Venant equations.

A strategy that is shared across many neural networks for dynamical predictions consists of first allowing the network to build its own representation of the spatio-temporal data and then propagating this latent space in time [2,3,9]. This procedure is based on the assumption that each individual propagation in time is governed by the same function. However, the propagator is either a convolutional layer or a single LSTM. Furthermore, there are differences in the way the system of interest is given as input to the network. For instance, Sanchez-Gonzales et al. use multiple graph networks that contain information about static or dynamic properties of the system [10], while Finn et al. employ a guided procedure where the network is presented with sensory data on actuators' live position and forces [3], and Ehrhardt et al. use a neural network which solely relies on visual input [2]. We followed the latter strategy, with the aim of extending the present work from simulations to experimental data.

3 The Saint-Venant Equations

We aim to solve the 2-dimensional Saint-Venant equations in non-conservative form with no coriolis or viscous forces (a non-linear coupled system). They successfully model wave propagation phenomena and are used in many fields, including avalanche [7] and urban flood modelling [8]. They are given by:

$$h_t + ((H + h)u)_x + ((H + h)v)_y = 0$$
$$u_t + uu_x + vu_y + gh_x - \nu(u_{xx} + u_{yy}) = 0$$
$$v_t + uv_x + vv_y + gh_y - \nu(v_{xx} + v_{yy}) = 0$$

Here, H is the reference water height and h is the deviation from this reference, u and v are the velocities in the x and y direction, respectively, g is the gravitational acceleration, ν is the kinematic viscosity (set to $10^{-6}\,\mathrm{m^2s^{-1}}$), and u_x is the partial derivative of u with respect to x. We used solid wall boundary conditions along the whole perimeter of the environment. The initial condition was given by a droplet (a localised 2-dimensional Gaussian), and we varied its location, as well as the propagation speed, within the dataset. The simulations were run using TriFlow [1], with the generation of each simulated frame taking, on average, 0.44 s on a CPU. The final images (Fig. 1) are given by a rendering model with azimuth of 45° and altitude of 20°. We ran 3,000 simulations, each with 100 images at a resolution of 128 × 128 and a sampling frequency of 100 Hz (300,000 images in total). During training, random data augmentation was implemented by random flips in both the horizontal and the vertical direction. For each network trained, we randomly split the 3,000 simulations into training (70%), validation (15%), and test (15%) sets. To assess the network's generalisation performance onto more challenging scenarios, we also simulated two new initial conditions: two droplets with random locations and a linear wave front.

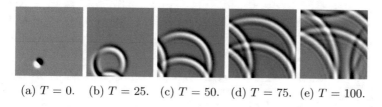

(a) $T = 0$.   (b) $T = 25$.   (c) $T = 50$.   (d) $T = 75$.   (e) $T = 100$.

Fig. 1. Example of surface wave propagation, generated using TriFlow [1]. T represents the number of time steps since the initial perturbation.

4   Long Term Predictor Network Architecture

Our neural network consists of an autoencoder structure with a propagator in its centre [2]. The data work-flow is as follows. Five states representing rendered wave patterns at 5 consecutive instants in time are given as the initial input. This is passed through the encoder, then repeatedly propagated in time by selecting one of three fully connected LSTMs (1000 units each), instead of just a single LSTM as applied in [2]. Finally, a decoder transforms the propagated latent vectors into 10 predicted future states. From these, the last 5 fields were reinserted again as input for further propagation. To handle this protocol seamlessly, we used three LSTMs in the propagator, with only one being active at a time, according to the type of prediction. One LSTM dealt with the initialisation of the network, one performed the self-propagation, and the last one was active during the reinsertion step. Preliminary results (in the Supplementary Material) appear to indicate that this modification is better than a single LSTM. Specifically, using only a single LSTM produced cyclical artefacts in time, with spikes appearing when predictions where reinserted as inputs. Such an effect was reduced when using the three LSTM structure. All three LSTMs shared the same hidden and cell states, ensuring complete information preservation (visualised in Fig. 2). The encoder has 4 convolutional layers (60, 120, 240 and 480 units, using a stride of 2, 7 × 7 and 3 × 3 kernels for the first and the following layers, respectively), and a fully connected layer to a latent space of 1000 nodes. The *tanh* non-linearity was used throughout. Regularisation was induced with dropout and batch normalisation layers. The architecture of the decoder mirrors that of the encoder, using deconvolutional layers. We used the Mean Square Error (MSE) between targets and predictions as the loss function, and ADAM as the optimization scheme. At training and test time, 20 and 80 frames were predicted, respectively, to test the network's extrapolation abilities. Using Pytorch and a NVIDIA GTX 1070, training each network for 50 epochs took approximately 7 h.

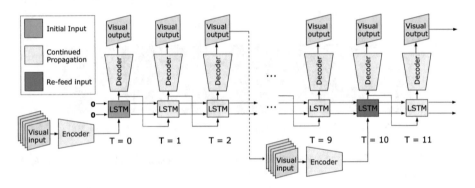

Fig. 2. Depiction of the internal structure and data flow of the prediction network with three LSTMs. Only one of the three LSTMs is used at any time.

5 Results

We trained 10 predictive networks, of which 6 converged successfully. The failure of the remaining 4 is likely due to the lack of sufficient data to train three large LSTMs (longer training did not help). By combining LSTM propagation and reinsertion of the resultant outputs, we predict 80 time steps into the future from an initial input of 5 images. We report not only the outcome of the most successful network, but also the average results across the entire ensemble of converged networks, to show the variability intrinsic in training for such a complex task.

5.1 Performance Assessment

A quantitative test of the network's predictive quality as a function of time steps from the last initial input is shown in Fig. 3. Specifically, we evaluated the image quality of the predictions using the Root Mean Squared Error (RMSE, Fig. 3a) and the Structural Similarity index (SSIM, Fig. 3b) between targets and predictions. The former aims to quantify the pixelwise spatio-temporal accuracy of the predictions, however it does not always correlate with perceptual similarity, which is instead captured by the latter [13]. Results are presented by the average (solid lines) and the standard deviation (shaded area) across all the networks that converged. Our best network achieved an average SSIM (RMSE) of 0.96 (0.04), 0.89 (0.09), 0.80 (0.11), 0.66 (0.16) and 0.56 (0.19) at time-steps 1, 20, 40, 60, and 80, respectively. One of the main challenges for the network is to extrapolate predictions to 80 frames after only being trained on 20: as expected, the SSIM decreases and the RMSE increases for longer term predictions. However, the lack of a sharp decrease in performance after $T = 20$ is promising. To contextualise the absolute accuracy values, we also plot RMSE and SSIM between the targets and the last frame of the input sequence ("last input" baseline). The purpose of such a measure is to estimate how well the network would perform without any physical knowledge, that is where its best guess at predicting the future is the last input it received. From $T = 60$, the SSIM suggests modest improvement with respect to this "last input" baseline, while visual inspection (Fig. 4) suggests that some predictions are closer to the target patterns than the last input frame.

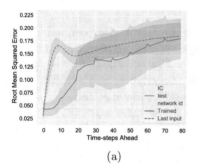 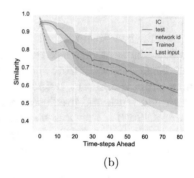

(a)   (b)

Fig. 3. RMSE (a) and SSIM (b) between targets and predictions (solid lines). For comparison, we show the same measures computed between the targets and the last input frames (dashed line). Lines and shadowed areas are the mean value and the standard deviation across the whole test dataset, respectively. Our best network achieved an average SSIM (RMSE) of 0.96 (0.04), 0.89 (0.09), 0.80 (0.11), 0.66 (0.16) and 0.56 (0.19) at time-steps 1, 20, 40, 60, and 80, respectively.

An example sequence of initial inputs, as well as target and predicted rendered wave patterns can be seen in Fig. 4. Five frames, taken at regular intervals, are shown to illustrate both short- and long-term prediction. Despite performance degradation over time, the network seems to sustain the relevant features of the propagating waves for long periods of time, almost until the end of the simulation for some features. It is worth remarking that the whole sequence is predicted by the neural network only, without any call to the numerical simulation software. Furthermore, to make correct predictions, the network not only

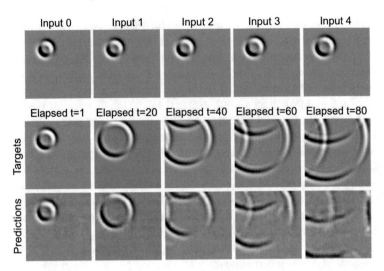

Fig. 4. Example initial inputs (top row), targets from numerical solver (middle row) and spatio-temporal pattern predicted by the trained network number 4 (bottom row): t represents the time-steps elapsed after the last initial input frame.

needs to infer how to update the current state, but is also required to perform an implicit visual estimate of relevant physical parameters, such as the viscosity and density, directly from the input data. Figure 5 shows another prediction example (from a different trained network), together with the intensity profile of the ground truth and predicted images, along a row situated near the boundary surface. Such a line was chosen to track the wave front during reflection, which is harder to model. It can be seen that the network is able to replicate the spatial pattern of the rendered wave amplitude (top row), albeit with lower accuracy around the reflection at the border (bottom row). Finally, we report that predicting and saving the entirety of the 80-frames time sequence took less than half a second. More example results can be found in the Supplementary Material (http://bit.ly/2P2Vf6g).

5.2   Generalisation Performance

On top of assessing the networks on left-out test data, we also tested them on two qualitatively different types of initial conditions: specifically, (i) two droplets and (ii) a linear propagating wave front. This allowed us to evaluate how well the trained networks captured the physics of the wave propagation. Results are shown in Figs. 6 and 7. The former shows the RMSE and the SSIM between targets and predictions for the two new types of initial conditions, averaged across all trained networks. For comparison, we included the same measures computed between the last original input and the current target frame. Figure 7 displays some example predictions. As shown by the error and accuracy curves corresponding to the "last input" baseline, the new scenarios are harder for the

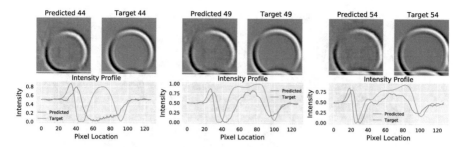

Fig. 5. Top: predicted two-dimensional spatial patterns from trained network number 3 vs targets from numerical solver. Bottom: spatial intensity profiles for predictions and targets along the horizontal line in the frames on top. This line approximately follows the wave front as it goes through reflection.

network to predict. The linear wave front is a particularly challenging test, since it involves both a new scenario and a new wave shape. Despite this, the results suggest that, although the predictions are not as accurate, the network can adapt to completely unseen data which varies significantly in spatial distribution from the training data. At the same time, the influence of the curved wave front – a prior created by the training data – is also clearly visible, especially for predictions further ahead in time. At the moment, we do not know exactly how strong this influence is.

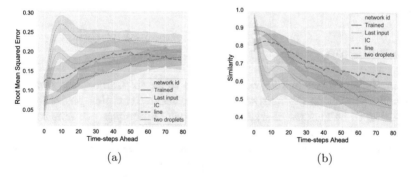

Fig. 6. RMSE (a) and SSIM (b) between targets and predictions (thick blue lines) for the two new initial conditions (each represented by a different dashed pattern). For comparison, we show the same measures computed between the targets and the last initial input frame (thin orange lines). Lines and shadowed areas are the mean values and the standard deviation, respectively, across all trained networks. For the linear wave front, our best network achieved an average SSIM (RMSE) of 0.83 (0.12), 0.79 (0.15), 0.69 (0.19), 0.65 (0.20) and 0.64 (0.18) at time-steps 1, 20, 40, 60, and 80, respectively. For the two droplets, the same values are 0.91 (0.07), 0.80 (0.11), 0.69 (0.14), 0.51 (0.19), 0.46 (0.20).

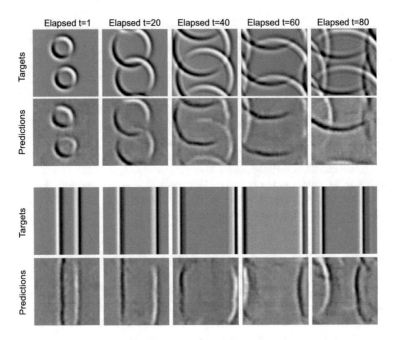

Fig. 7. Example targets from the numerical solver and spatio-temporal patterns predicted by the trained network nb. 3. The first and last two rows show predictions for the initial condition with two droplets and the linear wave front, respectively. The time-steps for each prediction corresponds to time-steps elapsed after the last initial input frame.

6   Discussion

The results presented in this work suggest that our trained network can perform state prediction of environments governed by complex physical laws. We believe this to be the first instance where such an analysis pertains to a system described by the Saint-Venant equations. Thus, no alternative state-of-the-art exists, though possible future extensions might involve comparison with optical flow methods. From 5 frames of visual input, the network is able to predict up to 80 frames into the future with reasonable accuracy, although with a lower quality than the numerical solver, especially for later time steps. Nevertheless, given that only the first 20 frames were back-propagated during training, the network exhibits a good extrapolation performance. The prediction of 80 frames using the neural network takes less than half a second, promoting the idea of using these techniques for real-time simulation and control. Given sufficiently general training, similar networks could be employed as an approximate simulation tool to perform initial parameter sweeps, or quickly iterate over designs. Results on the generalisation to two different unseen initial conditions are also promising. However, there are still some limitations. First, a minority of the networks we trained did not reach convergence and there was some variability in performance between networks. Secondly, some portions of the rendered wave amplitude pattern are lost over time. Finally, when tested on new scenarios, the influence of the prior learned during training is evident. For future work, we will seek to: (a) increase prediction accuracy (and thus the generalisation performance); (b) investigate architectures and benchmarks for generalisation (e.g. multiple sources, wave front shapes and boundary conditions); (c) stabilise the training by assessing different cost functions; (d) analyse the latent space representation learned by the network; (e) extend the approach to data from real-world wave propagation.

References

1. Cellier, N.: Locie/triflow: Triflow v0.5.0., May 2018. https://doi.org/10.5281/zenodo.1239703
2. Ehrhardt, S., Monszpart, A., Mitra, N.J., Vedaldi, A.: Learning a physical long-term predictor. ArXiv e-prints arXiv:1703.00247, March 2017
3. Finn, C., Goodfellow, I., Levine, S.: Unsupervised learning for physical interaction through video prediction. In: NeurIPS, pp. 64–72 (2016)
4. Guevara, T.L., Gutmann, M.U., Taylor, N.K., Ramamoorthy, S., Subr, K.: Adaptable pouring: teaching robots not to spill using fast but approximate fluid simulation. ArXiv e-prints arXiv:1708.01465v2, August 2017
5. Kim, B., Azevedo, V.C., Thuerey, N., Kim, T., Gross, M., Solenthaler, B.: Deep fluids: a generative network for parameterized fluid simulations. ArXiv e-prints arXiv:1806.02071, June 2018
6. Long, Y., She, X., Mukhopadhyay, S.: HybridNet: integrating model-based and data-driven learning to predict evolution of dynamical systems. ArXiv e-prints arXiv:1806.07439, June 2018
7. Mangeney-Castelnau, A., Vilotte, J.P., Bristeau, M.O., Perthame, B., Bouchut, F., Simeoni, C., Yerneni, S.: Numerical modeling of avalanches based on Saint Venant equations using a kinetic scheme. J. Geophys. Res. Solid Earth 108(B11) (2003)
8. Oezgen, I., Zhao, J., Liang, D., Hinkelmann, R.: Urban flood modeling using shallow water equations with depth-dependent anisotropic porosity. J. Hydrol. 541, 1165–1184 (2016)
9. Pathak, J., Lu, Z., Hunt, B.R., Girvan, M., Ott, E.: Using machine learning to replicate chaotic attractors and calculate lyapunov exponents from data. Chaos Interdisc. J. Nonlinear Sci. 27(12), 121102 (2017)
10. Sanchez-Gonzalez, A., Heess, N., Springenberg, J.T., Merel, J., Riedmiller, M., Hadsell, R., Battaglia, P.: Graph networks as learnable physics engines for inference and control. ArXiv e-prints arXiv:1806.01242, June 2018
11. Schenck, C., Fox, D.: SPNets: differentiable fluid dynamics for deep neural networks. In: Billard, A., Dragan, A., Peters, J., Morimoto, J. (eds.) Proceedings of The 2nd Conference on Robot Learning. Proceedings of Machine Learning Research, PMLR, 29–31 October 2018, vol. 87, pp. 317–335 (2018)
12. Tompson, J., Schlachter, K., Sprechmann, P., Perlin, K.: Accelerating Eulerian fluid simulation with convolutional networks. ArXiv e-prints arXiv:1607.03597v6, June 2017

13. Wang, Z., Bovik, A.C., Sheikh, H.R., Simoncelli, E.P.: Image quality assessment: from error visibility to structural similarity. IEEE Trans. Image Process. **13**(4), 600–612 (2004)
14. Wiewel, S., Becher, M., Thuerey, N.: Latent-space physics: towards learning the temporal evolution of fluid flow. ArXiv e-prints arXiv:1802.10123v2, June 2018
15. Yang, C., Yang, X., Xiao, X.: Data-driven projection method in fluid simulation. Comput. Animation Virtual Worlds **27**, 415–424 (2016)

A Semi-supervised Deep Rule-Based Approach for Remote Sensing Scene Classification

Xiaowei Gu[1,2] and Plamen P. Angelov[1,2,3(✉)]

[1] School of Computing and Communications, Lancaster University,
Lancaster LA1 4WA, UK
`p.angelov@lancaster.ac.uk`
[2] Lancaster Intelligent, Robotic and Autonomous Systems Centre (LIRA),
Lancaster University, Lancaster, UK
[3] Technical University, 1000 Sofia, Bulgaria

Abstract. This paper proposes a new approach that is based on the recently introduced semi-supervised deep rule-based classifier for remote sensing scene classification. The proposed approach employs a pre-trained deep convoluational neural network as the feature descriptor to extract high-level discriminative semantic features from the sub-regions of the remote sensing images. This approach is able to self-organize a set of prototype-based IF...THEN rules from few labeled training images through an efficient supervised initialization process, and continuously self-updates the rule base with the unlabeled images in an unsupervised, autonomous, transparent and human-interpretable manner. Highly accurate classification on the unlabeled images is performed at the end of the learning process. Numerical examples demonstrate that the proposed approach is a strong alternative to the state-of-the-art ones.

Keywords: Deep rule-based · Remote sensing scene classification ·
Semi-supervised learning

1 Introduction

The fast development of remote sensing techniques in the past decades results in a very large volume of high-resolution remote sensing images. These images are a data source of great importance for us to observe the ground surface of the Earth with detailed structures, and are instrumental for many real-world applications. Because of the drastically increasing number of remote sensing images and the very high complexity in terms of the semantic contents within these images, it is particularly difficult to label them manually. Therefore, the automatic classification of remote sensing scenes becomes a hotly studied problem.

The large majority of the existing approaches use fully supervised machine learning techniques for remote sensing scene classification [1–12]. Supervised

---

P. P. Angelov—Honorary Professor

© Springer Nature Switzerland AG 2020
L. Oneto et al. (Eds.): INNSBDDL 2019, INNS 1, pp. 257–266, 2020.
https://doi.org/10.1007/978-3-030-16841-4_27

approaches learn the classification model from the labeled images. In particular, as the state-of-the-art in the remote sensing domain, deep convoluational neural networks (DCNNs) require a large amount of labeled images for training [6,9,12]. In reality, labeled remote sensing images are scarce and expensive to obtain, but unlabeled images are plentiful. Supervised approaches, however, are unable to use the unlabeled images. On the other hand, semi-supervised machine learning approaches [13–16] consider both the labeled and unlabeled images for classification, and, thus, they are able to utilize information from the unlabeled images to a greater extent. Nonetheless, there are only very few published works applying semi-supervised techniques to remote sensing scene classification [17–19].

Semi-supervised deep rule-based (SSDRB) approach [20] is introduced as an semi-supervised learning extension of the deep rule-based (DRB) approach [11, 21] for image classification. In comparison with alternative approaches [13–19], the SSDRB classifier is able to perform semi-supervised learning in a self-organizing, fully transparent and human-interpretable manner thanks to its prototype-based nature. By exploiting the idea of "pseudo labeling", the SSDRB classifier can learn from unlabeled images offline and identify new classes actively without human expertise involvement. It further supports online learning on a sample-by-sample or chunk-by-chunk basis.

In this paper, a new SSDRB-based approach is proposed for remote sensing scene classification. The proposed approach extends the idea of our previous works [11,22] by conducting semi-supervised learning on the local regions of the remote sensing images. Using a DCNN-based feature descriptor [23] to extract high-level semantic features from the images locally, the SSDRB classifier is able to self-organize a set of prototype-based IF...THEN rules [24] from the local regions of, both, labeled and unlabeled images, and classify the unlabeled images based on the score of confidence obtained from each sub-region locally.

2   The Proposed Approach

The general architecture of the SSDRB classifier used in this paper is given by Fig. 1. As one can see from this figure, the SSDRB classifier is composed of the following components:

1. segmentation layer;
2. mean subtraction layer;
3. feature descriptor layer;
4. IF...THEN rule-based layer;
5. decision-maker,

and, their functionalities are described in the following five subsections, respectively.

2.1   Segmentation Layer

This layer crops each remote sensing image into five different sub-images [3], namely, ① upper-left corner; ② upper-right corner; ③ lower-left corner; ④ lower-right corner; ⑤ center with the required size (224×224 pixels) by the feature

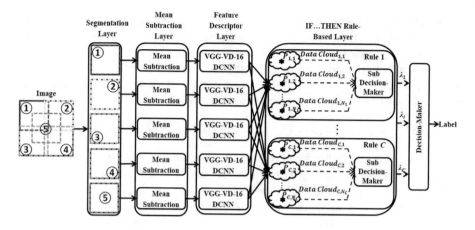

Fig. 1. The general architecture of SSDRB classifier.

descriptor [23]. In comparison with using the whole image for classification, segmenting the image in this way enables the feature descriptor to extract the semantic features locally, and, thus, improve the generalization ability of the proposed approach.

2.2 Mean Subtraction Layer

This layer pre-processes the segments by mean subtraction for feature extraction. This operation centers the three channels (R, G, B) of the segments around the zero mean, which helps the feature descriptor to perform faster since gradients act uniformly for each channel.

2.3 Feature Descriptor Layer

This layer is used for extracting high-level semantic feature vectors from the sub-regions of the images for the learning process. In this work, we use a pretrained VGG-VD-16 DCNN model [23] for feature extraction, and the 1×4096 dimensional activations from the first fully connected layer are used as the feature vector of each segment.

2.4 IF...THEN Rule-Based Layer

This layer is a massively parallel ensemble of zero-order IF...THEN rules of AnYa type, which is the "core" of the SSDRB classifier. The SSDRB classifier is, firstly, trained in a fully supervised manner with the labeled training set [20,21]. Assuming that there are the C known classes based on the labeled remote sensing images, C IF...THEN rules are initialized in parallel from the segments of labeled images of the corresponding classes (one rule per class).

After the supervised initialization process, the identified C IF...THEN rules are continuously updated based on the segments of the unlabeled images in a self-organizing manner. The SSDRB classifier is also able to perform active learning without human interference [20], but, we only consider the semi-supervised learning in this paper. Once the whole learning process is finished, one can obtain a set of IF...THEN rules in the following form [11, 20, 21] $(c = 1, 2, 3, ..., C)$:

$$IF\,(\mathbf{s} \sim \mathbf{P}_{c,1})\,OR\,(\mathbf{s} \sim \mathbf{P}_{c,2})\,OR...OR\,(\mathbf{s} \sim \mathbf{P}_{c,N_c})\,THEN\,(class\,c) \qquad (1)$$

where "$\sim$" denotes similarity, which can also be seen as a fuzzy degree of membership; $\mathbf{s}$ is a segment of a particular image, and $\boldsymbol{x}$ is the corresponding feature vector extracted by the DCNN model; $\mathbf{P}_{c,i}$ stands for the ith visual prototype of the cth class, and $\boldsymbol{p}_{c,i}$ is the corresponding feature vector; $i = 1, 2, ..., N_c$; N_c is the number of identified prototypes from the segments of images of the cth class.

2.5   Decision-Maker

During the validation process, for each segment of the unlabeled remote sensing images, the C IF...THEN rules will generate a vector of scores of confidence:

$$\boldsymbol{\lambda}(\mathbf{s}_{k,j}) = [\lambda_1(\mathbf{s}_{k,j}), \lambda_2(\mathbf{s}_{k,j}), ..., \lambda_C(\mathbf{s}_{k,j})] \qquad (2)$$

where $\mathbf{s}_{k,j}$ denotes the $j^{th}(j = 1, 2, ..., K_o; K_o = 5)$ segment of the kth unlabeled image, $\mathbf{I}_k$; $\lambda_c(\mathbf{s}_{k,j})$ stands for the score of confidence given by the cth IF...THEN rule using the following equation:

$$\lambda_c(\mathbf{s}_{k,j}) = \max_{i=1,2,...,N_c} \left(e^{-||\boldsymbol{x}_{k,j} - \boldsymbol{p}_{c,i}||^2}\right) \qquad (3)$$

where $\boldsymbol{x}_{k,j}$ corresponds to the feature vector of the visual prototype, $\mathbf{P}_{c,i}$.

The label of the remote sensing image is decided based on the vectors of scores of confidence calculated from all the segments:

$$Label(\mathbf{I}_k) \leftarrow class\,c^*;\quad c^* = \underset{c=1,2,...,C}{\operatorname{argmax}} \left(\frac{1}{K_o} \sum_{j=1}^{K_o} \lambda_c(\mathbf{s}_{k,j})\right) \qquad (4)$$

Due to the limited space of this paper, we skip the details of the supervised and semi-supervised learning processes of the SSDRB classifier. The detailed description of the algorithmic procedure can be found in [20] and Chapter 9 of the book [25]. One can download the open source software implemented on Matlab platform from the following web link with a detailed instruction provided in the Appendix C of the book [25]:

https://uk.mathworks.com/matlabcentral/fileexchange/69012-empirical-approach-to-machine-learning-software-package?s_tid=prof_contriblnk.

3   Numerical Examples

In this section, numerical examples based on benchmark problems in the remote sensing domain are presented to demonstrate the performance of the proposed approach.

3.1   Experimental Setup

In this paper, we consider the following widely used benchmark datasets:

1. Singapore dataset;
2. WHU-RS dataset;
3. RSSCN7 dataset.

Singapore Dataset [26] is a recently introduced benchmark image set for remote sensing scene classification. This image set consists of 1086 images of size 256×256 pixels with nine categories: *(i)* airplane, *(ii)* forest, *(iii)* harbor, *(iv)* industry, *(v)* meadow, *(vi)* overpass, *(vii)* residential, *(viii)* river, and *(ix)* runway. This image set is imbalanced, and the number of images in each class varies from 42 to 179. Examples of images of the nine classes are given in Fig. 2(a).

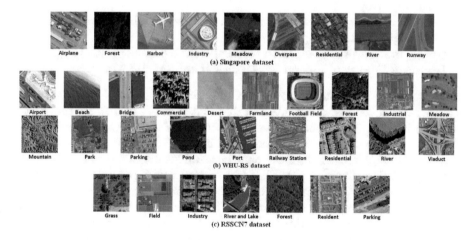

Fig. 2. Example images of the benchmark datasets.

WHU-RS Dataset [27] is a popular benchmark problem collected from Google Earth (Google Inc.). This image set consists of 950 images with size 600×600 pixels uniformly distributed in 19 scene classes, which include: *(i)* airport, *(ii)* beach, *(iii)* bridge, *(iv)* commercial, *(v)* desert, *(vi)* farmland, *(vii)* football field, *(viii)* forest, *(ix)* industrial, *(x)* meadow, *(xi)* mountain, *(xii)* park, *(xiii)* parking lot, *(xiv)* pond, *(xv)* port, *(xvi)* railway, *(xvii)* residential, *(xviii)* river, and *(xix)* viaduct. There are very high variations within the images in terms of illumination, scale, resolution, etc., which makes this a difficult classification problem. Examples of images of this problem are given in Fig. 2(b).

RSSCN7 Dataset [28] is a benchmark problem also collected from Google Earth. This image set is composed of images of seven different classes, which include *(i)* grassland, *(ii)* forest, *(iii)* farmland, *(iv)* parking lot, *(v)* residential region, *(vi)* industrial region, *(vii)* river and lake. Each class has 400 images of size 400×400 pixels. This dataset is a very challenging one due to the fact that the images of each class are sampled on four different scales (100 images per scale) with different imaging angles. Examples of images of this problem are given in Fig. 2(c).

In this paper, we use the offline semi-supervised learning strategy for the SSDRB classifier, and the user-controlled parameter required, namely, Ω_1, is set to $\Omega_1 = 1.2$. The reported experimental results are the average after five times Monte Carlo experiments. The images of the WHU-RS dataset have been re-scaled into the same size as the images of the RSSCN7 dataset, namely, 400×400 pixels, to avoid the loss of information during the segmentation operation.

We further involve the following seven popular approaches for comparison:

1. Deep rule-based classifier (DRB) [21];
2. Support vector machine classifier with linear kernel function (SVM) [29];
3. k-nearest neighbor classifier (kNN) [30];
4. AnchorGraphReg-based semi-supervised classifier with kernel weights (AnchorK) [16];
5. AnchorGraphReg-based semi-supervised classifier with Local Anchor Embedding weights (AnchorL) [16];
6. Greedy gradient Max-Cut based semi-supervised classifier (GGMC) [14];
7. Laplacian SVM semi-supervised classifier (LapSVM) [15, 17].

In the following numerical examples, the value of k for kNN is set to $k = 5$. The user-controlled parameter of AnchorK and AnchorL, s (number of the closest anchors) is set to $s = 3$, and the iteration number of Local Anchor Embedding (LAE) for AnchorL, is set to 10 as suggested in [16]. GGMC uses the KNN graph with $k = 5$. LapSVM employs the one-versus-all strategy, and it uses a radial basis function kernel with $\sigma = 10$. The two user-controlled parameters γ_I and γ_A are set to 1 and 10^{-6}, respectively; the number of neighbour, k, for computing the graph Laplacian is set to 15 as suggested in [15]. For the comparative algorithms, we follow the common practice by averaging the 1×4096 dimensional feature vectors of the five sub-regions to generate an overall image representation as the input feature vector [3].

3.2   Experimental Results on the Singapore Dataset

Firstly, we randomly select $L = 5, 8, 10, 12, 15$ images from each class of the Singapore dataset as the labeled training images, and use the remaining as the unlabeled ones to test the performance of the proposed approach. The classification accuracy on the unlabeled images obtained by the proposed approach is given in Table 1. We also report the results obtained by the seven comparative approaches in the same table.

Secondly, we follow the commonly used experimental protocol [26] by randomly selecting out 20% of images from each class as the labeled training images and using the remaining as the unlabeled ones, and report the classification performance of the eight approaches in Table 2. The state-of-the-art results reported by other approaches are also provided for a better comparison.

Table 1. Classification performance comparison on the Singapore dataset with different number of labeled training images.

Algorithm	L				
	5	8	10	12	15
The proposed	**0.9634**	**0.9661**	**0.9747**	**0.9746**	**0.9781**
DRB	0.9353	0.9339	0.9558	0.9568	0.9574
SVM	0.9155	0.9533	0.9606	0.9608	0.9689
kNN	0.9375	0.9381	0.9548	0.9576	0.9592
AnchorK	0.9591	0.9469	0.9538	0.9524	0.9546
AnchorL	0.9501	0.9314	0.9438	0.9489	0.9487
GGMC	0.8999	0.8828	0.9422	0.9286	0.9447
LapSVM	0.7858	0.9609	0.9679	0.9658	0.9588

Table 2. Classification performance comparison on the Singapore dataset under the commonly used experimental protocol.

Algorithm	Accuracy	Algorithm	Accuracy	Algorithm	Accuracy
The proposed	**0.9793**	DRB	0.9573	SVM	0.9643
kNN	0.9768	AnchorK	0.9678	AnchorL	0.9551
GGMC	0.9578	LapSVM	0.9293	TLFP [26]	0.9094
BoVW [31]	0.8741	VLAD [32]	0.8878	SPM [33]	0.8285

3.3   Experimental Results on the WHU-RS Dataset

In the following numerical example, we follow the commonly used experimental protocol [7] by randomly selecting out 40% of images from each class as the labeled training images and using the remaining as the unlabeled ones. The accuracy of the classification results obtained by the eight approaches are tabulated in Table 3, where the selected state-of-the-art results by other approaches are also reported.

Table 3. Classification performance comparison on the WHU-RS dataset under the commonly used experimental protocol.

Algorithm	Accuracy	Algorithm	Accuracy	Algorithm	Accuracy
The proposed	**0.9387**	DRB	0.9228	SVM	0.9291
kNN	0.9235	AnchorK	0.9013	AnchorL	0.9126
GGMC	0.9073	LapSVM	0.9305	BoVW (SIFT) [7]	0.7526
VLAD (SIFT) [7]	0.7637	SPM (CH) [7]	0.5595	CaffeNet [7]	0.9511
VGG-VD-16 [7]	0.9544	GoogLeNet [7]	0.9312	salM^{3}LBP-CLM [8]	0.9535

3.4   Experimental Results on the RSSCN7 Dataset

As a common practice [7], we randomly pick out 20% of images from each class, namely, 80 images per class as the labeled training images and use the remaining as the unlabeled ones, and conduct the experiments with the eight approaches. The experimental results are tabulated in Table 4. Similarly, the selected state-of-the-art results produced by other approaches are also reported in the same table.

Table 4. Classification performance comparison on the RSSCN7 dataset under the commonly used experimental protocol.

Algorithm	Accuracy	Algorithm	Accuracy	Algorithm	Accuracy
The proposed	**0.8670**	DRB	0.8422	SVM	0.8529
kNN	0.8532	AnchorK	0.8413	AnchorL	0.8465
GGMC	0.7705	LapSVM	0.8398	BoVW (SIFT) [7]	0.7633
VLAD (SIFT) [7]	0.7727	SPM (CH) [7]	0.6862	CaffeNet [7]	0.8557
VGG-VD-16 [7]	0.8398	GoogLeNet [7]	0.8255	DBNFS [28]	0.7119

3.5   Discussion

Tables 1, 2, 3 and 4 demonstrate that the proposed SSDRB approach consistently outperforms the seven comparative approaches. The accuracy of the classification results it produced is above or at least at the same level with the state-of-the-art approaches. In addition, our previous works [20] also show that the SSDRB classifier can achieve very high performance even with only one single labeled image per class, and is able to learn new classes actively without human expertise involvement.

Therefore, one can conclude that the proposed approach is a strong alliterative to the state-of-the-art ones for remote sensing scene classification problems.

4   Conclusion

In this paper, a semi-supervised deep rule-based (SSDRB) approach is proposed for remote sensing scene classification. By extracting the high-level features from the sub-regions of the images, the proposed approach is capable of capturing more discriminative semantic features from the images locally. After the supervised initialization process, the proposed approach self-organizes its system structure and self-updates its meta-parameters from the unlabeled images in a fully autonomous, unsupervised manner. Numerical examples on benchmark datasets demonstrate that the proposed approach can produce state-of-the-art classification results on the unlabeled remote sensing images surpassing the published alternatives.

References

1. Sheng, G., Yang, W., Xu, T., et al.: High-resolution satellite scene classification using a sparse coding based multiple feature combination. Int. J. Remote Sens. **33**(8), 2395–2412 (2012)
2. Cheriyadat, A.M.: Unsupervised feature learning for aerial scene classification. IEEE Trans. Geosci. Remote Sens. **52**(1), 439–451 (2014)
3. Hu, F., Xia, G.-S., Hu, J., et al.: Transferring deep convolutional neural networks for the scene classification of high-resolution remote sensing imagery. Remote Sens. **7**(11), 14680–14707 (2015)
4. Chen, S., Tian, Y.: Pyramid of spatial relatons for scene-level land use classification. IEEE Trans. Geosci. Remote Sens. **53**(4), 1947–1957 (2015)
5. Zhang, L., Zhang, L., Kumar, V.: Deep learning for remote sensing data. IEEE Geosci. Remote Sens. Mag. **4**(2), 22–40 (2016)
6. Scott, G.-J., England, M.R., Starms, W.A., et al.: Training deep convolutional neural networks for land-cover classification of high-resolution imagery. IEEE Geosci. Remote Sens. Lett. **14**(4), 549–553 (2017)
7. Xia, G.-S., Hu, J., Hu, F., et al.: AID: a benchmark dataset for performance evaluation of aerial scene classification. IEEE Trans. Geosci. Remote Sens. **55**(7), 3965–3981 (2017)
8. Bian, X., Chen, C., Tian, L., et al.: Fusing local and global features for high-resolution scene classification. IEEE J. Sel. Top. Appl. Earth Obs. Remote Sens. **10**(6), 2889–2901 (2017)
9. Li, Y., Zhang, H., Shen, Q.: Spectral-spatial classification of hyperspectral imagery with 3D convolutional neural network. Remote Sens. **9**(1), 67 (2017)
10. Liu, R., Bian, X., Sheng, Y.: Remote sensing image scene classification via multifeature fusion. In: Chinese Control and Decision Conference, pp. 3495–3500 (2018)
11. Gu, X., Angelov, P., Zhang, C., et al.: A massively parallel deep rule-based ensemble classifier for remote sensing scenes. IEEE Geosci. Remote Sens. Lett. **32**(11), 345–349 (2018)
12. Zhang, M., Li, W., Du, Q., et al.: Feature extraction for classification of hyperspectral and LiDAR data using patch-to-patch CNN. IEEE Trans. Cybern. (2018). https://doi.org/10.1109/TCYB.2018.2864670
13. Xiang, S., Nie, F., Zhang, C.: Semi-supervised classification via local spline regression. IEEE Trans. Pattern Anal. Mach. Intell. **15**(3), 2039–2053 (2010)

14. Wang, J., Jebara, T., Chang, S.-F.: Semi-supervised learning using greedy Max-Cut. J. Mach. Learn. Res. **14**, 771–800 (2013)
15. Belkin, M., Niyogi, P., Sindhwani, V.: Manifold regularization: a geometric framework for learning from labeled and unlabeled examples. J. Mach. Learn. Res. **7**, 2399–2434 (2006)
16. Liu, W., He, J., Chang, S.-F.: Large graph construction for scalable semi-supervised learning. In: International Conference on Machine Learning, pp. 679–689 (2010)
17. Gómez-Chova, L., Camps-Valls, G., Munoz-Mari, J., et al.: Semisupervised image classification with Laplacian support vector machines. IEEE Geosci. Remote Sens. Lett. **5**(3), 336–340 (2008)
18. Bruzzone, L., Chi, M., Marconcini, M.: A novel transductive SVM for semisupervised classification of remote-sensing images. IEEE Trans. Geosci. Remote Sens. **44**(11), 3363–3373 (2006)
19. Huo, L., Zhao, L., Tang, P.,: Semi-supervised deep rule-based approach for image classification. In: Workshop on Hyperspectral Image and Signal Processing: Evolution in Remote Sensing (WHISPERS), pp. 1–4 (2014)
20. Gu, X., Angelov, P.: Semi-supervised deep rule-based approach for image classification. Appl. Soft Comput. **68**, 53–68 (2018)
21. Angelov, P., Gu, X.: Deep rule-based classifier with human-level performance and characteristics. Inf. Sci. (Ny) **463–464**, 196–213 (2018)
22. Gu, X., Angelov, P.: A deep rule-based approach for satellite scene image analysis. In: IEEE International Conference on Systems, Man and Cybernetics (2018)
23. Simonyan, K., Zisserman, A.: Very deep convolutional networks for large-scale image recognition. In: International Conference on Learning Representations, pp. 1–14 (2015)
24. Angelov, P., Yager, P.: A new type of simplified fuzzy rule-based system. Int. J. Gen. Syst. **41**(2), 163–185 (2011)
25. Angelov, P., Gu, X.: Empirical Approach to Machine Learning. Springer International Publishing, Cham (2019)
26. Gan, J., Li, Q., Zhang, Z., et al.: Two-level feature representation for aerial scene classification. IEEE Geosci. Remote Sens. Lett. **13**(11), 1626–1639 (2016)
27. Xia, G., Yang, W., Delon. J., et al.: Structural high-resolution satellite image indexing. In: ISPRS, TC VII Symposium Part A: 100 Years ISPRS–Advancing Remote Sensing Science, pp. 298–303 (2010)
28. Zou, Q., Ni, L., Zhang, T., et al.: Deep learning based feature selection for remote sensing scene classification. IEEE Geosci. Remote Sens. Lett. **12**(11), 2321–2325 (2015)
29. Cristianin, N., Shawe-Taylo, J.: An Introduction to Support Vector Machines and Other Kernel-Based Learning Methods. Cambridge University Press, Cambridge (2000)
30. Cunningham, P., Delany, S.-J.: K-nearest neighbour classifiers. Mult. Classif. Syst. **34**(11), 1–17 (2017)
31. Yang, Y., Newsam, S.: Bag-of-visual-words and spatial extensions for land-use classification. In: International Conference on Advances in Geographic Information Systems, pp. 270–279 (2010)
32. Jégou, H., Douze, M., Schmid, C.: Aggregating local descriptors into a compact representation. In: IEEE Conference on Computer Vision and Pattern Recognition. pp. 3304–3311 (2010)
33. Lazebnik, S., Schmid, C., Ponce, J.,: Beyond bags of features: spatial pyramid matching for recognizing natural scene categories. In: IEEE Computer Society Conference on Computer Vision and Pattern Recognition, pp. 2169–2178 (2006)

Comparing the Estimations of Value-at-Risk Using Artificial Network and Other Methods for Business Sectors

Siu Cheung[1(✉)], Ziqi Chen[2], and Yanli Li[1]

[1] Department of Statistics and Actuarial Science, Faculty of Mathematics,
University of Waterloo, Waterloo, Canada
jameszhangxiao5@hotmail.com
[2] Department of Statistical and Actuarial Science, Faculty of Science,
Western University, London, Canada

Abstract. Previous studies on estimating Value-at-Risk mostly focus on the market index or specific portfolio, while few has been done on specific business sectors. In this paper, we compare the Value-at-Risk estimations from different methods, namely Artificial Neural Network model, extreme value theory-based method, and Monte Carlo simulation. We show that while non-parametric approaches such as Monte Carlo simulation performs better marginally, Artificial Neural Network has great potential for future development.

Keywords: Artificial Neural Network · Value-at-Risk · Risk management

1 Introduction

In financial market, we are always exposed to all kinds of risks, such as the threats from project failure, credit risk, or financial crisis among many others. It is always desirable to measure and manage the uncertainty we are facing, so that we could better minimize and hedge the potential risks. Recent researches point out that risk management is considered by financial executives as one of their most important goals [1].

Value-at-Risk (VaR) was first introduced and extensively developed as a risk management tool by J.P. Morgan [2], and now is arguably one of the most frequently used risk measures by both practitioners and researchers. VaR aims to estimate the quantile loss of a portfolio, under normal market conditions, in a given time period at a given confidence level. For a random variable X, the Value-at-Risk (VaR) of X with the confidence level $\alpha \in (0,1)$ is the smallest number y, such that the probability $Y := -X$ does not exceed y is at least $(1 - \alpha)$. Statistically, $VaR_\alpha(X)$ is the $(1 - \alpha)$-quantile of Y, or

$$VaR_\alpha(X) = \inf\{x \in \mathbb{R} : F_X(x) > \alpha\} = F_Y^{-1}(1 - \alpha).$$

© Springer Nature Switzerland AG 2020
L. Oneto et al. (Eds.): INNSBDDL 2019, INNS 1, pp. 267–275, 2020.
https://doi.org/10.1007/978-3-030-16841-4_28

Various methods for estimating VaR have been developed in previous studies. Most of these methods fall into two categories: nonparametric method such as the historical simulation, and the parametric method such as those based on extreme value theory. Historical simulation is based on the idea that the historical data can be used to forecast the future return without making assumption of any statistical parameters or distributions. However, problems arise when there are not enough historical observations. A natural extension of the historical simulation is the Monte Carlo simulation. Instead of merely collecting the historical return to estimate the empirical distribution of returns, Monte Carlo simulation adopts a random number generator to simulate the past information and then to generate the empirical distribution. Among the model-based approaches, GARCH filter proposed by Bollerslev in 1986 [3] is one of the most popular one. Inspired by the stylized facts of the financial market, GARCH model is widely used for modelling the market return, so we can further use it to estimate the VaR in practice. In recent years, researchers in risk management enjoy the development of machine learning techniques, and a lot of efforts have been put on applying the machine learning techniques in estimating VaR. Given that the financial market is increasingly complicated, Artificial Neural Networks (ANNs) stands out as one of the popular choices [5]. For example, ANN method was implemented by Locarek-junge and Prinzler when estimating the VaR of a US-Dollar portfolio [6].

Notice that previous studies overlooked a critical aspect. Most existing researches either only focus on estimating VaR of a specific portfolio or the index of the whole market. Few attentions have been paid to individual business sectors. The whole market can be separated into different business sectors, each of which includes relatively similar companies. The behavior of business sectors can be very different from each other. For example, it is reasonable to believe that the risk from IT companies should be very different from the risks from medical companies. Therefore, one major contribution of this paper is to investigate the VaR estimation by business sectors. We also compare the performance of three commonly used VaR estimation methods, namely Monte Carlo simulation, EVT-based method, and ANN.

The remainder of this paper is organized as follows: Sect. 2 discusses the methodology we use. Section 3 presents and compares the results of VaR estimation for each of the 7 business sectors in S&P500 stock index. Section 4 is the conclusion.

2   Methodology

In this section, we discuss three different methods to estimate the Value-at-Risk, namely Monte Carlo simulation, EVT (Extreme Value Theory)-based method, and artificial neural network (ANN).

2.1 Monte Carlo Simulation

In real life, most financial time series are clearly non-stationary, which means there exits dependence, mean drifting or change points among series. A GARCH (p, q) (generalized auto-regressive conditional heteroskedasticity) model introduced by Bollerslev [3], is generally used to capture this long-range and non-linear dependence between the log returns of financial data. This model allows today's variance to be condition on previous lagged values.

Generally, a GARCH (1,1) model is sufficient to capture above features of financial series [4]. Under GARCH (1,1) model we assume that

$$X_t = \mu_t + \sigma_t Z_t, \tag{1}$$

where X_t is the log return at time t, μ_t is the mean of the returns, σ_t is the volatility at time t, and Z_t is iid innovation terms with zero mean and standard deviation of 1. From (1) we know that to estimate VaR at time $t+1$, $VaR(X_{t+1})$, we need to estimate μ_{t+1}, σ_{t+1}, and quantile of Z separately. While μ_{t+1} and σ_{t+1} can be obtained when fitting the GARCH model, the challenge lies in estimating the quantile of Z. Here we implement the Monte Carlo simulation.

First, we collect the 5-year close price of a certain business sector (from 16-Oct-13 to 29-Dec-17) and compute the corresponding log return. Then we set a rolling window of which size is constant and equal to 500 days length. This rolling window will move along the data by adding a newest observation and dropping the oldest one. This loop will stop until the last observation is reached. We apply the AR (1)-GARCH (1,1) filter to each of the rolling window, and obtained the predictions of the μ and σ for the following day. We denote them as $\hat{\mu}$ and $\hat{\sigma}$.

The next step is to estimate the distribution of Z by the empirical distribution of the backed-out residuals. We use Monte Carlo simulation to estimate this empirical distribution. We generate 10000 repeated samples from the estimated residuals estimated by the AR (1)-GARCH (1,1) filter. We denote the αth quantile of the empirical distribution of residuals as $\widehat{VaR}(Z)$.

Last, we combined the estimated quantile of Z with the predicted $\hat{\mu}$ and $\hat{\sigma}$ to obtain the estimated VaR of X_{t+1}:

$$\widehat{VaR}(X_{t+1}) = \hat{\mu} + \hat{\sigma} \times \widehat{VaR}(Z).$$

2.2 EVT Based Method

One problem with the Monte Carlo simulation is that it only includes past information. We cannot model potential risks that have not yet happened. A common approach is to combine the GARCH filter with extreme value theory. Following the similar approach as described in Monte Carlo simulation, we fit the log returns of stocks to AR (1)-GARCH (1,1) filter and obtained the estimated innovations $Z_t's$. The mean $\hat{\mu}$ and standard deviation $\hat{\sigma}$ of next day's return X_{t+1} is forecasted similarly as above. However, instead of estimating the quantile of Z by simulation, we fit the threshold exceedances to some specialized distributions. Suppose X is a random variable with distribution function F with upper end point x_F, and u is some threshold, then the conditional excess distribution is defined as

$$F_u = P(X - u \le x | X > u)$$

Then by the Pickands-Balkema-de Haan theorem [7] there exists a function $\beta(u)$ such that

$$\lim_{u \to x_F} \max_{0 \le x \le x_F - u} |F_u(x) - G_{\xi, \beta(u)}(x)| = 0.$$

In other words, the large values of these residuals follow some special heavy-tailed quantile distribution. One of the commonly used quantile distribution is Generalized Pareto Distribution [8], which has the following distribution function:

$$G_{\xi, \beta(u)}(x) = \begin{cases} 1 - \left(1 + \frac{\xi x}{\beta}\right) & \text{if } \xi \ne 0 \\ 1 - exp\left(\frac{-x}{\beta}\right) & \text{if } \xi = 0 \end{cases} \qquad (2)$$

where we select the threshold u as the 90 percent quantile of Z. We obtain the estimate the quantile from this Generalized Pareto Distribution, and denote it as $\widehat{VaR}^*(Z)$. Then the VaR of X_{t+1} can be estimated as

$$\widehat{VaR}^*(X_{t+1}) = \hat{\mu} + \hat{\sigma} \times \widehat{VaR}^*(Z)$$

2.3   Artificial Neural Network

Now we discuss using ANN to estimate the VaR. An Artificial New Network is a common mathematical model used in the machine learning for information

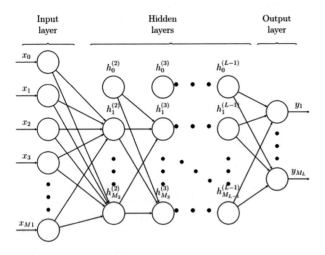

Fig. 1. Example of a common neural network architecture. [9]

processing. Neural network consists three sets of layers: input layers, output layers, and hidden layers that establish connections between input layers and output layers. A multi-layer neural network is illustrated in Fig. 1.

White [10] shows that multi-layer neural network are general-purpose, flexible, non-linear models that can approximate virtually any function to any desired degree of accuracy if given enough hidden neurons and enough data. To estimate VaR using ANN method, Chen et al. [6] suggest the following model:

$$r_t = \mu_t + \epsilon_t$$
$$\epsilon_t = z_t \sigma_t$$
$$\mu_t = \mu(\Omega_{t-1}; \theta)$$
$$\sigma_t^2 = \sigma^2(\Omega_{t-1}; \theta)$$

where r_t represents the log return at time t, μ_t is the conditional expected return, σ_t represents the conditional volatility of the portfolio change or innovation ε_t, Ω_{t-1} is the information set available at time t1, θ is a finite-dimensional vector of true parameter values, and z_t is an iid random variable satisfying $E(z_t|\Omega_{t-1}; \theta) = 0$ and $Var(z_t|\Omega_{t-1}; \theta) = 1$ [6].

The VaR at time t with alpha confidence level can then be estimated as:

$$VaR_\alpha^t = \mu_t + q_\alpha \times \sigma_t,$$

Where q_α is the αth quantile of Z_t. The expected return μ_t and the conditional volatility σ_t are estimated using ANN models built based on the following:

Each business sector's log returns are first calculated using the close price. Then the log returns are scaled using min-max normalization method to fit into neural network. The scaling of data is essential because otherwise a variable may have large impact on the prediction variable only because of its scale. Thus, using unscaled inputs may lead to meaningless results.

The scaled data is then split into a training set and a testing set. The training dataset is used to find the parameters of the best performed ANN model, and then the selected model is used to forecast VaR on the test set. In this study, training set consists the first 500 data records, and testing set consists the rest of the data. The rolling window size of ANN model is fixed as 100 days. We use rolling window method to keep the sample size constant as 100, by adding the 101th observation and dropping the 1st observation. This procedure is repeated until the last observation of the entire sample.

The ANN models in this study are one-step-ahead forecast model beginning from the 501th observations. Daily data from the 1st observation to 500th observation is used to estimate the parameters of ANN model. The best neurons combination is determined by comparing MSE across all testing combinations. In this study, we test a couple of layer-neutron combinations and find the fully connected ANN model with 7 hidden layer and 7 neutrons within each layer to perform the best.

After the optimal ANN models for fitting mean and mean-squared are determined, we start to produce VaR forecast on the testing set. The model based VaR for the testing dataset is then calculated based on the VaR formula above.

To further utilize the flexibility of an ANN model, we test an extended model specifically for Financial sector by adding three variables that are strongly related to the Financial sector: gold price, interest, and inflation. VaR is then estimated based on the same procedures to determine if the ANN model is improved.

3   Results

Figure 2 shows the estimated VaR in Utility sector of S&P 500 Market Index with Monte Carlo simulation, EVT-based method, and ANN model. As shown in the graph, the estimations from the three methods are very close to each other. In order to test and compare the performance of each VaR model, we need a better method to quantify their differences. Ehm et al. [11] suggest a piecewise linear scoring function which is consistent in ranking different estimations for the quantiles. The formula is as follows:

$$S_\alpha(x,y) = \begin{cases} \alpha|x - y| & x \leq y \\ (1 - \alpha)|x - y| & x \geq y \end{cases} \tag{3}$$

The intuition behind this score function is that it is always better to overestimate VaR rather than underestimate VaR. Therefore, there would be a large penalty on the underestimated VaR. The smaller the mean score is, the more accurate and preferable the method in estimating VaR. If the estimation is perfect, the score should be 0; otherwise, score is strictly positive.

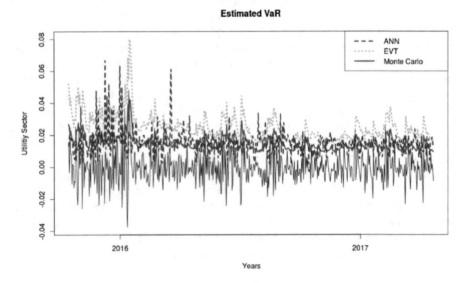

Fig. 2. The log returns of utility sector from 2015-10-13 to 2017-12-29, and the corresponding daily estimated value-at-risk from three methods discussed above.

Let x be the predicted value and y be the corresponding observed value, then $\alpha\%$ score can be calculated by the above formula. If the estimated value is less than the observed value, i.e. $x \leq y$, a higher weight $\alpha\%$ is assigned in calculation, otherwise a lower weight is assigned when $x \geq y$.

The results of mean scores of VaR estimation for 7 different business sectors which consist S&P 500 stock index by ANN model, Monte-Carlo simulation, and EVT-based method are presented in Table 1 below:

Table 1. Comparison of mean scores (at the level of 10^{-3}) of VaR estimations across different business sectors by ANN model, Monte-Carlo simulation, and EVT-based method.

Sectors	ANN	Monte Carlo	EVT
Financial	1.2	1.14	1.41
Health care	0.94	0.79	1.06
IT	1.14	0.89	1.18
Industrials	0.92	0.77	0.93
Materials	1.03	0.84	1.24
Telecom	1.09	1.06	1.28
Utilities	0.89	0.78	1.05

We also try to extend the ANN model by adding gold price, interest rate, and inflation rate as additional variables in predicting VaR in Financial sector. It shows the same score as the original model without the three extra variables. The result indicates that the three variables, which has strong correlation with Financial sector in our opinion, do not influence VaR estimation. One possible explanation could be correlations between the three variables and the log returns of Financial sector have already been exposed in the previous log return. Therefore, the results of the extended ANN model are omitted.

Overall, the Monte Carlo simulation has the smallest mean score in every business sector. The second smallest mean score is generated by ANN method. The mean scores generated by EVT-based model are the highest among those three methods. While we expect the EVT-based method to outperform the Monte Carlo simulation since the financial data is well known to be heavy-tailed. However, our results show that it actually preforms the worst among the three. One possible explanation is that the EVT-based method is parametric-based and has some very strong assumptions on the dynamic of the data. For example, GARCH (1,1) filter assumes today's squared volatility to be a linear combination of previous day's squared volatility and squared square [3], while the actual mechanism of stock market would definitely be far from this model assumption. Hence it is reasonable to see the parametric method not perform as well as nonparametric method. To overcome this problem, higher order of GARCH filters such as GARCH (2,2) or Exponential Arch (EGARCH) are recommended [12].

S. Cheung et al.

We further compare the VaR estimations across different business sectors. While the mean scores of these 7 different sectors of S&P 500 stock index are very close to each other, the financial sector has the highest mean scores. It which means that VaR estimations are not as accurate and reliable on financial market. In contrary, these VaR estimation methods perform relatively better in health care market and utility market. Our explanation is that the financial sector of S&P 500 stock index consists of 8 financial industries such as bank, insurance company and capital market [13]. Those industries are highly volatile, which increases the difficulty on estimating of VaR precisely. The medical industry, on the other hand, is relatively stable compared with financial industry. It results a better performance of VaR prediction.

We also find that in traditional sectors such as Industrials, Materials, and Health Care, ANN method is outperformed by Monte Carlo simulation. One possible explanation is that traditional sectors has lower volatility, and as a new method introduced in financial forecasting, ANN's lack of consistent statistical basis result in less stable estimation. On the other hand, ANN performs at a similar level as Monte Carlo simulation in emerging sectors such as Financial, IT, and Telecommunications.

4 Conclusion

In summary, the Monte Carlo method proves to be the best performed model in VaR estimation across all business sectors of S&P 500 in our study. The accuracy and efficiency of Monte Carlo method has established itself as one of the benchmark models in financial forecasting. The EVT-based method has the worst performance in VaR estimation across all business sectors of S&P 500. ANN method performs better that the EVT-based method in VaR estimation across all business sectors of S&P 500, while retaining a similar level of accuracy compared to Monte Carlo method in business sectors which are more volatile. In addition, the extended ANN model shows the flexibility of the ANN method and the possibility of including external variables in financial forecasting. Since the market is influenced by many factors, the more we understand about the market, the better the ANN method will perform in the future. Notice that recurrent neural network (RNN) is implemented more often for modeling time series than a traditional ANN. We invite future researchers to further integrate RNN with the prediction of VaR to reach the full potential of neural networks.

References

1. Rawls, S.W., Smithson, C.W.: Strategic risk management. J. Appl. Corp. Finan. **2**(4), 6–18 (1990)
2. Nadarajah, S., Chan, S.: Estimation methods for value at risk. In: Extreme Events in Finance, pp. 283–356 (2016)
3. Bollerslev, T.: Generalized autoregressive conditional heteroskedasticity. J. Econ. **31**(3), 307–327 (1986)

4. Brooks, C.: Introductory Econometrics for Finance. Cambridge University Press, Cambridge (2017)
5. Locarek-Junge, H., Prinzler, R.: Estimating value-at-risk using neural networks. In: Informationssysteme in der Finanzwirtschaft, pp. 385–397 (1998)
6. Chen, X., Lai, K.K., Yen, J.: A statistical neural network approach for value-at-risk analysis. In: International Joint Conference on Computational Sciences and Optimization, vol. 2, pp. 17–21. IEEE (2009)
7. Iii, J.P.: Statistical inference using extreme order statistics. Ann. Stat. 3(1), 119–131 (1975)
8. Hosking, J.R.M., Wallis, J.R.: Parameter and quantile estimation for the generalized pareto distribution. Technometrics 29(3), 339 (1987)
9. Hagan, M.T., Demuth, H.B., Beale, M.H., De Jesús, O.: Neural Network Design, vol. 20 (1996)
10. White, H.: Nonparametric estimation of conditional quantiles using neural networks. In: Computing Science and Statistics, pp. 190–199 (1992)
11. Ehm, W., Gneiting, T., Jordan, A., Krüger, F.: Of quantiles and expectiles: consistent scoring functions, Choquet representations and forecast rankings. J. Roy. Stat. Soc. Ser. B (Stat. Methodol.) 78(3), 505–562 (2016)
12. Nelson, D.B.: Conditional Heteroskedasticity in asset returns: a new approach. Econometrica 59(2), 347 (1991)
13. Kennon, J.: What Are the Sectors and Industries of the S&P 500? The Balance. https://www.thebalance.com/what-are-the-sectors-and-industries-of-the-sandp-500-3957507. Accessed 1 Nov 2018

Using Convolutional Neural Networks to Distinguish Different Sign Language Alphanumerics

Stephen Green[1]([✉]), Ivan Tyukin[1,2], and Alexander Gorban[1,2]

[1] Leicester University, University Road, Leicester, Leicestershire LE1 7RH, UK
slg46@le.ac.uk
[2] Lobachevsky University, Prospekt Gagarina, 23, Nizhnij Novgorod,
Nizhegorodskaya oblast', 603022 Nizhny Novgorod, Russia

Abstract. Using Convolutional Neural Networks (CNN)'s to create Deep Learning systems that turns Sign Language into text has been a vital tool in breaking communication barriers between deaf-mute people. Conventional research on this subject concerns training networks to recognize alphanumerical gestures and produce their textual equivalents.

A problem with current methods is that images are scarce, with little variation in available gestures, often skewed towards skin tones and hand sizes that makes a significant subset of gestures hard to detect. Current identification programs are only trained in a single language despite there being over two-hundred known variants so far. This presents a limitation for traditional exploitation for the state of current technologies such as CNN's, due to their large number of required parameters.

This work presents a technology that aims to resolve this issue by combining a pretrained legacy AI system for a generic object recognition task with a corrector method to uptrain the legacy network. As a result, a program is created that can receive finger spelling from multiple tactile languages and deduct the corresponding alphanumeric and its language which no other neural network has been able to replicate.

Keywords: Convolutional Neural Networks · Sign language · Deep Learning · Legacy AI

1 Introduction

Sign language classification using neural networks has been a goal of data scientists since the start of the decade. Many papers exist on the subject and using Convolutional Neural Networks, each gesture can be assigned classes for the network to predict making them behave like most other iterations of CNN's. The direction that Sign Language recognition has focused on prioritises the correct identification of singular letters and numbers in an assigned language over whole words. This process called Fingerspelling is very appealing to classify as there

© Springer Nature Switzerland AG 2020
L. Oneto et al. (Eds.): INNSBDDL 2019, INNS 1, pp. 276–285, 2020.
https://doi.org/10.1007/978-3-030-16841-4_29

are only the 26 letters of the alphabet and the ten numerical digits to categorise over the thousands of words that exist in any given sign language dictionary.

The implications of a technology that could read sign language and convert the results into text would be a huge leap forward for communication of the deaf and hard of hearing. A study conducted by the World Health Organisation [1] states that about 466 million people in the world suffer from hearing loss (with 34 million of these people being children) and 70 million of these people are adept in Sign Language. This number is split between over two hundred known variants of Sign Language practiced all over the world [2]. Some of the difficulties of Sign Language detection is due to similarities over other types of gesture recognition is the similarities between poses. Changes that would otherwise be ignored in full body detection such as the placement of a thumb or the angle the hand makes with the wrist can change the meaning of the gesture being conveyed. This is why the majority of papers focus on specialising in the alphanumerics of one sign language.

With this paper, the gesture sets for American Sign Language (practised by about half a million people in the United States) [3], British Sign Language (practised by about one hundred and fifty thousand people in the United Kingdom) [4] and Chinese (Pinyin) Sign Language are fed into the Inception neural network and an algorithm is created that can recognise all three languages. An error corrector is then appended to the results so any misclassification produced by Inception can be detected and amended before final output. This combination provides a neural network that is able to read multiple sign languages that is fast and performs well.

2   Previous Literature

The two most popular categories of gesture recognition are with glove based systems and vision based systems. Glove based methods [7] have the advantage of being able to record every slight motion that the user makes that provides the specifics of the flex motion the hand is making along its trajectory, overcoming the problem of movement based gestures that exist especially outside of sign language. They are also able to provide very accurate results, with a test on the 36 alphanumerics in American Sign Language giving an average accuracy of 92% [8]. The problem is that this method is cumbersome. Motion detection gloves are currently very expensive and are not recommended for everyday interactions with the hard of hearing. There is also a link that needs to be maintained between the gloves and the computer for the gestures to be captured effectively. There are examples of this being done wirelessly but this connection still needs to be maintained and therefore impractical for the required kind of simultaneous translation this project needs.

Vision based system often require devices such as the Microsoft Kinect which is able to track the positions of hand gestures relative to the human body with similar accuracy rates to gesture capture using gloves [9]. Both of these methods however are impractical for this project so instead all gestures in the dataset were filmed with a Samsung Galaxy S8 camera. The video files were then split into images and this is what has made up the dataset.

With one notable exception [10], The idea of a multi-language detection system has not been mentioned in the existing literature. The most likely reason is that current research dedicated to just one widely used language is already computationally expensive. Even the most basic alphabet fingerspelling recognition networks require large datasets to train on to receive good results and these sets only exist for a handful of known sign languages [11]. The appended error correction program is a continuation of our previous work [12] where Inception is run on the American Sign Language signs for 0–9. The results of the initial experiments was a correct classification rate of 82.4% on 10000 tested images. After the images were split into a training and testing set with a 4:1 ratio, the corrector was applied that was able to successfully remove misclassifications from the test set with very little change to the number of True Positive results that were falsely taken away.

3   Implementation

3.1   Pre-processing

For these experiments, Inception Version 3 was trained on 268800 RGB images covering 112 gestures filmed with multiple participants of varying age, gender and skin colour, with 2400 images for each respective gesture. Each image was taken with a Samsung Galaxy S8's front 8 megapixel camera with resolution 1920×1080, which are later resized to 299×299 when fed through Inception. Each image set represents the gesture taken at various angles and distances covering as many variations as possible where the camera is in front of the gesture at some angle. To mitigate the problem of background noise, each participant's gestures were taken in different locations so the only common element within each group was the gesture itself.

Conventional tracking software often requires the image to be within a certain distance from the camera for the sign to be detected while this method provide a plethora of images that can help detection regardless of which direction the camera is located when pointed at the sign. While the majority of static detection algorithms omit gestures that require movement (in this experiment, ASL J, ASL Z, BSL H and BSL J usually require motion), this work instead focuses on still images that are unique to the process of making the gesture and can be used to distinguish them from the rest of the set without the loss of an alphanumeric.

3.2   Architecture

The Inception model is a computationally efficient Convolutional Neural Network originally developed by Google in December 2015 to provide state of the art performance rates on the ImageNet Large Scale Visual Recognition Challenge [13]. Unlike similar learning algorithms like VGGNet and AlexNet which use deep, wide networks to obtain high performance rates at the cost of long computation times which made these procedures much more impractical for computers with small processing power and lack of access to high-end GPU's, Inception utilises repeated modules to replicate these model's success at a quicker rate, with 12 times less parameters and AlexNet and 36 times less parameters than VGGNet [14].

From a performance perspective, Inception-v3 on average performs better than the majority of it's alternatives producing a smaller error rate in the top 1% and 5% boundaries of 18.77% and 4.2% respectively [15]. This improves on previous versions of Inception through an RMSProp optimiser, additional convolution layers, a BatchNorm process that normalises the activation of every convolution layer and a label smoothing component that prevents overfitting on the largest logits [16]. As Inception was originally created to be trained on Imagenet, a database with over 15 million images in around 22000 categories which makes this the ideal system to sort the created set as there are 112 categories to differentiate between (Fig. 1).

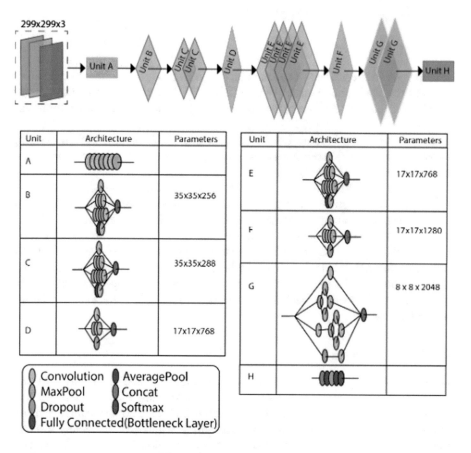

Fig. 1. Inception framework

The model starts with an $299 \times 299 \times 3$ input corresponding to a 3 colour (RGB) channel square image with a size of 299 pixels that is processed through stacked Inception modules with five regular convolution layers along with a max pooling function and a softmax output that produces the predicted label for the input [14]. Each module processes the correlation statistics of the layer before it and groups the clusters of units into filter banks [17]. This results in a large percentage of units concentrated in single regions which can be covered with a 1×1 convolution preserving the original dimensions of the structure with minimal loss of information while other structures are fed through 3×3 or 5×5 filters depending on how spread out the clusters are (the operations through these larger filters will increase as Inception progresses once the spatial concentration decreases such that dimensionality needs to be reduced to compensate). A fourth pooling path is also included for outliers that have little to no relation to the rest of the data. Small 1×1 convolution filters are attached to the larger layers reducing the number of necessary calculations and the successive stack of modules forms the Inception framework.

3.3   Training

The top layer of Inception is a softmax layer that takes a 2048-dimensional vector as an input called a bottleneck. With 112 labels, the sum of each of the weights and bias's necessary for Inception to be trained comes to 229,488 parameters that are learned in total [18].

Inception is then trained on the dataset of 268800 images with a learning rate of 0.01 and 50000 training steps in total with a training batch size of 100 images. In total, 26880 images are used for testing, 26880 images are used for training and 215,040 images are used for training. Once these steps have been carried out, 268800 images are given to the retrained Inception algorithm. The training, test, and validation steps takes 100 images for each batch size, with no further transformations performed due to the already large number of images available. For each image, the softmax layer takes the bottlenecks and produces a probability for each label that corresponds to the likelihood of that label correctly matching the given image according to what Inception has been trained on. The label with the highest respective probability is chosen as the final prediction made by Inception.

3.4   Post-processing

When each image has a predicted label attached toward it, a corrector can be appended to the current system that is able to detect incorrect labels and reassess them [19]. Each element is first sorted into with set $\mathcal{M}$ which contains each element with a correct label attached and $\mathcal{Y}$ which contains every element with an incorrect label attached along with their union $\mathcal{S} = \mathcal{M} \cup \mathcal{Y}$. The sets $\mathcal{M}^1 = \mathcal{M}, \mathcal{Y}^1 = \mathcal{Y} \,\&\, \mathcal{S}^1 = \mathcal{S}$ are then initialised, the number of clusters p is chosen and the filtering threshold θ is set.

Centering. The data is first centered, where $\mathcal{S}_c^i$ & $\mathcal{Y}_c^i$ are created by subtracting the mean $\overline{x}(\mathcal{S}^i)$ from the elements of each set respectively.

$$\mathcal{S}_c^i = \{\mathbf{x} \in \mathbb{R}^n | \mathbf{x} = \xi - \overline{\mathbf{x}}(\mathcal{S}^i), \xi \in \mathcal{S}^i\} \tag{1}$$

$$\mathcal{Y}_c^i = \{\mathbf{x} \in \mathbb{R}^n | \mathbf{x} = \xi - \overline{\mathbf{x}}(\mathcal{S}^i), \xi \in \mathcal{Y}^i\} \tag{2}$$

Regularization. The covariance matrix of $\mathcal{S}^i$ is calculated along with the corresponding eigenvalues and eigenvectors. All eigenvectors $h_1, h_2, \ldots, h_m$ that's eigenvectors $\lambda_1, \lambda_2, \ldots, \lambda_m$ that pass the Kaiser-Guttman test are combined into a single matrix H [20] before being multiplied by each element in $\mathcal{S}_c^i$ & $\mathcal{Y}_c^i$:

$$\mathcal{S}_r^1 = \{\mathbf{x} \in \mathbb{R}^n | \mathbf{x} = H^T \xi, \xi \in \mathcal{S}_c^1\} \tag{3}$$

$$\mathcal{Y}_r^1 = \{\mathbf{x} \in \mathbb{R}^n | \mathbf{x} = H^T \xi, \xi \in \mathcal{Y}_c^1\} \tag{4}$$

Whitening. The two sets then undergo a whitening coordinate transformation ensuring that the covariance matrix of the transformed data is the identity matrix:

$$\mathcal{S}_w^1 = \{\mathbf{x} \in \mathbb{R}^m | \mathbf{x} = Cov(\mathcal{S}_r^1)^{-\frac{1}{2}}\xi, \xi \in \mathcal{S}_r^1\} \tag{5}$$

$$\mathcal{Y}_w^1 = \{\mathbf{x} \in \mathbb{R}^m | \mathbf{x} = Cov(\mathcal{S}_r^1)^{-\frac{1}{2}}\xi, \xi \in \mathcal{Y}_r^1\} \tag{6}$$

Projection. Elements of $\mathcal{S}_w^1, \mathcal{Y}_w^1$ are then projected onto the unit sphere by scaling them to unit length $\mathbf{x} \mapsto \mathbf{x}/\|\mathbf{x}\|$

Clustering. The error set $\mathcal{Y}_w^1$ is partitioned into p clusters $\mathcal{Y}_{w,1}^1, \mathcal{Y}_{w,2}^1, \ldots, \mathcal{Y}_{w,p}^1$ with elements that are pairwise positively correlated.

Training. For each cluster $\mathcal{Y}_{w,j}^1, j = 1, \ldots, p$ and its complement $\mathcal{S}_w^1 \setminus \mathcal{Y}_{w,j}^1$ we construct the following separating hyperplanes:

$$h_j(\mathbf{x}) = \ell_j(\mathbf{x}) - c_j, \tag{7}$$

$$\ell_j(\mathbf{x}) = \left\langle \frac{\mathbf{w}_j}{\|\mathbf{w}_j\|}, \mathbf{x} \right\rangle, \quad c_j = \min_{\xi \in \mathcal{Y}_{w,j}^1} \left\langle \frac{\mathbf{w}_j}{\|\mathbf{w}_j\|}, \xi \right\rangle \tag{8}$$

$$\mathbf{w}_j = (Cov(\mathcal{S}_w^1 \setminus \mathcal{Y}_{w,j}^1) + Cov(\mathcal{Y}_{w,j}^1))^{-1} \times (\overline{\mathbf{x}}(\mathcal{Y}_{w,j}^1) - \overline{\mathbf{x}}(\mathcal{S}_w^1 \setminus \mathcal{Y}_{w,j}^1)). \tag{9}$$

The values of $\mathbf{w}_j$ that are greater than θ have their respective hyperplanes kept and a corresponding element $f_j(x)$ is created.

$$f_j(\mathbf{x}) = f\left(\langle \frac{WH^T(\mathbf{x} - \overline{\mathbf{x}}(\mathcal{S}^i)))}{|x|}, \frac{\mathbf{w}_j}{\|\mathbf{w}_j\|} \rangle - c_j \right) \tag{10}$$

The set of elements $\mathbf{x} \in s_w^i \setminus y_w^i$ where $h_j(\mathbf{x}) \geq 0$ is called $\mathcal{C}$ and the elements of the set $\mathcal{C}_j \cup \mathcal{Y}_{w,j}^i$ are projected orthogonally onto the hyperplane $h_j(\mathbf{x}) = l_j(\mathbf{x}) - c_j$ as:

$$\mathbf{x} \mapsto \left(I - \frac{\mathbf{w}_j \mathbf{w}_j^T}{||\mathbf{w}_j||^2} \right) \mathbf{x} + \frac{c_j \mathbf{w}_j}{||\mathbf{w}_j||} = P(w_j)\mathbf{x} + b(\mathbf{w}_j, c_j) \qquad (11)$$

This determines a hyperplane $h_{j,2}(\mathbf{x}) = \langle \mathbf{w}_{j,\,2}, \mathbf{x} \rangle - c_{j,\,2}$ whose values is less than 0 for all projections of $\mathcal{C}_j$ and greater than or equal to 0 for all projections of $y_{\mathbf{w},j}^i$. If no such planes exist, Linear Fisher Discriminant is used.

A second function is then created followed by an amalgamation that only produces a positive response when the output of the previous two functions is also positive.

$$f_j^\perp(\mathbf{x}) = f\left(\langle P(\mathbf{w}_j) \left(\frac{WH^T(\mathbf{x} - \bar{x}(S^1))}{||\mathbf{x}||} \right) + b(\mathbf{w}_j, c_j), \mathbf{w}_{j,\,2} \rangle - c_{j,\,2} \right) \qquad (12)$$

$$f_j^c(\mathbf{x}) = f(\text{Step}(f_j(\mathbf{x})) + \text{Step}(f_j^\perp(\mathbf{x})) - 2) \qquad (13)$$

Integration. For any $\mathbf{x}$ that produces a value of $f_j^c(\mathbf{x})$ that's greater than 0, the label can then be swapped accordingly to the next best matching label that Inception reported for that image.

Testing. The newly generated sets $\mathcal{M}^2$, $\mathcal{Y}^2$ & $\mathcal{S}^2$, the cluster number p and the filtering threshold θ can be carried over to another iteration of correcting and this continue until the desired results are achieved.

4   Results

After testing, Inception was able to correctly identify each gesture 66.7% of the time with the standard threshold $\theta = 0$. Reasons for this comparatively low initial result arise from similarities within the data, with small variations in gestures crossing over into different alphanumerics compounded with existing problems all tactile translators face. The spread of errors shows a fairly consistent variance of errors with no outliers 3 standard deviations from the average of 800 errors present in a set of 2400 gestures.

After each gesture was classified, a case was then added of a gesture not being present in the given image. This was done as a precaution for later testing where a wide lexicon of gestures could attempt to identify gestures where none are present. To circumvent this, a series of indicators were made as functions of θ in relation to traditional error types (elaborated in Table 1):

$$\text{True Positive Rate } (\theta) = \frac{TP(\theta)}{FP(\theta) + TP(\theta)} \qquad (14)$$

$$\text{Misclassification Rate } (\theta) = \frac{FP(\theta)}{FP(0)} \qquad (15)$$

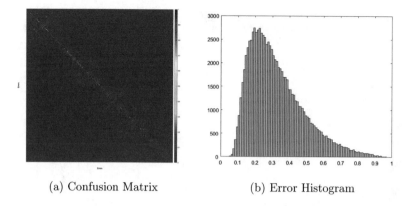

(a) Confusion Matrix   (b) Error Histogram

Fig. 2. Errors were often taken precedent by small margins compared to the true labels where the histogram of the highest rating for each mismatch and the Confusion Matrix of the whole set shows that very minor changes to the sorting procedure could provide drastic improvement.

In the situations described, the system has to make a decision regardless of the maximum score even when the vast majority of errors are only the victor through relatively small amounts as shown in Fig. 2b along with the normalised data points $\langle \frac{x_i}{||x_i||}, \frac{x_j}{||x_j||} \rangle$ with a dimensionality of 266 principle components (a reduction of over 90% from the original bottleneck length of 2048) where points labelled as errors are largely orthogonal to each other.

Table 1. Categorisation of error types, where (*) indicates results not accepted into recognition system

Presence of a gesture	System's response	Error type
Yes	Correctly classified	True positive
	Incorrectly classified	False positive
No	Not reported	False negative
	Reported	False positive*
	Not reported	True negative*

The number of clusters scaled exponentially from 1 unsorted cluster to 1000 clusters and for each value of p, k-means was run 20 times and the average True Positive rate was noted. For each cluster iteration, the corresponding separating hyperplaces h_i were created and if the values were above the chosen threshold θ then the features hyperplanes were retained. Figure 3 shows the results of these trials.

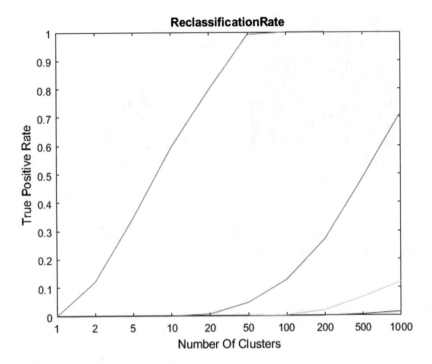

Fig. 3. Average True Positives rate in relation to number of clusters with thresholds of 0.0, 0.1, 0.2 and 0.3.

5   Conclusion and Future Work

This work presents the first algorithm capable of detecting multiple sign languages with reasonable accuracy that has the ability to self correct detected errors. The applications of this program are numerous, most notably to continue to break down the communication barriers between the audibly impaired and the rest of the world. The error correction process helps to mitigate accuracy loss common in huge datasets with large intra-class variations and furthering from this research, future variations can have language preference options so if the end user thinks they know what language is being signed then a priority status can be appended to each gesture from that language, vastly improving final results.

Future work involves expanding on current results as the final stages required large computation times to calculate the optimum threshold value and number of clusters to ensure that the maximum number of errors are detected while remaining computationally efficient. More languages can also be added to the current dataset as once the optimum parameters are established, new classes can be established without any drops in accuracy provided that the new classes are as diverse as the original three languages.

The research presented can then be made open source. Large sign language databases only exist for the most popular tectile languages while hundreds of other variations have no available image sets. By releasing the project to the public, contributions can be accepted worldwide in addition to its current state and then Inception can continue to train until even more languages are recognised.

References

1. World Health Organization: Deafness and hearing loss. http://www.who.int/mediacentre/factsheets/fs300/en/
2. Clarion UK. https://www.clarion-uk.com/know-many-sign-languages-world/
3. https://www.startasl.com/american-sign-language
4. British Deaf Association. https://bda.org.uk/help-resources/
5. http://www.washington.edu/news/2016/04/12/uw-undergraduate-team-wins-10000-lemelson-mit-student-prize-for-gloves-that-translate-sign-language/
6. https://www.microsoft.com/en-us/research/blog/kinect-sign-language-translator-part-1/
7. Mohandes, M., Aliyu, S., Deriche, M.: Prototype Arabic Sign language recognition using multi-sensor data fusion of two leap motion controllers. In: 2015 IEEE 12th International Multi-conference on Systems, Signals & Devices (SSD15), Mahdia, pp. 1–6 (2015)
8. Abhishek, K.S., Qubeley, L.C.F., Ho, D.: Glove-based hand gesture recognition sign language translator using capacitive touch sensor. In: 2016 IEEE International Conference on Electron Devices and Solid-State Circuits (EDSSC). IEEE (2016)
9. Yang, H.-D.: Sign language recognition with the Kinect sensor based on conditional random fields. Sensors (2014)
10. Kumar, V.K., Goudar, R.H., Desai, V.T.: Sign language unification: the need for next generation deaf education. Procedia Comput. Sci. **48**, 673–678 (2015). https://doi.org/10.1016/j.procs.2015.04.151
11. http://facundoq.github.io/unlp/sign_language_datasets/index.html
12. Efficiency of Shallow Cascades for Improving Deep Learning AI Systems. https://doi.org/10.1109/IJCNN.2018.8489266
13. [cs.CV] arXiv:1512.00567
14. https://medium.com/initialized-capital/we-need-to-go-deeper-a-practical-guide-to-tensorflow-and-inception-50e66281804f
15. https://medium.com/@sh.tsang/review-inception-v3-1st-runner-up-image-classification-in-ilsvrc-2015-17915421f77c
16. https://towardsdatascience.com/a-simple-guide-to-the-versions-of-the-inception-network-7fc52b863202
17. [cs.CV] arXiv:1409.4842
18. [cs.CV] arXiv:1805.06618
19. Tyukin, I.Y., Gorban, A.N., Green, S., Prokhorov, D.: Fast construction of correcting ensembles for legacy artificial intelligence systems: algorithms and a case study
20. Jackson, D.: Stopping rules in principal components analysis: a comparison of heuristic and statistical approaches. Ecology **74**(8), 2204–2214 (1993)

Mise en abyme with Artificial Intelligence: How to Predict the Accuracy of NN, Applied to Hyper-parameter Tuning

Giorgia Franchini[1], Mathilde Galinier[1,2(✉)], and Micaela Verucchi[1]

[1] Università degli studi di Modena e Reggio Emilia,
Via Università, 4, 41121 Modena, Italy
`mathildeemmanuelle.galinier@unimore.it`
[2] Marie Sklodowska-Curie fellow of the Istituto Nazionale di Alta Matematica,
Rome, Italy

Abstract. In the context of deep learning, the costliest phase from a computational point of view is the full training of the learning algorithm. However, this process is to be used a significant number of times during the design of a new artificial neural network, leading therefore to extremely expensive operations. Here, we propose a low-cost strategy to predict the accuracy of the algorithm, based only on its initial behaviour. To do so, we train the network of interest up to convergence several times, modifying its characteristics at each training. The initial and final accuracies observed during this beforehand process are stored in a database. We then make use of both curve fitting and Support Vector Machines techniques, the latter being trained on the created database, to predict the accuracy of the network, given its accuracy on the primary iterations of its learning. This approach can be of particular interest when the space of the characteristics of the network is notably large or when its full training is highly time-consuming. The results we obtained are promising and encouraged us to apply this strategy to a topical issue: hyper-parameter optimisation (HO). In particular, we focused on the HO of a convolutional neural network for the classification of the databases MNIST and CIFAR-10. By using our method of prediction, and an algorithm implemented by us for a probabilistic exploration of the hyper-parameter space, we were able to find the hyper-parameter settings corresponding to the optimal accuracies already known in literature, at a quite low-cost.

Keywords: Machine learning · Support Vector Machines ·
Curve fitting · Artificial neural network · Hyper-parameter optimisation

1 Introduction

During the last decades, machine learning algorithms and deep neural networks have shown remarkable potentials in numerous fields. However, despite their

© Springer Nature Switzerland AG 2020
L. Oneto et al. (Eds.): INNSBDDL 2019, INNS 1, pp. 286–295, 2020.
https://doi.org/10.1007/978-3-030-16841-4_30

success, algorithms of this kind may still be hard to design, and their performance usually highly depend upon the choice of numerous criteria (learning rate, optimiser, structure of the layers, etc.), called hyper-parameters. In practice, finding an optimal combination of those hyper-parameters can often make the difference between bad or average results and state-of-the-art performance. The search of this optimal setting can be performed by maximising the accuracy f of the given learning algorithm. However, this process is made particularly arduous in that the maximisation of such a function is usually very expensive.

On the basis of that observation, we decided to develop a strategy for predicting the value of the objective function f at low-cost, in an off-line fashion. Based on Support Vector Machine (SVM) and curve fitting, this method enables to obtain a prediction of the accuracy of an artificial neural network (NN), exploiting only the behaviour of the NN over the first epochs of its learning. To the best of our knowledge, no other article in the literature mentions approaches for a low-cost approximation of the accuracy of a NN, and in this respect, the ideas lying behind our method may constitute a breakthrough in the domain of artificial intelligence. The algorithm we propose can be of particular interest for a rather fast quality assessment of a new learning algorithm. Specially, it may facilitate hyper-parameters optimisation in an interesting way.

The Sect. 2 of this article describes the main steps of our method for predicting the accuracy of a given NN. Section 3 deals with an application of our method to a topical issue: hyper-parameter optimisation. Section 4 shows the results of the experiments on the MNIST and CIFAR-10 databases. Finally, the paper ends with concluding remarks in Sect. 5.

2   Methodology

A database is created beforehand, by training the NN algorithm of interest up to convergence, with different sets of randomly chosen hyper-parameters. The prediction accuracies of the NN algorithm are gathered throughout this process, and eventually constitute a database on which the hereinafter-described method is applied. Please note that a new database is to be created every time a new set of training/test samples is considered.

Between SVM and Curve Fitting

The goal is to be able to predict the final accuracy of the NN algorithm after convergence, based only on its behaviour over the first epochs of its learning. To do so, we have combined two techniques, well-known in the literature: Support Vector Machines (SVM) [10] and curve fitting [11].

We first present a theoretical background of the mentioned techniques, and then indicate how we used them in this work.

A SVM is a discriminative classifier formally defined by a separating hyper-plane. In other words, given labelled training data (supervised learning), the algorithm outputs an optimal hyper-plane which categorises new examples. SVM algorithms are characterised by the usage of kernels, the sparseness of the solution and the capacity control obtained by acting on a margin, or on the number of support vectors. SVM can be applied not only to classification problems but also to the case of regression. One of the most important ideas in SVM for regression is that presenting the solution by means of small subsets of training points gives enormous computational advantages.

We consider the well-known dual problem of the SVM algorithm [10], whose solution is given by:

$$f(\mathbf{x}) = \sum_{i=1}^{n_{SV}} (\alpha_i - \alpha_i^*) K(\mathbf{x_i}, \mathbf{x}) + b$$

where n_{SV} is the number of support vectors, α and $\alpha^* \in \mathbb{R}^{n_{SV}}$ are the multipliers of the support vectors and $b \in \mathbb{R}$ is a constant. $K(\mathbf{x_i}, \mathbf{x})$ is the kernel function, that we can choose to be linear, polynomial or Gaussian. In particular the Gaussian kernel form is

$$K(\mathbf{x_i}, \mathbf{x}) = \exp(-\gamma \|\mathbf{x_i} - \mathbf{x}\|^2), \text{ where } \gamma \text{ is a free positive parameter.}$$

Curve fitting is the process of constructing a curve that has the best fit to a series of data points, possibly subject to constraints. In other words, the goal of curve fitting is to find the appropriate parameters that, based on a chosen model, describe experimental data as precisely as possible. The values of such parameters are usually computed by the Least Squares (LS) method, which minimises the square of the error between the original data and the values predicted by the model. In this work, the fitting function is chosen to have the form:

$$g(x) = \alpha x^\beta \tag{1}$$

where α and β are the parameters to be computed by non-linear LS.

Our methodology is based on the following procedure: knowing the accuracies over the first epochs of its training process, the accuracy reached by an NN after convergence is predicted thanks to an SVM algorithm, beforehand trained on the created database. This value is considered as an approximation of the objective function f of interest, in the case it is neither greater than 1, nor smaller than the maximum of the initial observed accuracies. Otherwise, the final accuracy of the NN is predicted based on a curve fitting strategy: given the values of the accuracies of the first epochs, the corresponding curve is fitted thanks to the function (1). Once the parameters α and β are computed, the function g is used to predict the value of the accuracy after convergence of the algorithm. It is to be noted that the parameters are subject to the following constraints:

$$\begin{cases} \frac{acc\_max}{fin\_epoch} < \alpha \\ 0 < \beta < 1 \end{cases} \tag{2}$$

where acc_max is the maximum among the accuracies over the first epochs, and fin_epoch is the maximum number of epochs that are allowed to be seen by the learning algorithm. In particular, the first constraint forces the curve to reach an accuracy at the final epoch higher than the one already observed during the first epochs.

Combining those two different techniques for the prediction of the final accuracy makes the prediction process more efficient, this is why we decided not to use SVM only. Furthermore, the curve fitting method seemed to us appealing insofar as the fitting function can easily be constrained, as explained above. This heuristic choice was then validated by the experiments.

This method can be convenient from a computational point of view because, once the database is created and the SVM is trained, it allows to have new evaluations of the objective function in a short time. The time taken in creating the database must be evaluated; and obviously, the convenience of such a method increases as the complexity of the problem increases.

3   Application : From Prediction to Hyper-parameter Optimisation

Being able to predict the accuracy of an NN for a given hyper-parameter setting is particularly advantageous for finding the optimal combination of hyper-parameters. In this section, we aim to apply the hereinabove described method to the hyper-parameter optimisation (HO) of a Convolutional Neural Network (CNN). We first describe the state-of-the-art HO approaches. Secondly, we present the algorithm we designed to find the optimised hyper-parameters.

3.1   Related Works

Two of the most intuitive and widely spread methods are the grid search and the random search [5]. These techniques however are not well-suited in applications where a given set of hyper-parameters is costly to evaluate. As a result, Sequential Model-Based Optimisation (SMBO) [1] algorithms have been employed in many settings when the performance evaluation of a model is expensive. They approximate the black-box objective function f that is to be maximised by a surrogate function, cheaper to evaluate. At each iteration of the algorithm, the new point where the surrogate is to be evaluated is chosen by maximising a chosen criterion. Several SMBO algorithms have been proposed in the literature, and differ in the criteria by which they optimise the surrogate, and in the way they model the surrogate given the observation history. Two of the most famous SMBO approaches are the Bayesian Optimisation approach [3,4] and the Tree-structured Parzen Estimator strategy [2].

More recently, new hyper-parameter optimisation methods based on reinforcement learning have emerged [6–9]. The goal for most of them was to find the neural network (NN) or CNN architectures that are likely to yield to an optimised performance. Thus, they were seeking the appropriate architectural hyper-parameters, as the number of layers or the structure of each convolutional layer, but many other hyper-parameters such as the learning rate and regularisation parameters are manually chosen in the end. In any case, while all the above-mentioned strategies aim to evaluate the expensive objective function f (which, in the case of a NN or CNN is the prediction accuracy) as seldom as possible, to the best of our knowledge very few algorithms offer a method to reduce the evaluation cost of f.

3.2   Our Method Applied to HO

Here, we assume that we are given a CNN, and we seek its optimal hyper-parameters. Basing the optimisation of these hyper-parameters on our above-mentioned method actually implies the choice of other parameters, such as the type of kernel function, or the form of the fitting function (1). However, it is to be observed that those new parameters can be much more easily chosen. For instance, SVM for regression is yet a very well known method in literature and offers a good robustness in its results with respect to its hyper-parameters. On the contrary, this is definitely not the case for NN in general, which makes the choice of their parameters much harder.

We focus on the optimisation of the learning rate, the optimiser and the mini-batch size leading to the best accuracy, but this method could be extended to other hyper-parameters (layer number, number of neurons in each layer, activation function).

The procedure we propose consists in the following steps. First, a database is created, as above-mentioned, and an SVM algorithm is trained on it. We chose to create this database using 10% of the possible settings of hyper-parameters, randomly selected with uniform probability. Although time-consuming, this preliminary step may lead to a less expensive hyper-parameter optimisation process than the ones already known in the literature.

In our HO approach, each hyper-parameter i of interest is represented as a vector V_i containing values in a range chosen by the experimenter. To this vector corresponds a second vector of probabilities P_i, initialised with uniform distribution. At each iteration of the exploration process, a set of hyper-parameters is randomly chosen, based on the probabilities in the corresponding vectors P_i. The CNN is parameterised with the selected hyper-parameters, and trained on a few epochs. Based on its behaviour at the beginning of its learning, and thanks to the methodology detailed in Sect. 2, the accuracy of the CNN after convergence of its learning is predicted, this value being considered as a reward. If the reward $r^{(t)}$ at the iteration t is higher than the reward $r^{(t-1)}$ at the previous iteration, the probabilities in P corresponding to the selected hyper-parameters and their neighbourhood in the vector are increased, while the other ones are penalised. On the contrary, if $r^{(t)} < r^{(t-1)}$ the probabilities of the selected hyper-parameters are penalised and the other ones are increased. Thus, the hyper-parameters are weighted by probabilities that are modified throughout the exploration process, until one value for each hyper-parameters reaches a probability greater than some threshold t_i. Then, our exploration algorithm is considered to have converged.

Finally, the NN is parameterised with the sets of hyper-parameters corresponding to the 10 highest predicted final accuracies, and brought to convergence. The setting leading to the best observed final accuracy is defined as the optimal hyper-parameter setting.

4   Experiments

In order to evaluate and tune our method, we performed different tests. We decided to consider a spreadly used NN, namely CNN for classification. We picked two well-known datasets, MNIST[1] and CIFAR-10[2], and we designed two different networks, capable of classifying images in one of 10 classes. For the MNIST dataset, the network is composed of an input layer, two sequences of convolutional and max-pooling layers, a fully connected layer and an output layer. We make use of Rectified Linear Unit (ReLU) activations and dropout technique. On the other hand, for the CIFAR dataset, the CNN is composed of an input layer, four sequences of convolutional and max-pooling layers, four fully connected layers and an output layer, exploiting ReLU activations, batch normalisation, and dropout. For the training process of the two networks, we chose to insert a form of regularisation: the early stopping. This technique is used to avoid overfitting when training a learner with an iterative method, such as gradient descent. In particular, if the training of the network is not providing better results after a determined amount of epochs, the learning process is stopped because it is considered to be stuck in a minimum and may lead to overfitting. As already mentioned, we first needed a database to train our SVM on. With the underlying idea of this paper, we won a project that granted us the use of CINECA resources on the Marconi cluster[3]. This grant enabled us to perform all our tests and generate the databases for MNIST and CIFAR. The database consists in a table in which each row corresponds to a full training of a network. We picked learning rate, optimiser and batch size as hyper-parameters and we stored them along with the accuracies observed at every epoch to create the database. Actually, the number of fully trained network needed was quite small. In our case, we needed only 44 examples: we used 35 to train the SVM and 9 to test it. Once we collected the database, we were able to understand which SVM configuration was more suitable for our problem. We tried with three different kernels: linear, polynomial and Gaussian, and we picked the last one because it outperformed the other two in terms of loss (Mean Squared Error - MSE).

In Table 1 the predictions of the three different kernels are reported. In this case, predictions are made on the test set and are based on three epochs, for the CIFAR dataset, the MSE losses are reported in Table 1, where is shown that the Gaussian one is the kernel that dominates the other two.

We also made some tests to understand which were the most important features to feed the SVM predictor with and finally chose, given the results, to feed the method only with the epoch accuracies. Regarding the curve fitting, we also tried several functions in order to find the one that produces the best fitting (1). Figure 1 shows a graphical example of the prediction made by the SVM and the curve fitting. In the chart, the full dots represent the real values of the accuracies, epoch by epoch; the star at the final epoch is the SVM prediction,

---

[1] http://yann.lecun.com/exdb/mnist/.

[2] https://www.cs.toronto.edu/~kriz/cifar.html.

[3] https://www.cineca.it/en/content/marconi.

Table 1. Given the accuracies of the NN over its first three epochs, the final accuracies are predicted using the aforementioned method. Three different SVM kernels are investigated: linear, polynomial and gaussian. The nine first lines show the predicted final accuracies of nine different test cases, the last line indicates the MSE losses corresponding to each method.

	Ground-truth	Linear	Polynomial	Gaussian
0	0.7129	0.816350	0.854324	0.713435
1	0.1871	0.197152	0.207737	0.175775
2	0.1000	0.200523	0.204122	0.200774
3	0.6820	0.704606	0.599408	0.781832
4	0.4340	0.381075	0.301048	0.456969
5	0.6369	0.572801	0.489510	0.638199
6	0.7315	0.747803	0.682245	0.805067
7	0.1000	0.190538	0.203211	0.175411
8	0.4783	0.410684	0.325788	0.491291
MSE	0	0.04136	0.1138	0.0320

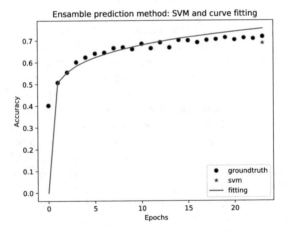

Fig. 1. Prediction of curve fitting and SVM with three epoches, compared with the real accuracies.

while the line is curve obtained with the fitting. For both SVM and curve fitting, only the first three epochs accuracies are used to predict the accuracy at the final epoch.

Figure 2 shows the results of our method applied to the prediction of the final accuracy of the CNN for the CIFAR-10 dataset. In this case, we use only the accuracies of the first two or four epochs of training to predict the final one. The orange line represents the prediction made by our method, the blue one the ground truth (*i.e.* the network is fully trained up to convergence with the

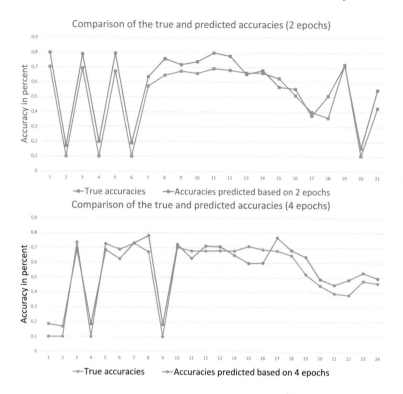

Fig. 2. Comparison of the predicted final accuracy with SVM and curve fitting and the ground truth, feeding the method with only two training epochs (on the left) and four (on the right).

same hyper-parameters). As it can be seen, the method can effectively provide a satisfactory prediction of the final behaviour of the network after only few epochs. Encouraged by those results, we applied our strategy to a challenging topical case: to tune automatically the hyper-parameters of a network, with the procedure explained in Sect. 3. We performed this further test using the two first epochs to predict the final one. After 200 iterations, we were able to find the parameters that lead to the best final accuracy, confirmed by a real full convergence. The hyper-parameters provided by our method are: the learning rate is equal to 0.0008425, the mini-batch size is 128, and the chosen optimiser is ADAM.

The source code can be found at https://git.hipert.unimore.it/mverucchi/optics.

5 Conclusion

We proposed and implemented a novel approach to predict the final behaviour of a learning process. This new method exploits both SVM and curve fitting to foresee the resulting accuracy of a long method, using only some initial steps. We applied this technique to a CNN, in order to quickly understand if the training of the network will end up in a good or bad manner. The results show that the predictions achieved with our technique are quite similar to the ground-truth, and confirm that this strategy can be of particular interest in the hyper-parameter optimisation domain. Further, we will focus on a more complete procedure to automatically tune hyper-parameters, such as the number of layers or the activation function, of a network exploiting our SVM-curve-fitting predictor.

Acknowledgements. The research leading to these results has received funding from the European Union's Horizon 2020 Programme under the CLASS Project (https://class-project.eu/), grant agreement n 780622.

This work was also partially supported by INdAM-GNCS (Research Projects 2018). Furthermore, it was partially supported by INdAM Doctoral Programme in Mathematics and/or Applications Cofunded by Marie Sklodowska-Curie Actions (INdAM-DP-COFUND-2015) whose grant number is 713485.

References

1. Hutter, F., Hoos, H., Leyton-Brown, K.: Sequential model-based optimization for general algorithm configuration. In: LION-5 2011. Extended version as UBC Technical report TR-2010-10 (2011)
2. Bergstra, J.S., Bardenet, R., Bengio, Y., Kégl, B.: Algorithms for hyper-parameter optimization. In: NIPS (2011)
3. Shahriari, B., Swersky, K., Wang, Z., Adams, R.P., de Freitas, N.: Taking the human out of the loop: a review of bayesian optimization. Proc. IEEE **104**(1), 148–175 (2016)
4. Mockus, J., Tiesis, V., Zilinskas, A.: The application of Bayesian methods for seeking the extremum. In: Dixon, L.C.W., Szego, G.P. (eds.) Towards Global Optimization. volume 2, pp. 117–129. North Holland, New York (1978)
5. Bergstra, J., Bengio, Y.: Random search for hyper-parameter optimization. J. Mach. Learn. Res. **13**(1), 281–305 (2012)
6. Zoph, B., Le, Q.V.: Neural architecture search with reinforcement learning. In: International Conference on Learning Representations, Toulon, France, pp. 1–16 (2017)
7. Baker, B., Gupta, O., Naik, N., Raskar, R.: Designing neural network architectures using reinforcement learning. In: International Conference on Learning Representations, pp. 1–18 (2017)
8. Zhong, Z., Yan, J., Wei, W., Shao, J., Liu, C.-L.: Practical block-wise neural network architecture generation. In: Conference on Computer Vision and Pattern Recognition, Salt Lake City, Utah, USA (2018). arXiv preprint:1708.05552
9. Cai, H., Chen, T., Zhang, W., Yu, Y., Wang, J.: Efficient architecture search by network transformation. In: AAAI Conference on Artificial Intelligence, New Orleans, Louisiana, USA, pp. 2787–2794 (2018)

10. Chapelle, O., Vapnik, V.: Model selection for support vector machines. In: Advances in Neural Information Processing Systems, vol. 12 (1999)
11. Arlinghaus, S.L.: PHB Practical Handbook of Curve Fitting. CRC Press, Boca Raton (1994)

Asynchronous Stochastic Variational Inference

Saad Mohamad[1(✉)], Abdelhamid Bouchachia[1],
and Moamar Sayed-Mouchaweh[2]

[1] Department of Computing, Bournemouth University, Poole, UK
saad.mohamad@outlook.com
[2] Department of Informatics and Automatics, Ecole des Mines, Douai, France

Abstract. Stochastic variational inference (SVI) employs stochastic optimization to scale up Bayesian computation to massive data. Since SVI is at its core a stochastic gradient-based algorithm, horizontal parallelism can be harnessed to allow larger scale inference. We propose a lock-free parallel implementation for SVI which allows distributed computations over multiple slaves in an asynchronous style. We show that our implementation leads to linear speed-up while guaranteeing an asymptotic ergodic convergence rate $O(1/\sqrt{T})$ while the number of slaves is bounded by $\sqrt{T}$ (T is the total number of iterations). The implementation is done in a high-performance computing environment using message passing interface for python (MPI4py). The empirical evaluation shows that our parallel SVI is lossless, performing comparably well to its counterpart serial SVI with linear speed-up.

1 Introduction

Probabilistic models with latent variables have grown into a backbone in many modern machine learning applications such as text analysis, computer vision, time series analysis, network modelling, and others. The main challenge in such models is to compute the posterior distribution over some hidden variables encoding hidden structure in the observed data. Generally, computing the posterior is intractable and approximation is required. Markov chain Monte Carlo (MCMC) sampling has been the dominant paradigm for posterior computation. It constructs a Markov chain on the hidden variables whose stationary distribution is the desired posterior. Hence, the approximation is based on sampling for a long time to (hopefully) collect samples from the posterior [1].

Recently, variational inference (VI) has emerged as a deterministic alternative approach to Markov chain Monte Carlo (MCMC) sampling. In general, VI tends to be faster than MCMC which makes it more suitable for problems with large data sets. VI turns the inference problem to an optimization problem by

A. Bouchachia was supported by the European Commission under the Horizon 2020 Grant 687691 related to the project: *PROTEUS: Scalable Online Machine Learning for Predictive Analytics and Real-Time Interactive Visualization.*

© Springer Nature Switzerland AG 2020
L. Oneto et al. (Eds.): INNSBDDL 2019, INNS 1, pp. 296–308, 2020.
https://doi.org/10.1007/978-3-030-16841-4_31

positing a simpler family of distributions and finding the member of the family that is closest to the true posterior distribution [2]. Such optimization problem is a non-convex one which requires sophisticated tools to tackle. Stochastic optimisation has been applied to VI in order to cope with massive data [3]. While VI requires repeatedly iterating over the whole data set before updating the variational parameters (parameters of the variational objective), stochastic VI (SVI) updates the parameters every time a data example is processed. Therefore, by the end of one pass through the dataset, the parameters will have been updated multiple times. Hence, the parameters converge faster with less computational resources. The idea of SVI is to move the variational parameters at each iteration in the direction of a noisy estimate of the variational objective's natural gradient based on a couple of examples [3]. Following these gradients with certain conditions on the (decreasing) learning rate schedule, SVI provably converges to a local optimum [4].

Although stochastic optimization improves the performance of VI, its serial employment prevents scaling up the inference. Since SVI is basically a stochastic gradient-based optimisation algorithm, horizontal parallelism is straightforward. That is, computing stochastic gradients of a batch of data samples can be done locally in parallel (on multi-core machines) given that the parameters update is synchronised. However, such synchronisation limits the scalability by since slaves need to send their stochastic gradients to the master prior to each parameter update. Hence, synchronous methods suffer from the curse of the last reducer; that is, a single slow slave can dramatically slow down the whole performance. Thus, asynchronous parallel optimization is an interesting alternative provided it maintains comparable convergence rate to its synchronous counterpart. Indeed, asynchronous parallel stochastic gradient-based optimisation algorithms have recently received broad attention [5–9].

Authors in [6] show that for smooth stochastic convex problems the asynchronisation effects are asymptotically negligible and order-optimal convergence results can be achieved. Since the SVI objective function is non-convex, we are interested in the asynchronous parallel stochastic gradient algorithm (ASYSG) for smooth non-convex optimization [10]. A recent study [11] breaks the usual convexity assumption taken by [6]. Nonetheless, theoretical guarantees (convergence and speed-up) for many recent successes of ASYSG are reported. In this paper, we use the ASYSG algorithm proposed in [6] to come up with an asynchronous SVI (ASYSVI) algorithm for a wide family of Bayesian models. We also adapt the theoretical studies of ASYSG for smooth non-convex optimization from [11] to explain ASYSVIs' convergence and speed-up properties. We also propose a novel contribution to linearly speed up SVI by distributing its stochastic natural gradient computations in an asynchronous way while guaranteeing an ergodic convergence rate $O(1/\sqrt{T})$ under some assumptions. Latent Dirichlet allocation is used as a case study to empirically evaluate ASYSVI.

The rest of the paper is structured as follows. We describe our ASYSVI algorithm and its convergence analysis in Sect. 2. Latent Dirichlet allocation case study is derived in Sect. 3. Empirical evaluation is presented in Sect. 4 and the paper concludes with a discussion in Sect. 5. We also attached an appendix where

---

Algorithm 1. ASYSVI-Master

1: **Input:** number of iteration T and step-size $\{\rho_t\}_{t=0,\ldots,T-1}$
2: **initialize:** $\boldsymbol{\lambda}^0$ randomly and t to 0
3: **while** $(t < T)$ **do**
4:     Aggregate M stochastic natural gradients from the slaves: $\hat{\nabla}\mathcal{L}_1(\boldsymbol{\lambda}^{t-\tau_{t,1}}), \ldots, \hat{\nabla}\mathcal{L}_M(\boldsymbol{\lambda}^{t-\tau_{t,M}})$
5:     Average the M stochastic natural gradients. $G_M^t = \sum_m \hat{\nabla}\mathcal{L}_m(\boldsymbol{\lambda}^{t-\tau_{t,m}})$
6:     Update the current estimate of the global variational parameter. $\boldsymbol{\lambda}^{t+1} = \boldsymbol{\lambda}^t + \rho_t G_M^t$
7:     $t = t + 1$
8: **end while**

Algorithm 2. ASYSVI-Slave

1: **Input:** data size D
2: **while** (True) **do**
3:     Sample a data point $\boldsymbol{x}_i$ uniformly from the data set
4:     Pull a global variational parameter $\boldsymbol{\lambda}^*$ from the master
5:     Compute the local variational parameters $\phi_i^*(\boldsymbol{\lambda}^*)$ corresponding to the data point $\boldsymbol{x}_i$ and the global variational parameter $\boldsymbol{\lambda}^*$, $\phi_i^*(\boldsymbol{\lambda}^*) = \arg\max_{\phi_i} \mathcal{L}_i(\boldsymbol{\lambda}^*, \phi_i)$
6:     Compute the stochastic natural gradient with respect to the global parameter $\boldsymbol{\lambda}$, $\boldsymbol{g}_i(\boldsymbol{\lambda}) = \boldsymbol{\alpha} + D E_{\phi_i(\boldsymbol{\lambda})}[t(\boldsymbol{x}_i, \boldsymbol{z}_i)] - \boldsymbol{\lambda}$
7:     Push $\boldsymbol{g}_i(\boldsymbol{\lambda}^*)$ to the master
8: **end while**

---

a background on variational and stochastic variational inference is provided in Sect. A. Related work is discussed in Sect. B.

2   Asynchronous Stochastic Variational Inference

ASYSVI is presented analogously to ASYSG in [11] but in the context of VI. The architecture of the computer network on which ASYSVI is run is known as the *star-shaped* network. In this network, a master machine maintains the global variational parameter $\boldsymbol{\lambda}$, whereas other machines serve as slaves which independently and simultaneously compute the local variational parameters ϕ and stochastic gradients of ELBO $\mathcal{L}(\boldsymbol{\lambda})$. The slaves only communicate with the master to exchange information in which they access the state of the global variational parameter and provide the master with the stochastic gradients. These gradients are computed with respect to $\boldsymbol{\lambda}$ based on few data points acquired from distributed sources. The master aggregates predefined amounts of stochastic gradients from slaves nonchalantly about the sources of the collected stochastic gradients. Then, it updates its current global variational parameter. The update step is performed as an atomic operation where slaves cannot read the value of the global variational parameter during this step. However, vertical parallelism can be achieved by adopting the ASYSG algorithm proposed in [5]. Furthermore, a hybrid horizontal-vertical parallelism could be achieved by combining the mechanism used in [12] with ASYSVI (more details in Sect. 5).

The key difference between ASYSVI and the synchronous parallel SVI is that ASYSVI does not lock the slaves until the master's update step is done. That is, the slaves might compute some stochastic gradients based on early value of the global variational parameter. By allowing delayed and asynchronous updates, one might expect slower convergence if any. In the next section, we apply the study of [11] on SVI to show that the effect of stochastic gradients delay will vanish asymptotically. The algorithms of ASYSVI-mater and ASYSVI-salve are shown in Algorithms 1 and 2. We denote by $\tau_{t,m}$ the delays between the current iteration t and the one when the slave pulled the global variational parameter at which it computed the stochastic gradient.

2.1   Convergence Analysis

Following [6,11], we take the same assumptions, but replace the gradient with the natural gradient:

- Unbiased gradient: the expectation of the stochastic natural gradient of Eq. (16) is equivalent to the natural gradient of Eq. (13):

$$\hat{\nabla}\mathcal{L}(\boldsymbol{\lambda}) = E[\hat{\nabla}\mathcal{L}_i(\boldsymbol{\lambda})] \tag{1}$$

  where $\hat{\nabla}$ denotes natural gradient. This assumption already holds in SVI problems for the family of models shown in Sect. A.
- Bounded variance: the variance of the stochastic natural gradient is bounded for all $\lambda \in \mathcal{G}$, $E[||\hat{\nabla}\mathcal{L}_i(\boldsymbol{\lambda}) - \hat{\nabla}\mathcal{L}(\boldsymbol{\lambda})||^2] \leq \sigma^2$. By applying SVI natural gradient, we obtain:

$$E[||nE_{\phi_i(\boldsymbol{\lambda})}[t(\boldsymbol{x}_i, \boldsymbol{z}_i)] - \sum_{i=1}^{n} E_{\phi_i(\boldsymbol{\lambda})}[t(\boldsymbol{x}_i, \boldsymbol{z}_i)]||^2] \leq \sigma^2 \tag{2}$$

- Lipschitz-continuous gradient: the natural gradient is L-Lipschitz-continuous for all $\lambda \in \mathcal{G}$ an $\lambda' \in \mathcal{G}$, $||\hat{\nabla}\mathcal{L}(\boldsymbol{\lambda}) - \hat{\nabla}\mathcal{L}(\boldsymbol{\lambda'})|| \leq L||\lambda - \lambda'||$. By applying SVI natural gradient, we end up with the following formulation:

$$||\sum_{i=1}^{n} E_{\phi_i(\boldsymbol{\lambda})}[t(\boldsymbol{x}_i, \boldsymbol{z}_i)] - \lambda - \sum_{i=1}^{n} E_{\phi_i(\boldsymbol{\lambda'})}[t(\boldsymbol{x}_i, \boldsymbol{z}_i)] + \lambda'|| \leq L||\lambda - \lambda'|| \tag{3}$$

- Bounded delay: All delay variables $\tau_{t,m}$ are bounded:

$$\max_{t,m} \tau_{t,m} \leq B \tag{4}$$

In addition to these assumptions, authors [6,11] assume that each slave receives a stream of independent data points. Although this assumption might not be satisfied strictly in practice, we follow the same assumption for analysis purpose. Thus, the same theoretical results obtained by [11] can be applied for ASYSVI, namely, an ergodic convergence rate $O(1/\sqrt{MT})$ provided that T is greater than $O(B^2)$. The results also show that, since the number of slaves is proportional to B, the ergodic convergence rate is achieved as long as the number of salves is bounded by $O(\sqrt{T/M})$. Note that $O(1/\sqrt{MT})$ is consistent with the serial stochastic gradient (SG) and the stochastic variational inference (SVI). Thus, ASYSG and ASYSVI allow for a linear speed-up if $B \leq O(\sqrt{T/M})$.

3 Case Study: Latent Dirichlet Allocation

Latent Dirichlet allocation (LDA) is an instance of the family of models described in Sect. A where the global, local, observed variables and their distributions are set as follows:

- the global variables $\{\beta\}_{k=1}^{K}$ are the topics in LDA. A topic is a distribution over the vocabulary, where the probability of a word w in topic k is denoted by $\beta_{k,w}$. Hence, the prior distribution of β is a Dirichlet distribution $p(\beta) = \prod_k Dir(\beta_k; \eta)$
- the local variables are the topic proportions $\{\theta_d\}_{d=1}^{D}$ and the topic assignments $\{\{z_{d,w}\}_{d=1}^{D}\}_{w=1}^{W}$ which index the topic that generates the observations. Each document is associated with a topic proportion which is a distribution over topics, $p(\theta) = \prod_d Dir(\theta_d; \alpha)$. The assignments $\{\{z_{d,w}\}_{d=1}^{D}\}_{w=1}^{W}$ are indices, generated by θ_d, that couple topics with words, $p(z_d|\theta) = \prod_w \theta_{d,z_{d,w}}$
- the observations x_d are the words of the documents which are assumed to be drawn from topics β selected by indices z_d, $p(x_d|z_d, \beta) = \prod_w \beta_{z_{d,w}, x_{d,w}}$

In LDA, documents are represented as random mixtures over latent topics, where each topic is characterized by a distribution over words [13]. LDA assumes the following generative process:

1 Draw topics $\beta_k \sim Dir(\eta, ..., \eta)$ for $k \in \{1, ..., K\}$
2 Draw topic proportions $\theta_d \sim Dir(\alpha, ..., \alpha)$ for $d \in \{1, ..., D\}$
   2.1 Draw topic assignments $z_{d,w} \sim Mult(\theta_d)$ for $w \in \{1, ..., W\}$
      2.1.1 Draw word $x_{d,w} \sim Mult(\beta_{z_{d,w}})$

According to Sect. A, each variational distribution is assumed to come from the same family of the true one. Hence, $q(\beta_k|\lambda_k) = Dir(\lambda_k)$, $q(\theta_d|\gamma_d) = Dir(\gamma_d)$ and $q(z_{d,w}|\phi_{d,w}) = Mult(\phi_{d,w})$. To compute the stochastic natural gradient g_i in Algorithm 2 for LDA, we need to find the sufficient statistic $t(.)$ presented in Eq. (7). By writing the likelihood of LDA in the form of Eq. (7), we obtain $t(x_d, z_d) = \sum_{w=1}^{W} \mathbf{I}_{z_{d,w}, x_{d,w}}$, where $\mathbf{I}_{i,j}$ is equal to 1 for entry (i, j) and 0 for all the rest. Hence, the stochastic natural gradient $g_i(\lambda_k)$ can be written as follows:

$$g_i(\lambda_k) = \eta + D \sum_{w=1}^{W} \phi_{i,w}^k \mathbf{I}_{k,x_{i,w}} - \lambda_k \tag{5}$$

Details on how to compute the local variational parameters $\phi_i^*(\lambda^*)$ in Algorithm 2 can be found in [3].

Having computed the elements needed to run ASYSVI's Algorithms 1 and 2, we move to the convergence analysis. Since the data is assumed to be subsampled uniformly, the unbiased gradient assumption holds for LDA. We can always find a constant variable to bound the variance. At the worst case, the variance of the stochastic natural gradient of LDA can be bounded by $DW(max_{i,w}(\phi_{i,w}^k)^2 - min_{i',w'}(\phi_{i',w'}^k)^2)$, $\forall k$. Therefore, it can be bounded by $O((DW)^2)$. It is clear that the Lipschitz-continuous gradient can be satisfied for any class of the family models proposed in Sect. A and hence, for LDA. Finally, the bounded delay can be guaranteed through the implementation. Therefore, ASYSVI of LDA can converge since the aforementioned assumptions can be satisfied.

Table 1. Parameters settings

Data sets	Enron emails				NYTimes news articles				Wikipedia articles			
batch	16	64	256	1024	16	64	256	1024	16	64	256	1024
κ	0.7	0.7	0.5	0.5	0.7	0.7	0.5	0.5	0.7	0.7	0.5	0.5
τ_0	1024	24	24	1	1024	24	24	1	1024	1024	1024	1024
perplexity	5919	5348	5264	4771	11989	10156	9015	5501	1446	1390	1355	1332

4 Experimental Results

In the following, we demonstrate the usefulness of distributing the computation of SVI, mainly the speed-up advantages of ASYSVI. For this purpose, we compare the speed-up of ASYSVI LDA against serial SVI LDA (online LDA [14]). The two versions are evaluated on three datasets consisting of very large document collections. We also evaluate ASYSVI LDA in the streaming setting where new documents arrive in the form of stream. The implementation is available in PROTEUS SOLMA Library[1]. The evaluation is done using held-out perplexity as a measure of model fit. Perplexity is defined as the geometric mean of the inverse marginal probability of each word in the held-out set of documents [13]. To validate the speed-up properties, following [11], we compute the running time speed-up (TSP):

$$TSP = \frac{\text{running time of SVI-LDA}}{\text{running time of DPSVI-LDA}}$$

The running time of both models is taken when they achieve the same final held-out perplexity.

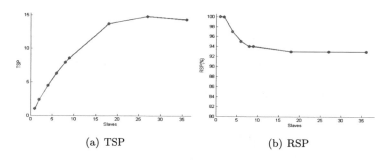

(a) TSP   (b) RSP

Fig. 1. Comparing ASYSVI LDA to online LDA on *Enron* dataset

Datasets: we perform all comparisons and evaluations on three corpora of documents. The first two corpora are available on [15]. The third corpus was used in [14].

---

[1] https://github.com/proteus-h2020/proteus-solma/tree/master/src/main/scala/eu/proteus/solma/asvi.

302   S. Mohamad et al.

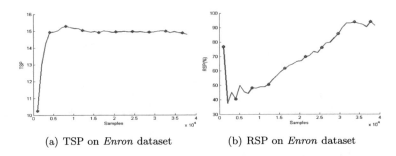

(a) TSP on *Enron* dataset (b) RSP on *Enron* dataset

Fig. 2. TSP and RSP with respect to streaming samples on *Enron* dataset

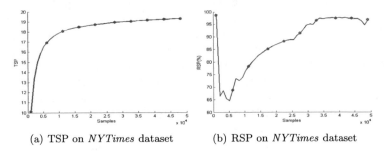

(a) TSP on *NYTimes* dataset (b) RSP on *NYTimes* dataset

Fig. 3. TSP and RSP with respect to streaming samples on *NYTimes* dataset

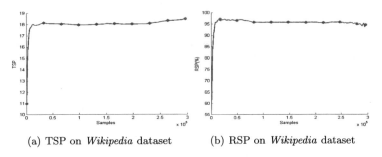

(a) TSP on *Wikipedia* dataset (b) RSP on *Wikipedia* dataset

Fig. 4. TSP and RSP with respect to streaming samples on *Wikipedia* dataset

- *Enron* emails: contains $39,861$ email messages from about 150 users. Data is pre-processed by removing all words not in a vocabulary dictionary of $28,102$ words.
- *NYTimes* news dataset: contains $300,000$ news articles from *New York Times*. Data is pre-processed by removing stopwords (not in a dictionary of $102,660$ words).
- *Wikipedia* articles: contains $1M$ documents downloaded from *Wikipedia*. Data is proceeded before usage by removing all words not in a vocabulary dictionary of $7,700$ words.

Settings the Parameters: In all experiments, α and η are fixed at 0.01 and the number of topics $K = 50$. We evaluated a range of settings of the learning parameters, κ, τ_0, and *batch* on all corpora. The parameters κ and τ_0, defined in [14], control the learning step-size ρ_t. We use 29,861 emails from *Enron* data, 50,000 news articles from *NYTimes* data and 300,000 documents from *Wikipedia* data as training sets. We also reserve 5,000 documents as a validation set and another 5,000 documents as a testing set. The online LDA is run (one time per corpus) on the training sets for $\kappa \in \{0.5, 0.7, 0.9\}$, $\tau_0 \in \{1, 24, 256, 1024\}$, and *batch* $\in \{16, 64, 256, 1024\}$. Table 1 summarises the best settings of each *batch* along with the resulting perplexity on the test set for each corpus.

Comparing Serial Online LDA and Asynchronous LDA: for each dataset, we set the parameters setting that give the best performance (least perplexity). ASYSVI LDA is then compared against serial SVI LDA using the same parameters setting.

The empirical results shown in this paper are obtained from a python implementation on high-performance computing (HPC) environment using message passing interface (MPI) for python (MPI4py). The cluster consists of 10 nodes, excluding the head node, with each node is a four-core processor. We run AYSVI LDA on *Enron* dataset for number of workers $nW \in \{2, 4, 6, 8, 9, 18, 27, 36\}$, B is set to 5. The number of employed nodes is equal to nW as long as nW is less than 9. As nW becomes higher than the available nodes, the processors' cores of nodes are employed as slaves until all the cores of all the nodes are used i.e., $9 \times 4 = 36$. Since the batch size is fixed to 1024, each slave processes a batch of data of size $S = 1024/M$ per iteration, where M is fixed to 36. Thus, the gradient computed by each slave will be multiplied by D/S. Hence, line number 6 of Algorithm 2 becomes, $g_i(\boldsymbol{\lambda}) = \boldsymbol{\alpha} + (D/S)E_{\phi_i(\boldsymbol{\lambda})}[t(\boldsymbol{x}_i, \boldsymbol{z}_i)] - \boldsymbol{\lambda}$.

Figure 1 summarises the total speed-up (i.e., TSP measured at the end of the algorithm) as well as the ratio of serial LDA perplexity to parallel LDA (RSP) on the test set for *Enron* dataset. It shows TSP and RSP results with respect to the number of slaves. It is clear that as long as each node is assigned one slave, the speed-up is linear which demonstrates the convergence analysis done in Sect. 2. Linear speed-up slowly converts to sub-linear as solo machines host more than one slave. The main reason of such behaviour is the communication delay caused by the increase of the network traffic. Hence, TSP is affected by the hardware. The communication cost starts affecting the speed-up when it becomes comparable to the local computation. Hence, increasing the local computation by increasing the batch size can be adopted to soften the communication effect. However, this decreases the convergence rate and increase local memory load. Hence, a balanced trade-off should be considered. RSP in Fig. 1 shows that although the speed of online LDA has been increased up to 15 times, performance is not seriously affected. We also evaluate TSP and RSP on *NYTimes* and *Wikipedia* for $nW = 27$. The processing speed of Online LDA on *NYTimes* has been increased 19.29 times, $TSP = 19.29$, with slight loss of performance, $RSP = 0.97$. For *Wikipedia*, $TSP = 18.58$ and $RSP = 0.94$.

Figures 2, 3 and 4 present TSP and RSP with respect to streaming samples from *Enron*, *NYTimes* and *Wikipedia* datasets. These figures show the performance of ASYSVI in a true online setting where the algorithm continually collects samples from the hard driver for the case of *Enron* and *NYTimes* or by downloading online in the case of *Wikipedia*. The perplexity is obtained online on the coming batches before being used to update the model parameters. The plots in the figures are slightly softened using a low-pass filter in order to make them easy to read. These plots show that the speed-up becomes invariant as more samples are processed. The poor speed-up in the beginning is normally caused by initialization and loading process. It can be noticed that the performance of ASYSVI LDA suffers at the beginning then it becomes comparable to online LDA after certain number of iterations. This behaviour can be explained by the convergence condition shown in Sect. 2 (T is greater than $O(B^2)$). Thus, as the number of iterations increases, the convergence of ASYSVI LDA is guaranteed and its performance becomes comparable to that of online LDA. Hence, RSP approaches 1.

5   Conclusion and Discussion

We have introduced ASYSVI, an asynchronous parallel implementations for SVI on computer cluster. ASYSVI leads to linear speed-up, while guaranteeing an asymptotic convergence rate given some assumptions involving the number of the slaves and iterations. Empirical results using latent Dirichlet allocation topic model as a case study have demonstrated the advantages of ASYSVI over SVI, particularly with respect to the key issue of speeding-up the computation while maintaining comparable performance to SVI.

In future work, vertical parallelism can be adopted along with the proposed horizontal one leading to a hybrid horizontal-vertical parallelism. In such case, multi-core processors will be used for the vertical parallelism, while horizontal parallelism is achieved on a multi-node machine. Another avenue of interest is to derive an algorithm for streaming, distributed, asynchronous inference where the number of instances is not known. Moreover, it is interesting to apply ASYSVI on very large scale problems and particularly on other models of the family discussed in Sect. A and studying the effect of the statistical properties of those models.

A   Background

We derive the model family studied in this paper and review SVI following the same pattern as in [3].

Model Family. Our family of models consists of three random variables: observations $x = x_{1:n}$, local hidden variables $z = z_{1:n}$, global hidden variables β and fixed parameters α. The model assumes that the distribution of the n pairs of

(x_i, z_i) is conditionally independent given β. Further, their distribution and the prior distribution of β are in an exponential family:

$$p(\beta, x, z | \alpha) = p(\beta | \alpha) \prod_{i=1}^{n} p(z_i, x_i | \beta), \tag{6}$$

$$p(z_i, x_i | \beta) = h(x_i, z_i) \exp\left(\beta^T t(x_i, z_i) - a(\beta)\right), \tag{7}$$

$$p(\beta | \alpha) = h(\beta) \exp\left(\alpha^T t(\beta) - a(\alpha)\right) \tag{8}$$

Here, we overload the notation for the base measures $h(.)$, sufficient statistics $t(.)$ and log normalizer $a(.)$. While the proposed approach is generic, we assume a conjugacy relationship between (x_i, z_i) and β. That is, the distribution $p(\beta | x, z)$ is in the same family as the prior $p(\beta | \alpha)$.

Note that this innocent looking family of models includes (but is not limited to) latent Dirichlet allocation [13], Bayesian Gaussian mixture, probabilistic matrix factorization, hidden Markov models, hierarchical linear and probit regression, and many Bayesian non-parametric models.

Mean-Field Variational Inference. Variational inference (VI) approximates intractable posterior $p(\beta, z | x)$ by positing a family of simple distributions $q(\beta, z)$ and find the member of the family that is closest to the posterior (closeness is measured with KL divergence). The resulting optimization problem is equivalent maximizing the evidence lower bound (ELBO):

$$\mathcal{L}(q) = E_q[\log p(x, z, \beta)] - E_q[\log p(z\beta)] \leq \log p(x) \tag{9}$$

Mean-field is the simplest family as it allows the distribution over hidden variables to factorize:

$$q(\beta, z) = q(\beta | \lambda) \prod_{i=1}^{n} p(z_i | \phi_i) \tag{10}$$

Each variational distribution is assumed to come from the same family of the true one. Mean-field VI optimizes the new ELBO with respect to the local and global variational parameters ϕ and λ:

$$\mathcal{L}(\lambda, \phi) = E_q\left[\log \frac{p(\beta)}{q(\beta)}\right] + \sum_{i=1}^{n} E_q\left[\log \frac{p(x_i, z_i | \beta)}{q(z_i)}\right] \tag{11}$$

It iteratively updates each variational parameter holding the others fixed. With the assumptions taken so far, each update has a closed form solution. The local parameters are a function of the global ones:

$$\phi(\lambda_t) = \arg\max_{\phi} \mathcal{L}(\lambda_t, \phi) \tag{12}$$

The global parameters summarise the dataset (clusters in Bayesian Gaussian mixture, topics in LDA):

$$\mathcal{L}(\lambda) = \max_{\phi} \mathcal{L}(\lambda, \phi) \tag{13}$$

To find optimal $\boldsymbol{\lambda}$ given fixed $\boldsymbol{\phi}$, we compute the natural gradient of $\mathcal{L}(\boldsymbol{\lambda})$ and set it to zero by setting:

$$\boldsymbol{\lambda}^* = \boldsymbol{\alpha} + \sum_{i=1}^{n} E_{\phi_i(\boldsymbol{\lambda}_t)}[t(\boldsymbol{x}_i, \boldsymbol{z}_i)] \tag{14}$$

Thus, the new optimal global parameters are $\boldsymbol{\lambda}_{t+1} = \boldsymbol{\lambda}^*$. The algorithm works by iterating between computing the optimal local parameters given the global ones (Eq. (12)) and vice versa (Eq. (14)).

Stochastic Variational Inference. Rather than analysing all the data to compute $\boldsymbol{\lambda}^*$ at each iteration, stochastic optimization can be used. Assuming that the data is uniformity at random selected from the dataset, an unbiased noisy estimator of $\mathcal{L}(\boldsymbol{\lambda}, \boldsymbol{\phi})$ can be developed based on a single data point:

$$\mathcal{L}_i(\boldsymbol{\lambda}, \boldsymbol{\phi}_i) = E_q\left[\log \frac{p(\boldsymbol{\beta})}{q(\boldsymbol{\beta})}\right] + nE_q\left[\log \frac{p(\boldsymbol{x}_i, \boldsymbol{z}_i|\boldsymbol{\beta})}{q(\boldsymbol{z}_i)}\right] \tag{15}$$

The unbiased stochastic approximation of the ELBO as a function of $\boldsymbol{\lambda}$ can be written as follows:

$$\mathcal{L}_i(\boldsymbol{\lambda}) = \max_{\phi_i} \mathcal{L}_i(\boldsymbol{\lambda}, \boldsymbol{\phi}_i) \tag{16}$$

Following the same step in the previous section, we obtain a noisy unbiased estimate of Eq. (14):

$$\hat{\boldsymbol{\lambda}} = \boldsymbol{\alpha} + nE_{\phi_i(\boldsymbol{\lambda}_t)}[t(\boldsymbol{x}_i, \boldsymbol{z}_i)] \tag{17}$$

Iteratively, we move the global parameters a step-size ρ_t in the direction of the noisy natural gradient:

$$\boldsymbol{\lambda}_{t+1} = (1 - \rho_t)\boldsymbol{\lambda}_t + \rho_t \hat{\boldsymbol{\lambda}} \tag{18}$$

With certain conditions on ρ_t, the algorithm converges ($\sum_{t=1}^{\infty} \rho_t = \infty$, $\sum_{t=1}^{\infty} \rho_t^2 < \infty$) [4].

B   Related Work

Few work has been proposed to scale VI to large datasets. We can distinguish two major classes. The first class is based on the Bayesian filtering approach [16,17]. That is, the sequential nature of Bayes theorem is exploited to recursively update an approximation of the posterior. Particularly, VI is used between the updates to approximate the posterior which becomes the prior of the next step. Author in [16] uses forgetting factors to decay the contribution of old data in favour of a new better one. The algorithm proposed in [17] considers a sequence of data batches and iterates over the data in each batch until convergence. Relying on a master-slave architecture, the computation of the batches posterior is done in a distributed and asynchronous manner. That is, the algorithm applies VI by performing asynchronous Bayesian updates to the posterior as data batches arrive continuously.

The second class of work is based on optimization [3,12,18]. As we already discussed, SVI proposed by [3] employs stochastic optimization to scale up Bayesian computation to massive data. SVI is inherently serial and requires the model parameters to fit in the memory of a single processor. Authors in [18] present a VI based inference algorithm that runs in parallel on data divided across several slaves. However at each iteration, the slaves are synchronized to combine their obtained parameters. Such synchronisation limits the scalability and decreases the speed of the update to that of the slowest slave. To avoid bulk synchronization, authors in [12] propose an asynchronous and lock-free update. In this update, vertical parallelism is adopted, where each processor asynchronously updates a subset of the parameters based on a subset of attributes. In contrast, we adopt horizontal parallelism update based on few (mini-batched or single) data points acquired from distributed sources. The update steps are aggregated to form the global update. Our proposed approach can make use of the mechanism proposed by [12] to achieve a hybrid horizontal-vertical parallelism. On the contrary to [12], our approach is not customised for LDA and can be applied to any of the model family presented in Sect. A.

References

1. Andrieu, C., De Freitas, N., Doucet, A., Jordan, M.I.: An introduction to MCMC for machine learning. Mach. Learn. **50**(1–2), 5–43 (2003)
2. Wainwright, M.J., Jordan, M.I.: Graphical models, exponential families, and variational inference. Found. Trends® Mach. Learn. **1**(1–2), 1–305 (2008)
3. Hoffman, M.D., Blei, D.M., Wang, C., Paisley, J.: Stochastic variational inference. J. Mach. Learn. Res. **14**(1), 1303–1347 (2013)
4. Robbins, H., Monro, S.: A stochastic approximation method. Ann. Math. Stat. 400–407 (1951)
5. Recht, B., Re, C., Wright, S., Niu, F.: Hogwild: a lock-free approach to parallelizing stochastic gradient descent. In: Advances in Neural Information Processing Systems, pp. 693–701 (2011)
6. Agarwal, A., Duchi, J.C.: Distributed delayed stochastic optimization. In: Advances in Neural Information Processing Systems, pp. 873–881 (2011)
7. Zhang, R., Kwok, J.T.: Asynchronous distributed ADMM for consensus optimization. In: ICML, pp. 1701–1709 (2014)
8. Feyzmahdavian, H.R., Aytekin, A., Johansson, M.: An asynchronous mini-batch algorithm for regularized stochastic optimization. IEEE Trans. Autom. Control **61**(12), 3740–3754 (2016)
9. Mania, H., Pan, X., Papailiopoulos, D., Recht, B., Ramchandran, K., Jordan, M.I.: Perturbed iterate analysis for asynchronous stochastic optimization. arXiv preprint arXiv:1507.06970 (2015)
10. Bertsekas, D.P., Tsitsiklis, J.N.: Parallel and Distributed Computation: Numerical Methods, vol. 23. Prentice Hall, Englewood Cliffs (1989)
11. Lian, X., Huang, Y., Li, Y., Liu, J.: Asynchronous parallel stochastic gradient for nonconvex optimization. In: Advances in Neural Information Processing Systems, pp. 2737–2745 (2015)
12. Raman, P., Zhang, J., Yu, H.-F., Ji, S., Vishwanathan, S.V.N.: Extreme stochastic variational inference: distributed and asynchronous. arXiv preprint arXiv:1605.09499 (2016)

13. Blei, D.M., Ng, A.Y., Jordan, M.I.: Latent dirichlet allocation. J. Mach. Learn. Res. **3**, 993–1022 (2003)
14. Hoffman, M., Bach, F.R., Blei, D.M.: Online learning for latent dirichlet allocation. In: Advances in Neural Information Processing Systems, pp. 856–864 (2010)
15. Lichman, M.: UCI Machine Learning Repository (2013)
16. Honkela, A., Valpola, H.: On-line variational Bayesian learning. In: 4th International Symposium on Independent Component Analysis and Blind Signal Separation, pp. 803–808 (2003)
17. Broderick, T., Boyd, N., Wibisono, A., Wilson, A.C., Jordan, M.I.: Streaming variational bayes. In: Advances in Neural Information Processing Systems, pp. 1727–1735 (2013)
18. Neiswanger, W., Wang, C., Xing, E.: Embarrassingly parallel variational inference in nonconjugate models. arXiv preprint arXiv:1510.04163 (2015)

Probabilistic Bounds for Binary Classification of Large Data Sets

Věra Kůrková[1] and Marcello Sanguineti[2(✉)]

[1] Institute of Computer Science, Czech Academy of Sciences, Prague, Czech Republic
vera@cs.cas.cz
[2] Department of Computer Science, Bioengineering, Robotics, and Systems
Engineering (DIBRIS), University of Genoa, Genoa, Italy
marcello.sanguineti@unige.it

Abstract. A probabilistic model for classification of task relevance is investigated. Correlations between randomly-chosen functions and network input-output functions are estimated. Impact of large data sets is analyzed from the point of view of the concentration of measure phenomenon. The Azuma-Hoeffding Inequality is exploited, which can be applied also when the naive Bayes assumption is not satisfied (i.e., when assignments of class labels to feature vectors are not independent).

Keywords: Binary classification ·
Approximation by feedforward networks · Concentration of measure ·
Azuma-Hoeffding inequality

1 Introduction

It has long been known that many widely used classes of feedforward networks can exactly compute any function on a finite domain. In particular, shallow (i.e., one-hidden-layer) networks with computational units satisfying mild conditions can compute any finite mapping [9]. Thus shallow networks of many types can exactly perform any classification task on any finite set of data. Nevertheless, arguments proving the universal representation property of various classes of networks assume that the number of units is as large as the size of the domain of functions to be computed [9]. Thus for large sets of data to be classified, networks guaranteed by the universality-type arguments might not be suitable for efficient performance.

The number of all binary classification tasks on a given domain grows exponentially with its size. Even for domains of moderate sizes, it quickly reaches the estimated number 10^{80} of atoms in the Universe (see, e.g., [14]), which is smaller than the number 2^{267} of all binary classifiers on a domain of size 267. However, in real tasks typically most functions from the huge sets of all possible binary classifiers on a given domain are not likely to model any case of a practical interest. Relevance of such functions for a choice of a neural network class suitable for efficient computation is very low or negligible.

© Springer Nature Switzerland AG 2020
L. Oneto et al. (Eds.): INNSBDDL 2019, INNS 1, pp. 309–319, 2020.
https://doi.org/10.1007/978-3-030-16841-4_32

In [11], we introduced a probabilistic approach modeling a prior knowledge that a function is likely to occur in a given application. We assumed therein that elements of a finite domain represent vectors of features, measurements, or observations for which some knowledge is available about probabilities that a presence of each of these features implies a property described by one of the classes. For example in some medical applications, domains are formed by vectors of symptom values (such as blood pressure, temperature, cholesterol, etc.) and functions on such domains represent classifications of these vectors to ill/healthy. It is unlikely that patients with small values of some features should be diagnosed as ill.

In machine learning, typically the data are assumed to be independent and identically distributed (i.i.d.) [4,18]. Under this assumption, stochastic separation theorems for classification were derived in [7,8]. Also in [11,12], we assumed independence of probabilities that each of the vectors belongs to one of the classes. So we assumed that the probability modeling relevance of tasks on the set of all binary classifiers on a domain formed by m feature vectors can be expressed as a product of m independent random variables. In [11,12], we analyzed consequences of the concentration of measure phenomenon [13] (which holds for "sufficiently large" values of m) on correlations between network units and functions to be computed. To derive probabilistic bounds we used the Chernoff-Hoeffding Bound on sums of independent random variables not necessarily identically distributed.

However, independence of random variables is a strong assumption. The hypothesis that a probability distribution can be expressed as a product probability is called the "naive Bayes assumption" [16]. It supposes that class labels are assigned to vectors of feature values independently of each other. Often in real tasks, this is not satisfied. For example in the above mentioned case of medical applications, if a patient with certain values of symptoms is likely to be diagnosed as ill, another one with larger values of these symptoms is even more likely to be ill. While many tools for investigation of product probabilities are available, for the case of dependent variables sophisticated tools based on theory of martingales are applicable, which only hold under various special assumptions [6].

In this paper, we investigate classes of binary classification tasks characterized by general probability distributions, which might not be expressible as product distributions. We analyze effects of increasing sizes of data sets on probabilities that most randomly-chosen functions are correlated with some input-output functions of feedforward networks. To characterize such correlations for tasks modeled by probability distributions that do not satisfy the independence assumption, we exploit the Hoeffding-Azuma Inequality, which applies to expected values of functions of random variables under the "average bounded differences condition" [6]. The inequality states that probabilities of large deviations of a function of several random variables satisfying this condition from its expected value are decreasing exponentially fast with the amounts of the deviations [3,6]). Using this inequality we prove that on large sets of data, inner products of randomly-chosen classifiers with any fixed input-output function are sharply concentrated around the mean value of these inner products.

The paper is organized as follows. In Sect. 2, we introduce notations and basic concepts on feedforward networks and approximation. In Sect. 3, the probabilistic model is introduced. Section 4 investigates effects of increasing sizes of sets of data to be classified. The deviation of inner products of randomly-chosen functions with any given function from their expected value is estimated without the assumption of independence. In Sect. 5, consequences of probabilistic results for approximation by neural networks are analyzed. Section 6 is a brief discussion.

2   Feedforward Networks

A *feedforward network with a single linear output* computes input-output functions from the set

$$\operatorname{span} G := \left\{ \sum_{i=1}^{n} w_i g_i \,\middle|\, w_i \in \mathbb{R},\ g_i \in G,\ n \in \mathbb{N} \right\},$$

where G, called a *dictionary*, is a parameterized family of functions. In networks with one hidden layer (called *shallow*), G is formed by functions computable by a given type of computational units. In networks with several hidden layers (called *deep*), it is formed by combinations and compositions of functions representing units from lower layers (see, e.g., [2, 15]).

For binary classification tasks, *one-hidden-layer networks with single threshold output units* are used. Such networks with n units compute functions from sets

$$\operatorname{sgn} \operatorname{span}_n G := \left\{ \operatorname{sgn} \sum_{i=1}^{n} w_i g_i \,\middle|\, w_i \in \mathbb{R},\ g_i \in G \right\},$$

where sgn denotes the *signum function* defined as

$$\operatorname{sgn}(t) := -1 \text{ for } t < 0 \quad \text{and} \quad \operatorname{sgn}(t) := 1 \quad \text{for } t \geq 0.$$

Dictionaries are parameterized families of functions of the form

$$G_\phi(X, Y) := \{ \phi(\cdot, y) : X \to \mathbb{R} \,|\, y \in Y \},$$

where $\phi : X \times Y \to \mathbb{R}$ is a function of two variables: an input vector $x \in X \subseteq \mathbb{R}^d$ and a parameter vector $y \in Y \subseteq \mathbb{R}^s$. When the set of parameters is the whole $\mathbb{R}^s$, we write shortly $G_\phi(X)$.

For a domain $X \subset \mathbb{R}^d$ we denote by

$$\mathcal{F}(X) := \{ f \,|\, f : X \to \mathbb{R} \}$$

the *set of all real-valued functions on* X and by

$$\mathcal{B}(X) := \{ f \,|\, f : X \to \{-1, 1\} \}$$

the *set of all functions on* X *with values in* $\{-1, 1\}$.

In practical applications, the domain $X \subset \mathbb{R}^d$ is finite but its size card X and/or the input dimension d can be quite large. It is easy to see that when card $X = m$ and $X = \{x_1, \ldots, x_m\}$ is a linear ordering of X, then the mapping $\iota : \mathcal{F}(X) \to \mathbb{R}^m$ defined as $\iota(f) := (f(x_1), \ldots, f(x_m))$ is an isomorphism. So, on $\mathcal{F}(X)$ we have the Euclidean inner product defined as $\langle f, g \rangle := \sum_{u \in X} f(u)g(u)$ and the Euclidean norm $\|f\| := \sqrt{\langle f, f \rangle}$.

In the analysis of approximation capabilities of neural networks, it has to be taken into account that the approximation error $\|f - \operatorname{span}_n G\|$ in any norm $\|.\|$ can be made arbitrarily large by multiplying f by a scalar. Indeed, for every $c > 0$ one has

$$\|cf - \operatorname{span}_n G\| = c\|f - \operatorname{span}_n G\|.$$

Thus approximation errors have to be studied either in sets of normalized functions or in sets of functions of a given fixed norm. Thus it is convenient to consider binary-valued functions with the range $\{-1, 1\}$ instead of $\{0, 1\}$. All functions in $\mathcal{B}(X)$ have norms equal to $\sqrt{\operatorname{card} X}$.

The following proposition [10, Proposition 3.1] shows that when functions to be approximated are binary-valued, then lower bounds on approximation errors by networks with single linear output units can be obtained from lower bounds holding for networks with single threshold output units.

Proposition 1. Let $X \subset \mathbb{R}^d$ be finite, $f \in \mathcal{B}(X)$, and $h \in \mathcal{F}(X)$. Then $\|f - h\|_2 \geq \frac{1}{2} \|f - \operatorname{sgn}(h)\|_2$.

When two functions f, h have the same norms, then their Euclidean distance $\|f - h\|_2$ depends on the angle $\arccos(\langle f/\|f\|_2, h/\|h\|_2 \rangle)$, since $\|f - h\|_2 = 2(\|f\|^2 - \langle f, h \rangle)$. So a function $f \in \mathcal{B}(X)$ is well approximated by a function $h \in \mathcal{B}(X)$ when the inner product $\langle f, h \rangle$ is large, i.e., when f and h are *correlated*. On the other hand, when f is nearly orthogonal to all input-output functions computable by a class of networks, then it cannot be well-approximated by such networks.

3   Probabilistic Model of Relevance of Tasks

Typically in real classification tasks, most binary-valued functions on a given domain are not likely to represent any task of interest. A prior knowledge about relevance of these tasks can be expressed by a probability measure on the set of all functions on the domain.

We model the relevance of tasks for a given application by a discrete probability measure ρ on the set $\mathcal{B}(X)$ of all $\{-1, 1\}$-valued functions on the domain $X = \{x_1, \ldots x_m\} \subset \mathbb{R}^d$. A function $f \in \mathcal{B}(X)$ can be represented as a vector $(f(x_1), \ldots, f(x_m)) \in \{-1, 1\}^m \subset \mathbb{R}^m$. Hence, ρ can be considered as a probability distribution on the set $\{-1, 1\}^m$. For every f randomly chosen in $\mathcal{B}(X)$ according to ρ, we have a set of random variables

$$Y_1 := f(x_1), \ldots, Y_m := f(x_m).$$

Without any prior knowledge, one has to assume that the probability ρ is uniform. In such a case, each randomly-chosen function has the same probability $1/2^m$. Typically in real applications, probabilities modeling relevance of functions are far from being uniform. Often, they are non zero only on relatively small subsets of $\mathcal{B}(X)$. The simplest special case is when the probability ρ on the set $\mathcal{B}(X) \simeq \{-1, 1\}^m$ is a *product probability*, which can be expressed as

$$\rho(f) := \prod_{i=1}^{m} \rho_i(f), \tag{1}$$

where for each $i = 1, \ldots, m$, ρ_i is a probability on $\{-1, 1\}$. It can be characterized by $p_i \in [0, 1]$ for which

$$\mathrm{Pr}\,(Y_i = 1) = p_i \quad \text{and} \quad \mathrm{Pr}\,(Y_i = -1) = 1 - p_i.$$

When ρ is a product distribution, then the random variables $Y_1 = f(x_1), \ldots,$ $Y_m = f(x_m)$ are independent. The assumption of independence is quite strong. In many real applications it does not hold. For example, let the set $X = \{x_1, \ldots x_m\}$ of data to be classified be formed by vectors $x_i \in \mathbb{R}^d$ representing some symptoms and the probability that vectors with larger values of some of these symptoms are classified as belonging to the class 1 cannot be smaller than the probability that vectors with smaller values of the same entries belong to the same class. Such cases can be formally described in terms of an ordering of the set X and the following condition on the probability distribution

$$x_i \leq x_j \quad \Rightarrow \quad \mathrm{Pr}(Y_i = 1) \leq \mathrm{Pr}(Y_j = 1).$$

4   Effects of Increasing Sizes of Data Set

In this section, we investigate effects of increasing sizes of sets of data to be classified. When such set X is large, then the set $\mathcal{F}(X)$ of all functions on X is isometric to a high-dimensional Euclidean space and the set $\mathcal{B}(X)$ of all binary classifiers on X is isomorphic to a high-dimensional Hamming cube. High-dimensional geometry manifests in rather counter-intuitive properties. Various concentration of measure phenomena occur [3], [6, p. 140]. They imply that a function of a large number of random variables tends to concentrate its values in a relatively narrow range. This holds under certain combinations of conditions on independence of the random variables and on smoothness of the function. We show that inner products satisfy such conditions even when the random variables are not independent.

In the special case when the random variables are independent (but not necessarily identically distributed), the Chernoff-Hoeffding Bound [5, Theorem 1.11] implies sharp concentration of their sum around its mean value. In [11], we applied it to investigation of correlations of randomly-chosen classifiers according to product probabilities of large data sets.

Here, we investigate the more natural case when the probability cannot be expressed as a product and so the assumption of independence of random variables does not hold. Various generalizations of the Chernoff-Hoeffding Bound hold, which are formulated not merely for sums, but more generally, for "well-behaved" functions $\Phi(Y_1, \ldots, Y_m)$ of random variables $Y_1, \ldots, Y_m$, which satisfy certain "bounded differences conditions". They follow from the Azuma-Hoeffding Inequality and the theory of martingales (see, e.g., [1,3], [6, Theorem 5.1, p. 62], [6, p. 58]). We exploit one of these generalizations, called the *method of averaged bounded differences*. It applies to functions that satisfy the *averaged Lipschitz condition (ALC)* [6, p. 68]. We say that a function $\Phi : \mathbb{R}^m \to \mathbb{R}$ of m real variables satisfies the ALC with parameters $c_i, i = 1, \ldots, m$, with respect to the random variables $Y_1, \ldots, Y_m$ with values in $A_1, \ldots, A_m$, resp., if for every $i = 1 \ldots, m$ and every $a_i, \bar{a}_i \in A_i$ the following bound holds on conditioned expected values:

$$\left| E\left(\Phi \mid \mathbf{Y}_{i-1}, Y_i = a_i\right) - E\left(\Phi \mid \mathbf{Y}_{i-1}, Y_i = \bar{a}_i\right) \right| \leq c_i.$$

Intuitively, the ALC states that assigning some values to the first $i-1$ random variables, letting the i-th variable take two different values, and setting randomly (according to the given distribution) the values of the remaining variables, the difference between the two corresponding partial averages of Φ is uniformly bounded by c_i. The following bound [6, p. 68, Corollary 5.1] on deviation from expectation holds for functions satisfying the ALC, without the assumption of independence of the random variables.

Theorem 1 (Method of averaged bounded differences). *Let $\Phi : \mathbb{R}^m \to \mathbb{R}$ be a function of m random variables that satisfies the ALC condition with parameters $c_i, i = 1, \ldots, m$, with respect to the random variables Y_i, each taking values in A_i, $i = 1, \ldots, m$, and $c := \sum_{i=1}^{m} c_i^2$. Then*

$$\Pr\left(|\Phi - E(\Phi)| > t\right) \leq e^{-2t^2/c}.$$

Theorem 1 guarantees that the probabilities of large deviations of Φ from its expected value decrease exponentially fast with the amounts of the deviations. We apply Theorem 1 to inner products with a fixed function $h \in \mathcal{B}(X)$. By $h^\circ := h/\|h\|$ we denote the normalization of h.

Theorem 2. *Let $X = \{x_1, \ldots, x_m\} \subset \mathbb{R}^d$ and $h \in \mathcal{B}(X)$. Then the inner product of h with f randomly chosen from $\mathcal{B}(X)$ according to the probability ρ satisfies for every $\lambda > 0$*

$$(i)\ \Pr\left(|\langle f, h \rangle - \mu(h, \rho)| > m\lambda\right) \leq e^{-\frac{m\lambda^2}{2}};$$

$$(ii)\ \Pr\left(|\langle f^\circ, h^\circ \rangle - \frac{\mu(h, \rho)}{m}| > \lambda\right) \leq e^{-\frac{m\lambda^2}{2}}.$$

Proof. Let $Y_i := f(x_i)$, $i = 1, \ldots, m$, be random variables and define the function $\Phi_h = <\cdot, h>$ induced by the inner product with h, i.e.,

$$\Phi_h(Y_i, \ldots, Y_m) := \sum_{i=1}^{m} h(x_i) Y_i.$$

To verify that Φ_h satisfies the ALC with respect to $Y_1, \ldots, Y_m$, let $a_1, \ldots, a_i, \bar{a}_i \in \{-1, 1\}$, $\mathbf{a}_{i-1} := (a_1, \ldots, a_{i-1})$, and $\mathbf{Y}_{i-1} := (Y_1, \ldots, Y_{i-1})$. Then

$$\left| E\left(<f, h> | \mathbf{Y}_{i-1} = \mathbf{a}_{i-1}, Y_i = a_i\right) - E\left(<f, h> | \mathbf{Y}_{i-1} = \mathbf{a}_{i-1}, Y_i = \bar{a}_i\right) \right|$$

$$= \left| \sum_{j=1}^{i-1} a_j\, h(x_j) + a_i\, h(x_i) + E \sum_{j=i+1}^{m} f(x_j)\, h(x_j) + \right.$$

$$\left. - \sum_{j=1}^{i-1} a_j\, h(x_j) - \bar{a}_i\, h(x_i) - E \sum_{j=i+1}^{m} f(x_j)\, h(x_j) \right| = \left| (a_i - \bar{a}_i) h(x_i) \right| \leq 2.$$

So (i) follows from Theorem 1 with $t = m\lambda$ and (ii) is obtained via normalization. $\square$

Theorem 2 shows that on a large domain X, most inner products of functions randomly chosen in $\mathcal{B}(X)$ with any fixed function $h \in \mathcal{B}(X)$ are concentrated around their mean value. For example, setting $\lambda = m^{-1/4}$, we get the bound $e^{-\frac{m^{-1/2}}{2}}$, which decreases exponentially fast with increasing m. Thus for large sets of data and any given function h, correlations of most randomly chosen classifiers according to the given probability are sharply concentrated around the mean value of their inner products with h.

5 Suitability of Classes of Networks for Classification Tasks

In this section, we analyze consequences of Theorem 1 on correlations between randomly chosen classifiers and input-output functions of feedforward networks. We show that an observable performance of given type of networks on a randomized set of classification tasks on large sets of data is almost deterministic in the sense that either a given class of networks can approximate well almost all tasks or none of them.

For a probability measure ρ on $\mathcal{B}(X)$ and a function $h \in \mathcal{B}(X)$, we denote by

$$\mu(h, \rho) := E_\rho(\langle h, f \rangle \,|\, f \in \mathcal{B}(X))$$

the *mean value of inner products* of h with f randomly chosen from $\mathcal{B}(X)$ according to ρ. For a fixed ρ, the quantity $\mu(h, \rho)$ varies as a function of h. Let $X = \{x_1, \ldots, x_m\} \subset \mathbb{R}^d$, $h \in \mathcal{B}(X)$, and f be randomly chosen in $\mathcal{B}(X)$ according to ρ. For $i = 1, \ldots, m$, we define

$$h_\rho(x_i) := \mathrm{sgn}\left[E_\rho[f(x_i)] \right].$$

It follows immediately that

$$\mu(h_\rho, \rho) = \max_{h \in \mathcal{B}(X)} \mu(h, \rho).$$

For $S \subset \mathcal{B}(X)$ a set of input-output functions of a feedforward network with a threshold output we denote

$$\mu_S(\rho) := \max_{h \in S} \mu(h, \rho). \qquad (2)$$

The next corollary estimates the concentration of inner products around $\mu_S(\rho)/m$.

Corollary 1. *Let* $X = \{x_1, \ldots, x_m\} \subset \mathbb{R}^d$, $S \subset \mathcal{B}(X)$, *and* card $S = k$. *Then for every* $f \in \mathcal{B}(X)$ *randomly chosen according to* ρ *and every* $\lambda > 0$

$$\Pr\left(|<f^o, h^o>| \leq \frac{\mu_S(\rho)}{m} + \lambda \right) > 1 - k\, e^{-\frac{m\lambda^2}{2}}.$$

Proof. By Theorem 2 (ii), we get $\Pr\left(\left| \langle f^o, h^o \rangle - \frac{\mu(h,\rho)}{m} \right| > \lambda\ \forall h \in S \right) \leq ke^{-\frac{m\lambda^2}{2}}$. Hence,

$$\Pr\left(\left| \langle f^o, h^o \rangle - \frac{\mu(h,\rho)}{m} \right| \leq \lambda\ \forall h \in S \right) > 1 - ke^{-\frac{m\lambda^2}{2}}.$$

As $\left| \langle f^o, h^o \rangle - \frac{\mu(h,\rho)}{m} \right| \leq \lambda$ implies $|\langle f^o, h^o \rangle| \leq \frac{\mu(h,\rho)}{m} + \lambda$, we get

$$\Pr\left(|\langle f^o, h^o \rangle| \leq \frac{\mu(h,p)}{m} + \lambda\ \forall h \in S \right) > 1 - ke^{-\frac{m\lambda^2}{2}}.$$

By (2) for every $h \in S$, $\mu(h, \rho) \leq \mu_S(\rho)$ and so the statement holds.   □

To get some insights from our estimates, consider a fixed value of the deviation λ from the expected value of the inner product and assume that for an input-output function h, $\mu(h, \rho) = cm$, where $c \in (0, m]$. Then by Theorem 2 we get

$$\Pr\left(|\langle f^o, h^o \rangle - c| > \lambda \right) \leq e^{-\frac{m\lambda^2}{2}}.$$

Thus the larger the value of m, the sharper the concentration of inner products around c. If c is close to 1, then there us almost all randomly chosen classifiers are highly correlated with h. Thus a class of networks which can compute h or a function close to h is suitable for almost all classification tasks encountered in the application area described by ρ. If, instead, c is small, we have to look at other input-output functions h computable by the network class.

The worst situation is when for all h computable by the class, mean values of inner products of h with random functions selected according to ρ are small. By Corollary 1, if, e.g., $\mu_S(\rho) = 0$ then

$$\Pr\left(|<f^o, h^o>| \leq \lambda\right) > 1 - k\,e^{-\frac{m\lambda^2}{2}}.$$

So, Corollary 1 shows that when the set of data to be classified X is sufficiently large and for all functions h from the set S of input-output functions of a given class of feedforward networks with the threshold output the absolute values of the mean values $\mu(h, \rho)$ are small, then unless the size of S outweighs the factor $e^{-\frac{m\lambda^2}{2}}$, almost any randomly-chosen function with respect to ρ is nearly orthogonal to all elements of $\mathcal{B}(X)$.

So, our results imply that when all input-output functions computable by a given class of networks have small mean values of their inner products with randomly-chosen classifiers, then such class of networks can efficiently perform the classification task modeled by the probability distribution only when the size of the set of its input-output functions grows exponentially with the size of the set of data. In particular, this happens when there is no prior knowledge about the task and thus uniform distribution has to be assumed. On the other hand, if (i) a network can compute an input-output function with a relatively large mean value of its inner products with randomly-chosen functions with respect to the probability distribution and (ii) the set of data is "sufficiently large", then almost any randomly-chosen classifier can be approximated quite well by such network.

6   Discussion

To deal with an exponentially growing number of classification tasks on data sets of even moderate sizes, we investigated a probabilistic model describing relevance of binary classification tasks. Since in many real applications, assignments of class labels to feature vectors are not independent, we studied distributions that may not satisfy the naive Bayes assumption. We explored suitability of classes of feedforward networks for approximation of classification tasks in terms of correlations of randomly chosen functions according to given probability with network input-output functions. We proved that performance of a given class of networks on a randomized set of classification tasks on large sets of data is almost deterministic in the sense that either a given class of networks can approximate well almost all randomly selected tasks according to a given probability or none of them. We proved that correlations of randomly-chosen classifiers with any fixed input-output function are sharply concentrated around the mean value of these inner products. To derive our results we exploited the Azuma-Hoeffding Inequality on expected values of "sufficiently smooth" functions (not just sums) of several random variables. We showed that correlation formalized in terms of inner product satisfies the "average bounded differences condition".

Note that correlation plays also an important role in study of the cumulative behavior of a class of networks described by a concept of coherence [17].

It is defined as the maximum of absolute values of inner products of pairs of distinct input-output functions. Loosely speaking, it measures how much two input-output functions are correlated, i.e., similar. Although coherence only reflects the most extreme correlations, it is easy to calculate and it captures in a significant way the behavior of sets of functions. Since we assume a probabilistic framework, we consider expected values of inner products.

Extension of our results to regression task is subject of our future research. To this end, suitable versions of the Azuma-Hoeffding Inequality [6, Chapter 5] can be applied to random variables with more general than binary values.

Acknowledgments. V.K. was partially supported by the Czech Grant Foundation grant GA 18-23827S and by institutional support of the Institute of Computer Science RVO 67985807. M.S. was partially supported by a FFABR grant of the Italian Ministry of Education, University and Research (MIUR). He is Research Associate at INM (Institute for Marine Engineering) of CNR (National Research Council of Italy) under the Project PDGP 2018/20 DIT.AD016.001 "Technologies for Smart Communities" and he is a member of GNAMPA-INdAM (Gruppo Nazionale per l'Analisi Matematica, la Probabilità e le loro Applicazioni - Instituto Nazionale di Alta Matematica).

References

1. Azuma, K.: Weighted sums of certain dependent random variables. Tohoku Math. J. **19**, 357–367 (1967)
2. Bengio, Y., Courville, A.: Deep learning of representations. In: Bianchini, M., Maggini, M., Jain, L. (eds.) Handbook of Neural Information Processing. Springer, Heidelberg (2013)
3. Chung, F., Lui, L.: Concentration inequalities and martingale inequalities: a survey. Internet Math. **3**, 79–127 (2005)
4. Cucker, F., Smale, S.: On the mathematical foundations of learning. Bull. Am. Math. Soc. **39**, 1–49 (2002)
5. Doerr, B.: Analyzing randomized search heuristics: tools from probability theory. In: Theory of Randomized Search Heuristics - Foundations and Recent Developments, chap. 1, pp. 1–20. World Scientific Publishing (2011)
6. Dubhashi, D., Panconesi, A.: Concentration of Measure for the Analysis of Randomized Algorithms. Cambridge University Press, Cambridge (2009)
7. Gorban, A.N., Golubkov, A., Grechuk, B., Mirkes, E.M., Tyukin, I.Y.: Correction of AI systems by linear discriminants: probabilistic foundations. Inf. Sci. **466**, 303–322 (2018)
8. Gorban, A., Tyukin, I.: Stochastic separation theorems. Neural Netw. **94**, 255–259 (2017)
9. Ito, Y.: Finite mapping by neural networks and truth functions. Math. Sci. **17**, 69–77 (1992)
10. Kůrková, V., Sanguineti, M.: Probabilistic lower bounds for approximation by shallow perceptron networks. Neural Netw. **91**, 34–41 (2017)
11. Kůrková, V., Sanguineti, M.: Probabilistic bounds on complexity of networks computing binary classification tasks. In: Krajči, S. (ed.) Proceedings of ITAT 2018. CEUR Workshop Proceedings, vol. 2203, pp. 86–91 (2018)
12. Kůrková, V., Sanguineti, M.: Classification by sparse neural networks. IEEE Trans. Neural Netw. Learn. Syst. (2019). https://doi.org/10.1109/TNNLS.2018.2888517

13. Ledoux, M.: The Concentration of Measure Phenomenon. AMS, Providence (2001)
14. Lin, H., Tegmark, M., Rolnick, D.: Why does deep and cheap learning work so well? J. Stat. Phys. **168**, 1223–1247 (2017)
15. Mhaskar, H.N., Poggio, T.: Deep vs. shallow networks: an approximation theory perspective. Anal. Appl. **14**, 829–848 (2016)
16. Rennie, J., Shih, L., Teevan, J., Karger, D.: Tackling the poor assumptions of Naive Bayes classifiers. In: Proceedings of the 20th International Conference on Machine Learning (ICML 2003) (2003)
17. Tropp, A.: Greed is good: algorithmic results for sparse approximation. IEEE Trans. Inf. Theory **50**, 2231–2242 (2004)
18. Vapnik, V.: The Nature of Statistical Learning Theory. Springer, New York (1997)

Multikernel Activation Functions: Formulation and a Case Study

Simone Scardapane[1]([✉])[iD], Elena Nieddu[2], Donatella Firmani[2][iD], and Paolo Merialdo[2][iD]

[1] DIET Department, Sapienza University of Rome, Rome, Italy
simone.scardapane@uniroma1.it
[2] Department of Engineering, Roma Tre University, Rome, Italy
{elena.nieddu,donatella.firmani,paolo.merialdo}@uniroma3.com

Abstract. The design of activation functions is a growing research area in the field of neural networks. In particular, instead of using fixed point-wise functions (e.g., the rectified linear unit), several authors have proposed ways of learning these functions directly from the data in a non-parametric fashion. In this paper we focus on the kernel activation function (KAF), a recently proposed framework wherein each function is modeled as a one-dimensional kernel model, whose weights are adapted through standard backpropagation-based optimization. One drawback of KAFs is the need to select a single kernel function and its eventual hyper-parameters. To partially overcome this problem, we motivate an extension of the KAF model, in which multiple kernels are linearly combined at every neuron, inspired by the literature on multiple kernel learning. We provide an application of the resulting *multi-KAF* on a realistic use case, specifically handwritten Latin OCR, on a large dataset collected in the context of the 'In Codice Ratio' project. Results show that multi-KAFs can improve the accuracy of the convolutional networks previously developed for the task, with faster convergence, even with a smaller number of overall parameters.

Keywords: Activation function · Multikernel · OCR · Latin

1 Introduction

The recent successes in deep learning owe much to concurrent advancements in the design of activation functions, especially the introduction of the rectified linear unit (ReLU) [7], and its several variants [2,11,13]. Albeit the vast majority of applications consider fixed activation functions, a growing trend of research lately has focused on designing *flexible* ones, that are able to adapt their shape from the data through the use of additional trainable parameters. Common examples of this are parametric versions of ReLU [11], or the beta-Swish function [16].

The work of S. Scardapane was supported in part by Italian MIUR, "*Progetti di Ricerca di Rilevante Interesse Nazionale*", GAUChO project, under Grant 2015YPXH4W_004.

© Springer Nature Switzerland AG 2020
L. Oneto et al. (Eds.): INNSBDDL 2019, INNS 1, pp. 320–329, 2020.
https://doi.org/10.1007/978-3-030-16841-4_33

In the more extreme case, multiple authors have advocated for *non-parametric* formulations, in which the overall flexibility and number of parameters can be chosen freely by the user on the basis of one or more hyper-parameters. As a result, the trained functions can potentially approximate a much larger family of shapes. Different proposals, however, differ on the way in which each function is modeled, resulting in vastly different characteristics in terms of approximation, optimization, and simplicity of implementation. Examples of non-parametric activation functions are the maxout neuron [9,20], defined as the maximum over a fixed number of affine functions of its input, the Fourier activation function [3], defined as a linear combination of a predetermined trigonometric basis expansion, or the Hermitian-based expansion [19]. We refer to [17] for a fuller overview on the topic.

In this paper we focus on the recently proposed kernel activation function (KAF) [17], in which each (scalar) function is modeled as a one-dimensional kernel expansion, with the linear mixing coefficients adapted together with all the other parameters of the network during optimization. In [17] it was shown that KAFs can greatly simplify the design of neural networks, allowing to reach higher accuracies, sometimes with a smaller number of hidden layers. Linking neural networks with kernel methods also allows to leverage a large body of literature on the learning of kernel functions (e.g., kernel filters [14]), particularly with respect to their approximation capabilities. At the same time, compared to ReLUs, KAFs introduce a number of additional design choices, most notably the selection of which kernel function to use, and its eventual hyper-parameters (e.g., the bandwidth of the Gaussian kernel). Although in [17] and successive works we mostly focused on the Gaussian kernel, it is not guaranteed to be the optimal one in all applications.

Contribution of the Paper. To solve the kernel selection problem of KAFs, in this paper we propose an extension inspired to the theory of multiple kernel learning [1,8]. In the proposed *multi-KAF*, different kernels are linearly combined for every neuron through an additional set of mixing coefficients, adapted during training. In this way, the optimal kernel for each neuron (or a specific mixture of them) can be learned in a principled way during the optimization process. In addition, since in our KAF implementation the points where the kernels are evaluated are fixed, a large amount of computation can be shared between the different kernels, leading to a very small computational overhead overall, as we show in the following sections.

Case Study: In Codice Ratio. To show the usefulness of the proposed activation functions, we provide a realistic use case by applying them to the data from the 'In Codice Ratio' (ICR) project [4], whose aim is the automatic transcription of medieval manuscripts from the Vatican Secret Archive.[1] A key component of the project is an OCR tool applied to characters from a Latin handwritten text

---

[1] http://www.archiviosegretovaticano.va/.

(see Sect. 4 for additional details). In [5] we presented a convolutional neural network (CNN) for this task, which we applied to a dataset of 23 different Latin characters extracted from a sample selection of pages from the Vatican Register. In this paper we show that, using multi-KAFs, we can increase the accuracy of the CNN even while reducing the number of filters per layer.

Organization of the Paper. In Sects. 2 and 3 we describe standard activation functions for neural networks, KAFs [17], and the proposed multi-KAFs. The dataset from the ICR project is described in Sect. 4. We then perform a set of experiments in Sect. 5, before concluding in Sect. 6.

2 Preliminaries

2.1 Feedforward Neural Networks

Consider a generic feedforward NN layer, taking as input a vector $\mathbf{x} \in \mathbb{R}^d$ and producing in output a vector $\mathbf{y} \in \mathbb{R}^c$:

$$\mathbf{y} = g\left(\mathbf{Wx} + \mathbf{b}\right), \tag{1}$$

where $\{\mathbf{W}, \mathbf{b}\}$ are adaptable weight matrices, and $g(\cdot)$ is an element-wise activation function. Multiple layers can be stacked to obtain a complete NN. In the following we focus especially on the choice of $g(\cdot)$, but we note that everything extends immediately to more complex types of layer, including convolutional layers (wherein the matrix product is replaced by a convolutional operator), or recurrent layers [10].

Generally speaking, the activation functions $g(\cdot)$ for the hidden (not last) layers are chosen as simple operations, such as the rectified linear unit (ReLU), originally introduced in [7]:

$$g(s) = \max\{0, s\}, \tag{2}$$

where we use the letter s to denote a generic scalar input to the function, i.e., a single *activation* value. An NN is a generic composition of L such layers, denoted as $f(\mathbf{x})$, that is trained with a dataset of N training samples $\left\{\mathbf{x}^i, \mathbf{d}^i\right\}_{i=1}^N$. In the experimental section, in particular, we deal with multi-class classification with C classes, where the desired output $\mathbf{d}^i$ represents a one-hot encoding of the target class. We train it by minimizing a regularized cross-entropy cost:

$$\min\left\{-\sum_{i=1}^N \sum_{c=1}^C d_c^i \log\left(f_c(\mathbf{x}^i) + \lambda \cdot \|\mathbf{w}\|^2\right)\right\}, \tag{3}$$

where $\mathbf{w}$ is a weight vector collecting all the adaptable weights of the network, λ is a positive scalar, and we use a subscript to denote the c-th element of a vector.

2.2   Kernel Activation Functions

Differently from (2), a KAF can be adapted from the data. In particular, each activation function is modeled in terms of D expansions of the activation s with a kernel function κ:

$$g(s) = \sum_{i=1}^{D} \alpha_i \kappa\left(s, d_i\right),$$

(4)

where the scalars d_i form the so-called dictionary, while the scalars α_i are called the mixing coefficients. To make the training problem simpler, the dictionary is fixed beforehand (and not adapted) by sampling D values from the real line, uniformly around zero, and it is shared across the network, while a different set of mixing coefficients is adapted for every neuron. This makes implementation extremely efficient. The integer D is the key hyper-parameter of the model: a higher value of D increases the overall flexibility of each function, at the expense of adding additional mixing coefficients to be adapted.

Any kernel function from the literature can be used in (4), provided it respects the semi-definiteness property:

$$\sum_{i=1}^{D} \sum_{j=1}^{D} \alpha_i \alpha_j \kappa\left(d_i, d_j\right) \geq 0,$$

(5)

for any choice of the mixing coefficients and the dictionary. In practice, [17] and all subsequent papers only used the one-dimensional Gaussian kernel defined as:

$$\kappa(s, d_i) = \exp\left\{-\gamma\left(s - d_i\right)^2\right\},$$

(6)

where $\gamma > 0$ is a parameter of the kernel. The value of γ influences the 'locality' of each α_i with respect to the dictionary. In [17] we proposed the following rule-of-thumb, found empirically:

$$\gamma = \frac{1}{6\Delta^2},$$

(7)

where Δ is the distance between any two dictionary elements. However, note that neither the Gaussian kernel in (6) nor the rule-of-thumb in (7) are optimal in general. For example, simple smooth shapes (like slowly varying polynomials) could be more easily modeled via different types of kernels or much larger values of γ with a possibly smaller D. These are common problems also in the kernel literature. Leveraging it, in the next section we propose an extension of KAF to mitigate both problems.

3   Proposed Multiple Kernel Activation Functions

In order to mitigate the problems mentioned in the previous section, assume to have available a set of M *candidate* kernel functions $\kappa_1, \ldots, \kappa_M$. These can be entirely different functions or the same kernel with different choices of its parameters. There has been a vast research on how to successfully combine different kernels to obtain a new one, going under the name of multiple kernel learning (MKL) [1]. For the purpose of this paper we adopt a simple approach, in order to evaluate its feasibility. In particular, we build each KAF with a new kernel given by a linearly weighted sum of the constituents (base) kernels:

$$g(s) = \sum_{i=1}^{D} \alpha_i \left[\sum_{m=1}^{M} \mu_m \kappa_m (s, d_i) \right] = \sum_{i=1}^{D} \alpha_i \widetilde{\kappa} (s, d_i), \qquad (8)$$

where $\{\mu_m\}_{m=1}^{M}$ are an additional set of mixing coefficients. From the properties of reproducing kernel Hilbert spaces, it is straightforward to show that $\widetilde{\kappa}$ is a valid kernel if its constituents are also valid kernel functions. We call the resulting activation functions *multi-KAFs*. Note that such an approach only introduces $M - 1$ additional parameters for each neuron, where M is generally small. In addition, since in our implementation the dictionary is fixed, all kernels are evaluated on the same points, which can greatly simplify the implementation and allows to share a large part of the computation.

More in detail, for our experiments we consider an implementation with $M = 3$, where κ_1 is the Gaussian kernel in (6), κ_2 is chosen as the (isotropic) rational quadratic [6]:

$$\kappa_2(s, d_i) = 1 + \frac{(s - d_i)^2}{(s - d_i)^2 + c} \qquad (9)$$

with c being a parameter, and κ_3 is chosen as the polynomial kernel of order 2:

$$\kappa_3(s, d_i) = (1 + s d_i)^2. \qquad (10)$$

The rational quadratic is similar to the Gaussian, but it is sometimes preferred in practical applications [6], while the polynomial kernel allows to introduce a smoothly varying global trend to the function. We show a simple example of the mix of κ_1 and κ_3 in Fig. 1.

In order to simplify optimization, especially for deeper architectures, we apply the kernel ridge regression initialization procedure described in [17] to initialize all multi-KAFs to a known activation function, i.e., the exponential linear unit (ELU) [2]. For this purpose, denote by $\mathbf{t}$ the vector of ELU values computed on our dictionary points. We initialize all μ_j to $\frac{1}{3}$, and initialize the vector of mixing coefficients $\boldsymbol{\alpha}$ as:

$$\boldsymbol{\alpha} = \left(\widetilde{\mathbf{K}} + \varepsilon \mathbf{I} \right)^{-1} \mathbf{t}, \qquad (11)$$

where $\widetilde{\mathbf{K}} \in \mathbb{R}^{D \times D}$ is the kernel matrix computed between $\mathbf{t}$ and $\mathbf{d}$ using $\widetilde{\kappa}$, and $\varepsilon = 10^{-4}$. As a final remark, note that in (8) we considered an unrestricted linear combination of the constituting kernels. We can easily obtain more restricted formulations (which are sometimes found in the MKL literature [8]) by applying some nonlinear transformation to the mixing coefficients μ_m, e.g., a softmax function to obtain convex combinations. We leave such comparisons to a future work.

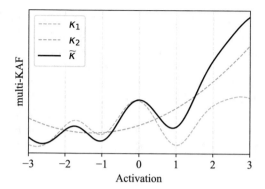

Fig. 1. An example of multi-KAF obtained from the mixture of a Gaussian kernel (green dashed line) and a polynomial kernel (red dashed line). In the example, we have $D = 15$ elements in the dictionary equispaced in $[-3.0, 3.0]$, mixing coefficients α_i sampled from the normal distribution, and $\mu_1 = 0.5$, $\mu_2 = 0.01$.

4   Case Study: In Codice Ratio

As stated in the introduction, we apply the proposed multi-KAF on a realistic case study taken from the ICR project. Apart from the details described below, we refer the reader to [4,5] for a fuller description of the project.

The overall goal of ICR is the transcription of a large portion of the Vatican Secret Archives, one of the largest existing historical libraries. In the first phase of the project, we collected and manually annotated a set of 23000 images representing 23 different types of characters (one of which is a special 'non-character' class). All characters were extracted from a sample of 30 (handwritten) pages of private correspondence of Pope Honorii III from the XIII century. Each character was then annotated using a crowdsourcing platform and 120 volunteer students.[2] A few examples taken from the dataset are shown in Fig. 2.

| (a) | (b) | (c) | (d) | (e) |

Fig. 2. Examples taken from the Latin OCR dataset. (a) and (d) are examples of the character 's'; (b) and (c) are examples of character 'd'; (e) is an example of a different version for the character 's' (considered as a separate class in the dataset).

In [5] we described the design and evaluation of a CNN for tackling the problem of automatically assigning a class to each character, represented as a

---

[2] The dataset is available on the web at http://www.dia.uniroma3.it/db/icr/.

56×56 black-and-white image. The final CNN had the following architecture: (i) a convolutive block with 42 filters of size 5×5; (ii) max-pooling with size 2×2; (iii) two additional series of convolutive blocks and max-pooling, this time with 28 filters per layer; (iv) a fully connected layer with 100 neurons, followed by (v) an output layer of 23 neurons. In the original implementation, all linear operations were preceded by dropout (with probability 50%), and followed by ReLU nonlinearities. We will use this dataset and architecture as baseline for our experiments. Note that this design was heavily fine-tuned, and it was not possible to increase the testing accuracy by simply adding more neurons/layers to the resulting CNN model.

5   Experimental Results

5.1   Experimental Setup

For our comparisons, we use the same CNN architecture described in Sect. 4, but we replace the ReLU functions by either KAFs or the proposed multi-KAFs, with the hyper-parameters described in the previous sections. In order to make the networks comparable in terms of parameters, for KAF and multi-KAF we decrease the number of filters and neurons in the linear layers by 10%. To stabilize training, we also replace dropout with a batch normalization step [12] before applying KAF-based nonlinearities.

We train the networks following a similar procedure as [5]. We use the Adam optimization algorithm on random mini-batches of 32 elements, with a small regularization factor $\lambda = 0.001$. After every 10 iterations of the optimization algorithm we evaluate the accuracy on a randomly held-out set of 2500 samples, taken from the original training set. Training is stopped whenever the validation accuracy has not improved for at least 250 iterations. The networks are then evaluated on a second independent held-out set of 2300 examples. All networks are implemented in PyTorch, and experiments are run on a CUDA backend using the Google Colaboratory platform.

5.2   Experimental Results

The results of the experiments, averaged over 5 different repetitions, are shown in Table 1, together with the number of trainable parameters of the different architectures. It can be seen that, while KAF fails to provide a meaningful improvement in this case, the multi-KAF architecture obtains a significant gain in testing accuracy, stably throughout the repetitions. Most notably, this gain is obtained with a strong decrease in the number of trainable parameters, which is around $16 \cdot 10^4$ for the KAF-based architectures, compared to $\approx 19 \cdot 10^4$ parameters for the baseline one.

This gain is accuracy is not only obtained with a smaller number of overall parameters, but also with a much faster rate of convergence. To see this, we plot in Fig. 3 the evolution of the loss function in (3) and the evolution of the accuracy on the validation portion of the dataset.

Table 1. Results of comparing the different activation functions on the OCR dataset described in Sect. 4. See the text for details on the experimental procedure.

Activation function	Testing accuracy [%]	Trainable parameters
ReLU	94.99 (± 0.1)	191871
KAF	94.00 (± 0.3)	158840
Proposed multi-KAF	96.31 (± 0.2)	159552

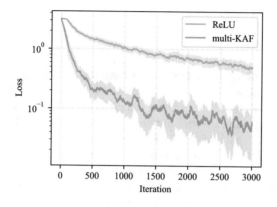

(a) Training loss

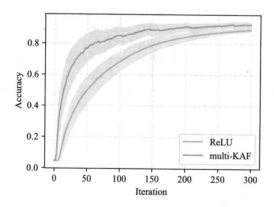

(b) Validation accuracy

Fig. 3. Loss and validation accuracy evolution for the baseline network (using ReLUs) and the proposed multi-KAF. Standard deviation for all curves is shown with a lighter color.

Finally, we also show the final mixing coefficients μ_m for 25 randomly sampled neurons of the first convolutive layer of the multi-KAF network, after training, in Fig. 4. Interestingly, different neurons require vastly different combinations of base kernels, reflected by light/dark colors in Fig. 4.

328 S. Scardapane et al.

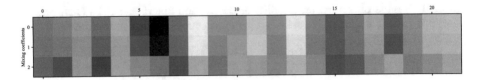

Fig. 4. Mixing coefficients of the multiple kernel after training, for the first convolutive layer and 25 randomly sampled filters. Light colors indicate stronger mixing coefficients.

6 Conclusions

In this paper we investigated a new non-parametric activation function for neural networks, which extends the recently proposed kernel activation function, by incorporating ideas from the field of multiple kernel learning to simplify the choice of the kernel function and further increase the expressiveness.

We evaluated the resulting multi-KAF on a benchmark dataset of Latin hand-written characters recognition, in the context of an ongoing real-world project. While in a sense these are only preliminary results, they point to the greater flexibility of such activation functions, coming with a faster rate of convergence and an overall smaller number of trainable parameters for the full architecture. We are currently in the process of collecting a larger dataset, considering a bigger amount of possible classes for the characters, in order to further evaluate the proposed architecture. Additional research directions will consider the application of multi-KAFs on different types of benchmarks, going beyond standard CNNs, particularly with respect to recurrent models, complex-valued kernels [18], and generative adversarial networks. Furthermore, we plan to test more extensively additional types of kernels, such as the periodic ones [6], in order to evaluate the scalability of multi-KAFs in the presence of a larger number of constituent kernels. Generalization bounds are also a promising research direction [15].

References

1. Aiolli, F., Donini, M.: EasyMKL: a scalable multiple kernel learning algorithm. Neurocomputing **169**, 215–224 (2015)
2. Clevert, D.A., Unterthiner, T., Hochreiter, S.: Fast and accurate deep network learning by exponential linear units (ELUs). In: Proceedings of the 2016 International Conference on Learning Representations, ICLR (2016)
3. Eisenach, C., Wang, Z., Liu, H.: Nonparametrically learning activation functions in deep neural nets. In: 5th International Conference for Learning Representations (Workshop Track) (2017)
4. Firmani, D., Maiorino, M., Merialdo, P., Nieddu, E.: Towards knowledge discovery from the Vatican Secret Archives. In Codice Ratio-episode 1: machine transcription of the manuscripts. In: Proceedings of the 24th ACM SIGKDD International Conference on Knowledge Discovery & Data Mining, pp. 263–272. ACM (2018)

5. Firmani, D., Merialdo, P., Nieddu, E., Scardapane, S.: In Codice Ratio: OCR of handwritten Latin documents using deep convolutional networks. In: 11th International Workshop on Artificial Intelligence for Cultural Heritage, AI*CH 2017, pp. 9–16 (2017)
6. Genton, M.G.: Classes of kernels for machine learning: a statistics perspective. J. Mach. Learn. Res. 2(Dec), 299–312 (2001)
7. Glorot, X., Bordes, A., Bengio, Y.: Deep sparse rectifier neural networks. In: Proceedings of the 14th International Conference on Artificial Intelligence and Statistics, AISTATS, p. 275 (2011)
8. Gönen, M., Alpaydın, E.: Multiple kernel learning algorithms. J. Mach. Learn. Res. 12(Jul), 2211–2268 (2011)
9. Goodfellow, I.J., Warde-Farley, D., Mirza, M., Courville, A., Bengio, Y.: Maxout networks. In: Proceedings of the 30th International Conference on Machine Learning, ICML (2013)
10. Goodfellow, I., Bengio, Y., Courville, A.: Deep Learning. MIT Press, Cambridge (2016)
11. He, K., Zhang, X., Ren, S., Sun, J.: Delving deep into rectifiers: Surpassing human-level performance on ImageNet classification. In: Proceedings of the IEEE International Conference on Computer Vision, ICCV, pp. 1026–1034 (2015)
12. Ioffe, S., Szegedy, C.: Batch normalization: Accelerating deep network training by reducing internal covariate shift. In: Proceedings of the 32nd International Conference on Machine Learning, ICML, pp. 448–456 (2015)
13. Jin, X., Xu, C., Feng, J., Wei, Y., Xiong, J., Yan, S.: Deep learning with S-shaped rectified linear activation units. In: Proceedings of the Thirtieth AAAI Conference on Artificial Intelligence (2016)
14. Liu, W., Principe, J.C., Haykin, S.: Kernel Adaptive Filtering: A Comprehensive Introduction. Wiley, Hoboken (2011)
15. Oneto, L., Navarin, N., Donini, M., Ridella, S., Sperduti, A., Aiolli, F., Anguita, D.: Learning with kernels: a local rademacher complexity-based analysis with application to graph kernels. IEEE Trans. Neural Netw. Learn. Syst. 29(10), 4660–4671 (2017)
16. Ramachandran, P., Zoph, B., Le, Q.V.: Searching for activation functions. arXiv preprint arXiv:1710.05941 (2017)
17. Scardapane, S., Van Vaerenbergh, S., Totaro, S., Uncini, A.: Kafnets: kernel-based non-parametric activation functions for neural networks. Neural Netw. (2018, in press)
18. Scardapane, S., Van Vaerenbergh, S., Hussain, A., Uncini, A.: Complex-valued neural networks with non-parametric activation functions. IEEE Trans. Emerg. Top. Comput. Intell. (2018, in press)
19. Siniscalchi, S.M., Salerno, V.M.: Adaptation to new microphones using artificial neural networks with trainable activation functions. IEEE Trans. Neural Netw. Learn. Syst. 28(8), 1959–1965 (2017)
20. Zhang, X., Trmal, J., Povey, D., Khudanpur, S.: Improving deep neural network acoustic models using generalized maxout networks. In: 2014 IEEE International Conference on Acoustics, Speech and Signal Processing, ICASSP, pp. 215–219. IEEE (2014)

Understanding Ancient Coin Images

Jessica Cooper and Ognjen Arandjelović[(✉)]

University of St Andrews, St Andrews, UK
jessicamarycooper@gmail.com, ognjen.arandjelovic@gmail.com

Abstract. In recent years, a range of problems within the broad umbrella of automatic, computer vision based analysis of ancient coins has been attracting an increasing amount of attention. Notwithstanding this research effort, the results achieved by the state of the art in the published literature remain poor and far from sufficiently well performing for any practical purpose. In the present paper we present a series of contributions which we believe will benefit the interested community. Firstly, we explain that the approach of visual matching of coins, universally adopted in all existing published papers on the topic, is not of practical interest because the number of ancient coin types exceeds by far the number of those types which have been imaged, be it in digital form (e.g. online) or otherwise (traditional film, in print, etc.). Rather, we argue that the focus should be on the understanding of the semantic content of coins. Hence, we describe a novel method which uses real-world multimodal input to extract and associate semantic concepts with the correct coin images and then using a novel convolutional neural network learn the appearance of these concepts. Empirical evidence on a real-world and by far the largest data set of ancient coins, we demonstrate highly promising results.

1 Introduction

Numismatics is the study of currency, including coins, paper money, and tokens. This discipline yields fascinating cultural and historical insights, and is a field of great interest to scholars, amateur collectors, and professional dealers alike. Important applications of machine learning in this field include theft detection, identification and classification of finds, and forgery prevention. The present work focuses on ancient coins.

Individual ancient coins of the same type can vary widely in appearance due to centering, wear, patination, and variance in artistic depiction of the same semantic elements. This poses a range of technical challenges and makes it difficult to reliably identify which concepts are depicted on a given coin using machine learning and computer vision based automatic techniques [4]. For this reason, ancient coins are typically identified and classified by professional dealers or scholars, which is a time consuming process demanding years of experience due to the specialist knowledge required. Human experts attribute a coin based

L. Oneto et al. (Eds.): INNSBDDL 2019, INNS 1, pp. 330–340, 2020.
https://doi.org/10.1007/978-3-030-16841-4_34

on its denomination, the ruler it was minted under and the time and place it was minted. A variety of different characteristics are commonly used for attribution, including:

- Physical characteristics such as weight, diameter, die alignment and colour,
- Obverse legend, which includes the name of the issuer, titles or other designations,
- Obverse motif, including the type of depiction (head or bust, conjunctional or facing, etc.), head adornments (bare, laureate, diademed, radiate, helmet, etc.), clothing (draperies, breastplates, robes, armour, etc.), and miscellaneous accessories (spears, shields, globes, etc.),
- Reverse legend, often related to the reverse motif,
- Reverse motif (primary interest herein), e.g. person (soldier, deity, etc.), place (harbour, fortress, etc.) or object (altar, wreath, animal, etc.), and
- Type and location of any mint markings.

1.1 Relevant Prior Work

Most existing algorithmic approaches to coin identification are local feature based which results in poor performance due to loss of spatial relationships between elements [4, 13, 14]. To overcome this limitation, approaches which divide a coin into segments have been proposed but these assume that coins are perfectly centred, accurately registered, and nearly circular in shape [7]. Recent work shows that existing approaches perform poorly on real world data because of the fragility of such assumptions [7]. Although the legend can be a valuable source of information, relying on it for identification is problematic because it is often significantly affected by wear, illumination, and minting flaws [5]. Lastly, numerous additional challenges emerge in and from the process of automatic data preparation, e.g. segmentation, normalization of scale, orientation, and colour [6].

In the context of the present work it is particularly important to note that all existing work in this area is inherently limited by the reliance on visual matching [2–4, 9, 17] and the assumption that the unknown query coin is one of a limited number of gallery types. However, this assumption is unrealistic as there are hundreds of thousands of different coin types [12]. Hence herein our idea is to explore the possibility of *understanding* the artistic content depicted on a coin, which could then be used for subsequent text based matching, allowing access to a far greater number of coin types without requiring them to be represented in the training data. In short, our ultimate aim is to attempt to a describe coin in the same way that a human expert would – by identifying the individual semantic elements depicted thereon.

2 Problem Specification, Constraints, and Context

Deep learning has proven successful in a range of image understanding tasks [16]. Recent pioneering work on its use on images of coins has demonstrated extremely promising results, outperforming more traditional approaches by an order of magnitude [15]. A major difference in the nature of the challenge we tackle here is that our data set is weakly supervised – that is, samples are labelled only at the image level by a corresponding largely unstructured textual description, even though the actual semantic elements of interest themselves occupy a relative small area of the image. This kind of data is much more abundant and easier to obtain than fully supervised data (pixel level labels), but poses a far greater challenge.

2.1   Challenge of Weak Supervision

Recall that our overarching goal is to identify the presence of various important semantic elements on an unknown query coin's reverse. Moreover, we are after a scalable approach – one which could be used with a large and automatically extracted list of elements. As this is the first attempt at solving the suggested task, we chose to use images which are relatively uncluttered, thus disentangling the separate problems of localizing the coin in an image and segmenting it out. Representative examples of images used can be seen in Fig. 1.

Fig. 1. Typical images used in the present work (obverses on the left, reverses on the right).

To facilitate the automatic extraction of salient semantic elements and the learning of their visual appearance, each coin image is associated with an unstructured text description of the coin, as provided by the professional dealer selling the coin; see Fig. 2. The information provided in this textual meta-data varies substantially – invariably it includes the descriptions of the coin's obverse and reverse, and the issuing authority (the person, usually emperor, on the obverse), but it may also feature catalogue references, minting date and place, provenance, etc.

We have already emphasised the practical need for a scalable methodology. Consequently, due to the nature of the available data (which itself is representative of the type of data which can be readily acquired in practice), at best it is possible to say whether a specific semantic element (as inferred from text; see next section) is present in an image or not, see Fig. 5. Due to the amount of human labour required it is not possible to perform finer labelling i.e. to specify where the element is, or its shape (there are far too many possible elements and the amount of data which would require labelling is excessive), necessitating weak supervision of the sort illustrated in Fig. 5.

```
83, Lot: 226. Estimate \$100. Sold for \$92.

MACRIANUS. 260-261 AD. Antoninianus (22mm, 3.40 gm). Samosata? mint. Radiate,
draped, and cuirassed bust  right / Apollo standing facing, head left, holding
laurel-branch and leaning on lyre; star in left  field. Cf. RIC V 6 (Antioch);
MIR 44, 1728k; cf. Cohen 2 (same). VF, light porosity. Rare.
```

Fig. 2. Typical unstructured text attribution of a coin.

2.2 Data Pre-processing and Clean-Up

As in most real-world applications unstructured, loosely standardized, and heterogeneous data, a major challenge in the broad domain of interest in the present work is posed by possibly erroneous labelling, idiosyncrasies, etc. Hence, robust pre-processing and clean-up of data is a crucial step for facilitating learning.

Image Based Pre-processing. We begin by cropping all images to the square bounding box which contains just the reverse of the corresponding coin, and isotropicaly resizing the result to the uniform scale of 300 × 300 pixels. The obverse of a coin typically depicts the bust of an emperor, whilst the reverse contains the semantic elements of interest herein.

Text Based Extraction of Semantics. In order to select which semantic elements to focus on, we use the unstructured text files to analyse the frequency of word occurrences and, specifically, focus on the most common concepts, see Figs. 3 and 4. Considering that the present work is the first attempt at addressing the problem at hand, the aforementioned decision was made in order to ensure that sufficient training data is available.

Fig. 3. Examples of depictions of reverse motif elements the present work focuses on.

Clean-up and Normalization of Text Data. To label the data we first clean the attribution text files to remove duplicate words and punctuation. Because the data included attributions in French, Spanish, German and English, we employ a translation API (googletrans) to generate translations of each keyword. Thus, when building the data set used to learn the concept of 'horse', images with associated text which contains the words 'horse', 'caballo', 'cheval' and 'pferd' are

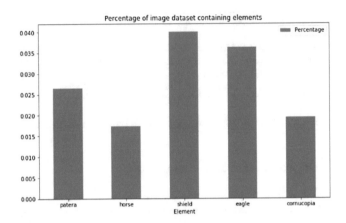

Fig. 4. Due to the great diversity of different semantic elements which appear on ancient coins only a relatively small number of coins in the corpus as a whole contain the specific semantic elements which we attempt to learn herein, raising class imbalance issues which we address using stratified sampling (see main text for thorough explanation).

also included as positive examples, making better use of available data. Plurals, synonyms and other words strongly associated with the concept in question (e.g. 'horseman'), and their translations are also included.

Randomization and Stratification. We shuffle the samples before building training, validation and test sets for each of the selected elements ('horse', 'cornucopia', 'patera', 'eagle' and 'shield'), with a ratio of 70% training set, 15% validation set and 15% test set. To address under-representation of positive examples, we use stratified sampling to ensure equal class representation.

2.3   Errors in Data

In the context of the present problem as in most large scale, real-world applications of automatic data analysis, even after preprocessing, the resulting data will contain a certain portion of erroneous data. Errors range in nature from incorrect label assignments to incorrectly prepared images.

To give a specific example, we observed a number of instances in which the keyword used to label whether or not a given coin image contains the element in question was actually referring to the obverse of the coin, rather than the reverse, with which we are concerned. This is most prevalent with shields, and leads to incorrect labelling when building the corresponding data sets. Our premise (later confirmed by empirical evidence) is that such incorrect labelling is not systematic in nature (c.f. RANSAC) and thus that the cumulative effect of a relatively small number of incorrect labels will be overwhelmed by the coherence (visual and otherwise) of correctly labelled data.

We also found that some of the original (unprocessed) images were unusually laid out in terms of the positioning of the corresponding coin's obverse and reverse, which thus end up distorted during preprocessing. However, these appear to be almost exclusively of the auction listing type described above, and as such would be labelled as negative samples and therefore cumulatively relatively unimportant.

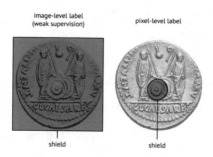

Fig. 5. The amount of labour which would be required to provide pixel-level labelling (right) of different concepts is prohibitive, imposing the much more challenging task of having to perform learning from weakly annotated data (left).

3   Proposed Framework

In this paper we describe a novel deep convolutional neural network loosely based on AlexNet [11] as a means of learning the appearance of semantic elements and thus determining whether an unknown query coin contains the said element or not. Although we do not use this information in the present work for other purposes than the analysis of results, we are also able to determine the location of the element when it is present by generating heatmaps using the occlusion technique [15].

3.1   Model Topology

As summarized in Fig. 6 and Table 1, our network consists of five convolutional layers (three of which are followed by a max pooling layer), after which the output is flattened and passed through three fully connected layers. The convolutional and pooling layers are responsible for detecting high level features in the input data, and the fully connected layers then use those features to generate a class prediction.

The convolutional layers enable location independent recognition of features in the data. The output of a convolutional layer is dependent on a number of hyperparameters which we specify for each layer. These are the stride, the number of kernels (also called the depth), the kernel size, and the padding. Using multiple kernels is necessary in order to learn a range of features. The complexity of the features we are able to detect depends on the number of convolutional

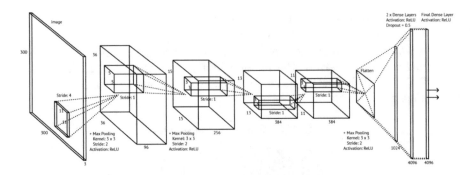

Fig. 6. Our model architecture: a global view.

Table 1. Our convolutional neural network topology: a summary.

Layer type	Kernel size	Stride	Kernel no	Activation
Convolutional	11 × 11	4	96	ReLU
Max pooling	3 × 3	2		
Convolutional	5 × 5	1	256	ReLU
Max pooling	3 × 3	2		
Convolutional	3 × 3	1	384	ReLU
Convolutional	3 × 3	1	384	ReLU
Convolutional	3 × 3	1	256	ReLU
Max pooling layer	3 × 3	2		
Flatten				
	Outputs	Dropout		
Dense	4096	0.5		ReLU
Dense	4096	0.5		ReLU
Dense	2	0.5		ReLU

layers present in a network – for example, one can reasonably expect to detect only simple, low-level features such as edges from the original 'raw' pixel inputs in the first layer, and then from those edges learn simple shapes in the second layer, and then more complex, higher level features from those simple shapes, and so on.

The pooling layers can be thought of as downsampling – the aim is to reduce the size of the output whilst retaining the most important information in each kernel. Pooling also decreases the complexity of the network by decreasing the number of parameters we need to learn. It also helps to make the network less sensitive to small variations, translations and distortions in the image input. Our design uses max-pooling, where the highest value in each kernel forms one element in the pooling layer output.

The dense layers (also called fully-connected layers) transform the high level features learned by the previous layers into a class prediction. To avoid overfitting, the dense layers in our model employ dropout, a regularisation technique which combats overfitting by randomly removing nodes at each iteration. The number of nodes that are removed is determined by a hyperparameter which, guided by previous work on CNNs, we set to 0.5.

We use the batch size of 24, and the maximum number of epochs of 200 which we experimentally found to be more than sufficient, with training invariably ending well before that number is reached. At each epoch, we check if the loss is lower than 0.001, or if there have been 30 epochs in a row with no improvement in the loss. If either of the conditions is fulfilled, training is terminated. This is done partly to avoid models overfitting, and partly to save time if a model or a certain set of hyperparameters are clearly not performing well.

The model is trained using adaptive moment estimation (Adam) optimization, a form of gradient descent, the learning rate and momentum being computed for each weight individually [10]. Because we are performing classification (in that each sample either 'contains element' or 'does not contain element') we use cross entropy loss [8]. The rectified linear unit (ReLU) [1] is employed at the activation function to avoid problems associated with vanishing gradients.

4 Experiments

Experiments discussed in the present section were performed using Asus H110M-Plus motherboard powered by Intel Core i5 6500 3.2 GHz Quad Core processor and NVidia GTX 1060 6 GB GPU, 8 GB RAM. Data comprised images and the associated textual descriptions of 100,000 auction lots kindly provided to us by the Ancient Coin Search auction aggregator https://www.acsearch.info/ (Fig. 1).

4.1 Results and Discussion

As readily seen from Table 2, our approach achieves a high level of accuracy across training, validation, and test data. The validation and test scores are not significantly lower than the training scores, which suggests little evidence of overfit and indicates a well designed architecture and sufficient data set. The models used to identify cornucopiae and paterae are particularly successful, which is

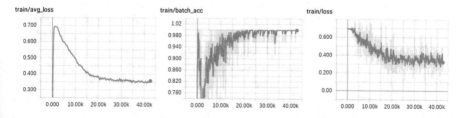

Fig. 7. Training a model to identify depictions of cornucopiae.

Table 2. Summary of experimental settings and results.

	Cornucopia	Patera	Shield	Eagle	Horse
Number of epochs	105	136	118	86	148
Training time (min)	58	30	82	51	106
Training accuracy	0.71	0.83	0.75	0.88	0.88
Validation accuracy	0.85	0.86	0.73	0.73	0.82
Validation precision	0.86	0.85	0.72	0.71	0.82
Validation recall	0.83	0.86	0.75	0.81	0.84
Validation F1	0.85	0.86	0.74	0.75	0.83
Test accuracy	0.84	0.84	0.72	0.73	0.82
Test precision	0.85	0.82	0.71	0.70	0.81
Test recall	0.83	0.87	0.74	0.81	0.82
Test F1	0.84	0.84	0.72	0.75	0.82

Fig. 8. Examples of automatically identified salient regions corresponding to a cornucopia, a patera, and a shield, respectively.

likely due to the low level of variance in artistic depictions of these elements – the orientation, position and style of shields, horses and eagles varies quite a bit, but cornucopiae and paterae are fairly constant in their depiction.

4.2 Learnt Salient Regions

We use the occlusion technique [15] to quantify the importance of different image regions in the context of the specific task at hand. In brief, the process involves synthetic occlusion by a uniform kernel and the quantification of the corresponding classification performance differential between unoccluded and occluded inputs, with a higher difference suggesting higher importance. Previous work on computer vision based ancient coin analysis has demonstrated the usefulness of this technique in the interpretation of empirical results [15]. In order to ensure robustness to relative size (semantic element to coin diameter), in the present work we adopt the use of three kernel sizes, 32×32, 48×48, and 64×64 pixels.

Typical examples of identified salient regions for different semantic elements are shown in Fig. 8. It can be readily seen that our algorithm picks up the most characteristic regions of the elements well – the winding pattern of cornucopiae, the elliptical shape of paterae, and the feather patterns of eagles. Since, as we noted before and as is apparent from Table 2, the performance of our algorithm is somewhat worse on the task of shield detection, we used our occlusion technique to examine the results visually in more detail. Having done so, our conclusion is reassuring – our hypothesis is that shields are inherently more challenging to detect as they exhibit significant variability in appearance and the style of depiction (shown from the front they appear circular, whereas shown from the side they assume an arc-like shape) and the least amount of characteristic detail (both circular and arc-like shapes are commonly found in more complex semantic elements shown on coins).

5 Summary and Conclusions

In this paper we made a series of important contributions to the field of computer vision based analysis of ancient coins. Firstly, we put forward the first argument against the use of visual matching of ancient coin images, having explained its practical lack of value. Instead, we argued that efforts should be directed towards the semantic understanding of coin images and described the first attempt at this challenging task. Specifically, we described a novel approach which combines unstructured text analysis and visual learning using a convolutional neural network, to create weak associations between semantic elements found on ancient coins and the corresponding images, and hence learn the appearance of the aforementioned elements. We demonstrated the effectiveness of the proposed approach using images of coins extracted from 100,000 auction lots, making the experiment the largest in the existing literature. In addition to a comprehensive statistical analysis, we presented the visualization of learnt concepts on specific instances of coins, showing that our algorithm is indeed creating the correct associations. We trust that our contributions will serve to direct future work and open avenues for new promising research.

References

1. Agarap, A.F.: Deep learning using rectified linear units (ReLU). arXiv:1803.08375 (2018)
2. Anwar, H., Zambanini, S., Kampel, M.: Supporting ancient coin classification by image-based reverse side symbol recognition. In: Proceedings of the International Conference on Computer Analysis of Images and Patterns, pp. 17–25 (2013)
3. Anwar, H., Zambanini, S., Kampel, M.: Coarse-grained ancient coin classification using image-based reverse side motif recognition. Mach. Vis. Appl. **26**(2), 295–304 (2015)
4. Arandjelović, O.: Automatic attribution of ancient Roman imperial coins. In: Proceedings of the IEEE Conference on Computer Vision and Pattern Recognition, pp. 1728–1734 (2010)

5. Arandjelović, O.: Reading ancient coins: automatically identifying denarii using obverse legend seeded retrieval. In: Proceedings of the European Conference on Computer Vision, vol. 4, pp. 317–330 (2012)

6. Conn, B., Arandjelović, O.: Towards computer vision based ancient coin recognition in the wild – automatic reliable image preprocessing and normalization. In: Proceedings of the IEEE International Joint Conference on Neural Networks, pp. 1457–1464 (2017)

7. Fare, C., Arandjelović, O.: Ancient Roman coin retrieval: a new dataset and a systematic examination of the effects of coin grade. In: Proceedings of the European Conference on Information Retrieval, pp. 410–423 (2017)

8. Janocha, K., Czarnecki, W.M.: On loss functions for deep neural networks in classification. arXiv:1702.05659 (2017)

9. Kampel, M., Zaharieva, M.: Recognizing ancient coins based on local features. In: Proceedings of the International Symposium on Visual Computing, vol. 1, pp. 11–22 (2008)

10. Kinga, D., Adam, J.B.: A method for stochastic optimization. In: Proceedings of the International Conference on Learning Representations, vol. 5 (2015)

11. Krizhevsky, A., Sutskever, I., Hinton, G.E.: ImageNet classification with deep convolutional neural networks. In: Advances in Neural Information Processing Systems, pp. 1097–1105 (2012)

12. Mattingly, H.: The Roman Imperial Coinage, vol. 7. Spink, London (1966)

13. Rieutort-Louis, W., Arandjelović, O.: Bo(V)W models for object recognition from video. In: Proceedings of the International Conference on Systems, Signals and Image Processing, pp. 89–92 (2015)

14. Rieutort-Louis, W., Arandjelović, O.: Description transition tables for object retrieval using unconstrained cluttered video acquired using a consumer level handheld mobile device. In: Proceedings of the IEEE International Joint Conference on Neural Networks, pp. 3030–3037 (2016)

15. Schlag, I., Arandjelović, O.: Ancient Roman coin recognition in the wild using deep learning based recognition of artistically depicted face profiles. In: Proceedings of the IEEE International Conference on Computer Vision, pp. 2898–2906 (2017)

16. Yue, X., Dimitriou, N., Arandjelović, O.: Colorectal cancer outcome prediction from H&E whole slide images using machine learning and automatically inferred phenotype profiles. In: Proceedings of the International Conference on Bioinformatics and Computational Biology (2019)

17. Zaharieva, M., Kampel, M., Zambanini, S.: Image based recognition of ancient coins. In: Proceedings of the International Conference on Computer Analysis of Images and Patterns, pp. 547–554 (2007)

Effects of Skip-Connection in ResNet and Batch-Normalization on Fisher Information Matrix

Yasutaka Furusho$^{(\boxtimes)}$ ⓘ and Kazushi Ikeda ⓘ

Nara Institute of Science and Technology, Nara 630-0192, Japan
{furusho.yasutaka.fm1,kazushi}@is.naist.jp

Abstract. Deep neural networks such as multi-layer perceptron (MLP) have intensively been studied and new techniques have been introduced for better generalization ability and faster convergence. One of the techniques is skip-connections between layers in the ResNet and another is the batch normalization (BN). To clarify effects of these techniques, we carried out the landscape analysis of the loss function for these networks. The landscape affects the convergence properties where the eigenvalues of the Fisher Information Matrix (FIM) plays an important role. Thus, we calculated the eigenvalues of the FIMs of the MLP, ResNet, and ResNet with BN by applying functional analysis to the networks with random weights, which of MLP was analyzed before in asymptotic case using the central limit theorem. Our results show that the MLP has eigenvalues that are independent of its depth, that the ResNet has eigenvalues that grow exponentially with its depth, and that the ResNet with BN has eigenvalues that grow sub-linear with its depth. These imply that the BN allows the ResNet to use larger learning rate and hence converges faster than the vanilla ResNet.

Keywords: ResNet · Batch-normalization · Fisher Information Matrix

1 Introduction

Deep neural networks have been changing the history of machine learning in performance [7,12] and the high performance is said to result from its exponential expressive power [2,10,11,14]. However, a standard feedforward network (MLP) cannot reduce its empirical risk even though it stacks more layers. In fact, an MLP with 56 layers has a larger empirical risk than one with 20 layers [4].

To overcome this problem, the ResNet incorporated skip-connections between layers [4]. In fact, the skip-connection enables even an extreme deep ResNet (1202 layers) to learn with a small empirical risk. This is because the ResNet with batch-normalization (BN) [5] copes with the shattered gradient problem in the MLP, that is, the correlation between gradients of data points decays exponentially and the spatial structure of gradients is obliterated as the network gets deeper [1]. In addition, the ResNet shows a much better property in

© Springer Nature Switzerland AG 2020
L. Oneto et al. (Eds.): INNSBDDL 2019, INNS 1, pp. 341–348, 2020.
https://doi.org/10.1007/978-3-030-16841-4_35

convergence when the BN is introduced [5]. Its better convergence properties result from smoothing loss landscape by the BN in the case of the linear MLP [13]. However, the mechanism in the case of the ResNet with non-linear ReLU activation is still unclear.

To clear the mechanism, we analyzed effects of skip-connections in the ResNet and the BN on its loss landscape, where the eigenvalues of the Fisher Information Matrix (FIM) play an important role on the convergence properties of the gradient descent method [6,9]. Karakida et al. [6] calculated the FIM of the MLP with random weights using the central limit theorem, that is, their analysis applies to only the asymptotic case of the MLP with infinite nodes.

We calculated the FIMs of the MLP, ResNet and ResNet with BN by applying functional analysis [1] to the networks with random weights. Our results explain why the ResNet shows a much better property in convergence when the BN is introduced.

2   Materials and Methods

We analyzed the FIMs of the MLP, ResNet, and ResNet with BN. This section formulates these networks and explain the FIM.

2.1   Dataset

We have a training set $S := \{x(n), y(n)\}_{n=1}^{N}$. Each example is a pair of input $x(n) \in \mathcal{X} \subset \mathbb{R}^D$ and output $y(n) \in \mathcal{Y} \subset \mathbb{R}$, which satisfy the following.

Assumption 1. *Input data is normalized as follows.*

$$\forall i \in [D], \quad \mathbb{E}_n[x_i] := \frac{1}{N} \sum_{n=1}^{N} x(n)_i = 0, \quad \mathbb{V}_n[x_i] := \frac{1}{N} \sum_{n=1}^{N} (x(n)_i - \mathbb{E}_n[x_i])^2 = 1 \tag{1}$$

where the expectation is taken with respect to empirical distribution of input $q(x)$.

We drop the index of data if context is clear.

2.2   Neural Networks

We consider the following L-layers neural networks $f : \mathcal{X} \times \mathcal{W} \to \mathcal{Y}$ with parameters (weight vector) $\theta \in \mathcal{W} \subset \mathbb{R}^{LD^2}$, which predict a corresponding output $y \in \mathcal{Y}$ for an input $x \in \mathcal{X}$. Let $h^0 = x$ and $\phi(\cdot) := \max(0, \cdot)$ is the ReLU function.

MLP:

$$u_i^l = \sum_{j=1}^{D} W_{ij}^l h_j^{l-1}, \quad h_i^l = \phi(u_i^l), \quad \hat{y} = 1^T h^L. \tag{2}$$

ResNet:

$$u_i^l = \phi(h_i^{l-1}), \qquad h_i^l = \sum_{j=1}^{D} W_{ij}^l u_j^l + h_i^{l-1}, \qquad \hat{y} = 1^T h^L. \tag{3}$$

ResNet with batch-normalization (BN):

$$u_i^l = \phi\left(\mathrm{BN}(h_i^{l-1})\right), \qquad \mathrm{BN}\left(h_i^l\right) = \frac{h_i^l - \mathbb{E}_n[h_i^l]}{\sqrt{\mathbb{V}_n[h_i^l]}},$$

$$h_i^l = \sum_{j=1}^{D} W_{ij}^l u_j^l + h_i^{l-1}, \qquad \hat{y} = 1^T h^L. \tag{4}$$

We initialize the weights by He-initialization [3]:

$$W_{ij}^l \sim \mathcal{N}\left(0, \frac{2}{D}\right). \tag{5}$$

We put the following assumption on the activations of the hidden units.

Assumption 2. *Hidden units are active with probability 0.5.*

2.3   Fisher Information Matrix (FIM)

The FIM of probability distribution $p(x, y; \theta)$ is defined as

$$F := \mathbb{E}_{x,y}\left[\nabla_\theta \log p(x, y; \theta)\nabla_\theta \log p(x, y; \theta)^T\right]. \tag{6}$$

The expectation is taken with respect to $p(x, y; \theta) = p(x)p(y|x; \theta)$ where $p(x)$ is a probability distribution of input and

$$p(y|x; \theta) = \mathcal{N}\left(f(x; \theta), 1\right) = \frac{1}{\sqrt{2\pi}} \exp\left(-\frac{(y - f(x; \theta))^2}{2}\right). \tag{7}$$

Then, the FIM can be written as

$$F = \mathbb{E}_x\left[\nabla_\theta f(x; \theta)\nabla_\theta f(x; \theta)^T\right]. \tag{8}$$

If we use the empirical distribution $q(x)$ as $p(x)$, the FIM can be written as

$$F = \mathbb{E}_n\left[\nabla_\theta f(x; \theta)\nabla_\theta f(x; \theta)^T\right]. \tag{9}$$

If loss function is mean squared error $E(\theta)$, the FIM is similar to the Hessian matrix around the local minima:

$$H := \nabla_\theta \nabla_\theta E(\theta) = F - \mathbb{E}_n\left[(y - f(x; \theta))\nabla_\theta \nabla_\theta f(x; \theta)\right] \tag{10}$$

where the expectation is taken with respect to empirical distribution of the pair of input and output $q(x, y)$.

Thanks to this property, the FIM is related to the convergence property of gradient descent. Gradient descent need to set learning rate smaller than $\frac{2}{\lambda_{max}}$ for convergence [6,9], where λ_{max} is the maximum eigenvalue of the FIM.

3   Results and Discussion

We calculated mean of the eigenvalue m_λ and the maximum eigenvalue λmax of the FIMs of the MLP, ResNet, and ResNet with BN with initialized random weights to clear how skip-connections and the BN affect its convergence property of gradient descent. We calculated these eigenvalues m_λ, λmax of expectation of the FIM with respect to parameter vector θ because initialized parameter is random variable (Table 1).

Table 1. Mean of the eigenvalue and the maximum eigenvalue of expectation of the FIM. Note that $H_{L+1} := \sum_{k=1}^{L+1} \frac{1}{k}$ is the harmonic number.

	Mean m_λ	Max $\lambda_{\max}$
MLP	$\frac{1}{2}$	
ResNet	2^{L-2}	$\frac{LD^2}{N} m_\lambda \leq \lambda_{\max} \leq LD^2 m_\lambda$
ResNet(BN)	$\frac{L+1}{L} \frac{H_{L+1}-1}{2}$	

Table 1 shows the three interesting results how the architecture of the networks affects its loss landscape and the convergence property of gradient descent. The first result is regarding the effect of the number of units in each hidden layer. Mean of the eigenvalue is independent of the number of the hidden units. However, the maximum eigenvalue grows with the square of it. This implies loss landscape is locally flat in most dimension but strongly steep in the other dimensions, which matches the theoretical result for the asymptotic case of the MLP with infinite units [6]. The second result is regarding the effect of skip-connections in the ResNet. Mean of the eigenvalue of the MLP is independent of its depth. However, skip-connections in the ResNet make mean of the eigenvalue grows exponentially with its depth. Therefore, gradient descent should set small learning rate for training the ResNet. The last result is regarding the effect of the BN. The BN changes this exponential dependence of the depth into the sub-linear dependence. This implies that the BN enables gradient descent to set larger learning rate and to converge faster than the vanilla ResNet, which explains its experimental fast convergence results [5].

In order to confirm our analysis is valid, we ran numerical experiments on the PCA-whitened MNIST dataset (see Appendix B for more details). We calculated the eigenvalues of expectation of the FIMs over 20 initialized parameters (Figs. 1, 2). Experimental results almost matched our theoretical results.

Moreover, in order to confirm the implication of our theoretical results: BN enables gradient descent to set larger learning rate than the vanilla ResNet, we trained the ResNet and the ResNet with BN by gradient descent (update parameters of only hidden layers 50 times) with different pair of the number of hidden units D, the number of hidden layers L, and the learning rate η and calculated its training loss (Figs. 3, 4). The loss are averaged over five trials. Red line means the theoretical values of learning rate $\frac{2}{\text{upper bound of } \lambda_{\max}}$ for convergence.

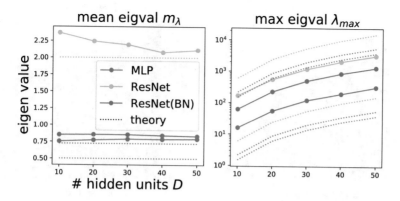

Fig. 1. The eigenvalues of 3-hidden layers networks with D hidden units.

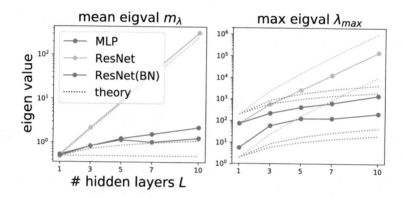

Fig. 2. The eigenvalues of L-hidden layers networks with 20 hidden units.

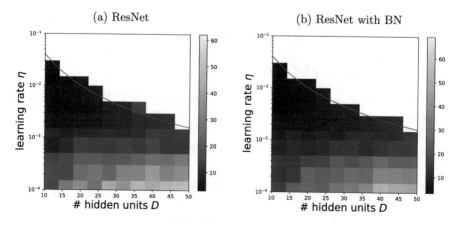

Fig. 3. Color map of training losses of 1-hidden layer networks with D-hidden units.

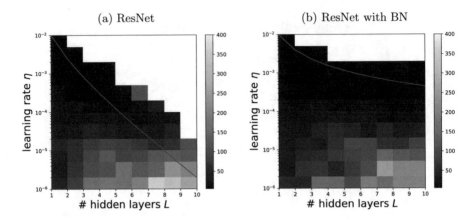

Fig. 4. Color map of training losses of L-hidden layers networks with 20-hidden units.

The white region means that loss is diverged (i.e., is larger than 1000). Experimental results almost matched the implication.

4   Conclusion

Deep neural networks have intensively been studied and new techniques have been introduced for better generalization ability and faster convergence. One of the techniques is skip-connections between layers in the ResNet and another is the batch normalization (BN). The ResNet shows a much better property in convergence when the BN is introduced. In order to clear its mechanism, we analyzed effects of skip-connections in the ResNet and the BN on its loss landscape. Our results show that the BN smooths the loss landscape of the ResNet, which enables gradient descent to set larger learning rate and to converge ResNet into minima faster than the vanilla ResNet.

Acknowledgements. This work was supported by JSPS KAKENHI Grant Number JP18J15055, JP18K19821, and NAIST Big Data Project.

Appendix A   Proof of Theoretical Results

This section sketches the proof of our main results of the MLP. This analysis procedure can be applied to the case of the ResNet and the ResNet with BN.

Mean of the eigenvalue m_λ and the maximum eigenvalue $\lambda_{\max}$ of the FIM can be written as error gradients, activations, and outputs of hidden units. This result is modification of [6].

Lemma 1. *Let* $\delta_i^l := \frac{\partial y}{\partial h_j^l}$ *is error gradient and* $\beta_i^l := \phi'(u_i^l)$ *is activation. Then,*

$$m_\lambda = \frac{1}{NP} \sum_{l=1}^{L} \sum_{i,j=1}^{D} \sum_{n=1}^{N} \delta_i^l(n)^2 \beta_i^l(n)^2 h_j^{l-1}(n)^2, \quad \frac{LD^2}{N} m_\lambda \leq \lambda_{max} \leq LD^2 m_\lambda. \quad (11)$$

We assume that $\mathbb{E}_{n,\theta}\big[\delta_i^{l^2}\beta_i^{l^2}h_j^{l-1^2}\big] = \mathbb{E}_{n,\theta}\big[\delta_i^{l^2}\big]\mathbb{E}_n\big[\beta_i^{l^2}\big]\mathbb{E}_{n,\theta}\big[h_j^{l-1^2}\big]$. Assumption 2 guarantees $\mathbb{E}_n\big[\beta_i^{l^2}\big] = \frac{1}{2}$. So, the remaining is to calculate $\mathbb{E}_{n,\theta}\big[\delta_i^{l^2}\big]$ and $\mathbb{E}_{n,\theta}\big[h_j^{l-1^2}\big]$.

In order to calculate these terms, we decompose h_j^l and δ_i^l into path weights of the network (Fig. 5), which is orthogonal elements, like Fourier expansion (Fig. 6). This result is extension of the result for the error gradient at input δ_i^0 [1] into the error gradients at general hidden units δi^l and outputs of hidden units h_j^l.

Lemma 2. *Let α be path indicating which edges of weight matrices are passed through and $A(n)_{l,i} := \beta_i^l$ be the matrix whose elements are state of units activation, active (1) or inactive (0), for n-th data. Then,*

$$h_j^l(n) = \sum_{\alpha \in [D]^l \times \{j\}} W_\alpha A(n)_\alpha x(n)_{\alpha_1} \tag{12}$$

where $W_\alpha := \prod_{k=1}^{|\alpha|-1} W_{\alpha_k,\alpha_{k+1}}^k$ is path weight and $A(n)_\alpha := \prod_{k=1}^{|\alpha|-1} A(n)_{k,\alpha_k}$ is state of path activation. These elements in sum are orthogonal because

$$\mathbb{E}_\theta\left[W_{\alpha 1} W_{\alpha 2}\right] = \begin{cases} \left(\frac{2}{D}\right)^{|\alpha|-1} & \text{if } \alpha 1 = \alpha 2 \\ 0 & \text{otherwise} \end{cases} . \tag{13}$$

Using Lemma 2, Assumptions 1 and 2,

$$\mathbb{E}_{n,\theta}\left[h_j^{l^2}\right] = \sum_{\alpha \in [D]^l \times \{j\}} \mathbb{E}_\theta\left[W_\alpha^2\right] \mathbb{E}_n\left[A_\alpha^2\right] \mathbb{E}_n\left[x_{\alpha_1}^2\right]$$

$$= \left(\frac{2}{D}\right)^l \sum_{\alpha \in [D]^l \times \{j\}} \mathbb{E}_n\left[A_\alpha^2\right] = 1. \tag{14}$$

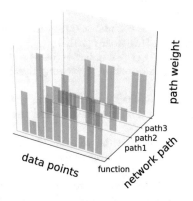

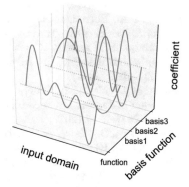

Fig. 5. Expansion of function (output or error gradient of hidden units) into path weights.

Fig. 6. Fourier expansion of function (time series) into basis functions (sin wave).

In the same way, we can calculate $\mathbb{E}_{n,\theta}\left[\delta_i^{l\,2}\right]$. Then, we can calculate mean of the eigenvalue and the maximum eigenvalue by substituting these values into Lemma 1.

Appendix B Experimental Settings

In our experiments, we prepared dataset for binary classification problem by subsampling 1000 examples whose class labels belong to 0 or 1 from MNIST dataset [8]. We set the value of output y whose class label is 0 to -1. Moreover, we preprocessed the input x of this dataset by the PCA-whitening in order to satisfy Assumption 1.

References

1. Balduzzi, D., Frean, M., Leary, L., Lewis, J., Ma, K.W.D., McWilliams, B.: The shattered gradients problem: if ResNets are the answer, then what is the question? In: International Conference on Machine Learning, pp. 342–350 (2017)
2. Bengio, Y., et al.: Learning deep architectures for AI. Found. Trends® Mach. Learn. **2**(1), 1–127 (2009)
3. He, K., Zhang, X., Ren, S., Sun, J.: Delving deep into rectifiers: surpassing human-level performance on ImageNet classification. In: Proceedings of the IEEE International Conference on Computer Vision, pp. 1026–1034 (2015)
4. He, K., Zhang, X., Ren, S., Sun, J.: Deep residual learning for image recognition. In: Proceedings of the IEEE Conference on Computer Vision and Pattern Recognition, pp. 770–778 (2016)
5. Ioffe, S., Szegedy, C.: Batch normalization: accelerating deep network training by reducing internal covariate shift. In: International Conference on Machine Learning, pp. 448–456 (2015)
6. Karakida, R., Akaho, S., Amari, S.: Universal statistics of fisher information in deep neural networks: mean field approach. arXiv preprint arXiv:1806.01316 (2018)
7. LeCun, Y., Bengio, Y., Hinton, G.: Deep learning. Nature **521**(7553), 436 (2015)
8. LeCun, Y., Bottou, L., Bengio, Y., Haffner, P.: Gradient-based learning applied to document recognition. Proc. IEEE **86**(11), 2278–2324 (1998)
9. LeCun, Y.A., Bottou, L., Orr, G.B., Müller, K.R.: Efficient backprop. In: Neural Networks: Tricks of the Trade, pp. 9–48. Springer, Heidelberg (2012)
10. Montufar, G.F., Pascanu, R., Cho, K., Bengio, Y.: On the number of linear regions of deep neural networks. In: Advances in Neural Information Processing Systems, pp. 2924–2932 (2014)
11. Raghu, M., Poole, B., Kleinberg, J., Ganguli, S., Sohl-Dickstein, J.: On the expressive power of deep neural networks. In: International Conference on Machine Learning, pp. 2847–2854 (2017)
12. Russakovsky, O., Deng, J., Su, H., Krause, J., Satheesh, S., Ma, S., Huang, Z., Karpathy, A., Khosla, A., Bernstein, M., et al.: ImageNet large scale visual recognition challenge. Int. J. Comput. Vis. **115**(3), 211–252 (2015)
13. Santurkar, S., Tsipras, D., Ilyas, A., Madry, A.: How does batch normalization help optimization? In: Advances in Neural Information Processing Systems, vol. 31 (2018)
14. Telgarsky, M.: Benefits of depth in neural networks. In: Conference on Learning Theory, pp. 1517–1539 (2016)

Skipping Two Layers in ResNet Makes the Generalization Gap Smaller than Skipping One or No Layer

Yasutaka Furusho[1]([✉]) [iD], Tongliang Liu[2] [iD], and Kazushi Ikeda[1] [iD]

[1] Nara Institute of Science and Technology, Nara 630-0192, Japan
{furusho.yasutaka.fm1,kazushi}@is.naist.jp
[2] The University of Sydney, Darlington, NSW 2008, Australia
tliang.liu@gmail.com

Abstract. The ResNet skipping two layers (ResNet2) is known to have a smaller expected risk than that skipping one layer (ResNet1) or no layer (MLP), however, the mechanism of the small expected risk is still unclear. The expected risk is divided into the three components, the generalization gap, the optimization error, and the sample expressivity, and the last two components are known to contribute the fast convergence of ResNet. We calculated the first component, the generalization gap, in the linear case, and show that ResNet2 has a smaller generalization gap than ResNet1 or MLP. Our numerical experiments confirmed the validity of our analysis and the applicability to the case with the ReLU activation function.

Keywords: Deep neural network · ResNet · Skip-connection · Generalization gap · Loss landscape

1 Introduction

Deep neural networks have been changing the history of machine learning in performance [10,16] and the high performance is said to result from its exponential expressive power [2,14,15,18]. However, a standard feedforward network cannot reduce its empirical risk even though it stacks more layers. In fact, an MLP with 56 layers has a larger empirical risk than one with 20 layers [6].

To overcome this problem, the ResNet incorporated skip-connections between layers [6]. The skip-connection enables even an extreme deep ResNet (1202 layers) to learn with a small empirical risk and a small expected risk. In addition, a ResNet skipping two layers (ResNet2) shows a better performance than a ResNet skipping one layer (ResNet1) or a standard feedforward network (MLP) [6,7,13].

The expected risk R is divided into the three components, the generalization gap ϵ_{gap}, the optimization error ϵ_{opt}, and the sample expressivity ϵ_{exp}, that is,

$$R(w_t) = \underbrace{R(w_t) - R_S(w_t)}_{\epsilon_{\text{gap}}} + \underbrace{R_S(w_t) - R_S(w_*)}_{\epsilon_{\text{opt}}} + \underbrace{R_S(w_*)}_{\epsilon_{\text{exp}}}, \tag{1}$$

© Springer Nature Switzerland AG 2020
L. Oneto et al. (Eds.): INNSBDDL 2019, INNS 1, pp. 349–358, 2020.
https://doi.org/10.1007/978-3-030-16841-4_36

where R_S is an empirical risk; w_t is an output of gradient descent at the t-th epoch and w_* is the empirical risk minimizer, both of which depend on the training set S. The last two components were analyzed and found that ResNet2 decreases the optimization error ϵ_{opt} easily compared with MLP and ResNet1 because the condition number of the Hessian matrix of ResNet2 is depth invariant unlike those of the others, which explode with its depth [11,13]. Moreover, ResNet2 has the perfect sample expressivity $\epsilon_{exp} = 0$ as soon as the number of its parameters exceeds the number of data in the training set [5]. These facts explain the fast convergence of ResNet but do not the small expected risk. Thus, we calculated the generalization gaps of ResNet2, ResNet1 and MLP and show the effects of the skip-connection on the generalization gap.

2   Materials and Methods

We analyzed the generalization gap of MLP, ResNet1, and ResNet2. This section formulates these neural networks and problem settings, and explains analysis methods for the generalization gap.

2.1   Linear Neural Networks

Since non-linear neural networks are difficult to analyze, we analyzed the generalization gap of linear neural networks as a step toward understanding those of the non-linear case. Actually, several studies explored loss landscapes and learning dynamics of the linear neural networks as the first step [1,4,5,8,17].

We consider the following L-layers neural networks $f : \mathcal{X} \times \mathcal{W} \to \mathcal{Y}$ with parameters (weights) $w \in \mathcal{W}$, which predict a corresponding output $y \in \mathcal{Y}$ for an input $x \in \mathcal{X}$:

$$\text{MLP}: \quad f(x,w) = \prod_{l=1}^{L} W^l x, \tag{2}$$

$$\text{ResNet1}: \quad f(x,w) = \prod_{l=1}^{L} \left(W^L + I \right) x, \tag{3}$$

$$\text{ResNet2}: \quad f(x,w) = \prod_{l=1}^{L/2} \left(W^{2l} W^{2l-1} \right) x \tag{4}$$

where $\mathcal{W}$ is a hypothesis space; $\mathcal{X}, \mathcal{Y}$ are input and output spaces respectively.

2.2   Problem Setting

We analyzed the generalization gap of the linear neural networks (Eqs. 2–4) in the following problem setting.

We have a training set $S := \{z^n\}_{n=1}^{N}$. Each training example is a pair of input and output $z^n := (x^n, y^n) \in \mathcal{X} \times \mathcal{Y} := \mathcal{Z}$ which is independently identically distributed from a probability distribution D.

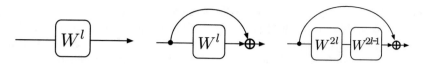

Fig. 1. Transformation of hidden layers in MLP, ResNet1, and ResNet2.

We put two assumptions on the training set to obtain special learning dynamics of the linear neural networks (Lemma 1). These special learning dynamics are key to our analysis.

Assumption 1. *The input correlation matrix* $\Sigma^{xx} := \sum_{n=1}^{N} x^n x^{nT}$ *is identity matrix.*

Assumption 1 is weak because practical data preprocessing methods like PCA-whitening satisfy this assumption.

Assumption 2. *The output-input correlation matrix* $\Sigma^{yx} := \sum_{n=1}^{N} y^n x^{nT}$ *is positive symmetric definite (PSD) with the eigenvalues greater than one.*

Assumption 2 can be relaxed as follows. We can extend our theoretical results into any rectangular output-input correlation matrix Σ^{yx} with the singular values greater than one (a weak version of Assumption 2) by modifying the skip-connections I to $R^{l+1}R^{l^T}$ where R^l is an orthonormal matrix; R^1, R^{L+1} are the right and left singular vectors of Σ^{yx}. For simplicity, we consider Assumption 2 in our analysis.

We train the linear neural networks (Eqs. 2–4) by gradient descent

$$w_{t+1} = w_t - \eta \nabla_w R_S(w_t) \tag{5}$$

to achieve a small empirical risk, measured by square loss

$$R_S(w) = \frac{1}{2} \sum_{n=1}^{N} \| f(x^n; w) - y^n \|^2. \tag{6}$$

2.3 Analysis Method for the Generalization Gap

The generalization gap of the linear neural networks depend on the two components; the speed of the parameter convergence by gradient descent and the flatness of loss landscape around the global minima. We analyzed these two components of MLP, ResNet1, and ResNet2 in next section.

This subsection explains the dependence between the generalization gap and the two components; the speed of the parameter convergence and the flatness of loss landscape around the global minima.

Generalization Gap and Algorithmic Stability: We used algorithmic stability [3] to derive the generalization gap of the linear neural networks.

A training algorithm $A : \mathcal{Z}^N \to \mathcal{W}$ receives the training set S and outputs a trained model $A(S)$. The algorithmic stability measures how much the removal of one example from the training set S affects the trained model $A(S)$ (Fig. 2).

Definition 1. [3] *The training algorithm A is pointwise hypothesis stable if the following holds*

$$\forall n \in [N], \; \mathbb{E}_{A,S}\left[|\ell(A(S), z^n) - \ell(A(S^n), z^n)|\right] \leq \epsilon_{\text{stab}} \tag{7}$$

where ℓ is a loss function and $S^n := S \setminus z^n$.

An output of a training algorithm A with a better stability (ϵ_{stab} is small) achieves a small generalization gap.

Theorem 1. [3] *If the training algorithm A is pointwise hypothesis stable, the following holds with probability at least $1 - \delta$:*

$$R(A(S)) - R_S(A(S)) \leq \sqrt{\frac{M^2 + 12MN\epsilon_{\text{stab}}}{2N\delta}} \tag{8}$$

where M is upper-bound of the loss function.

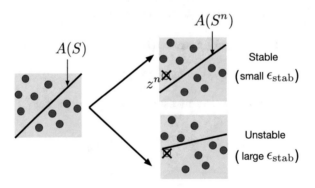

Fig. 2. The algorithmic stability measures how much the removal of one example from the training set S affects the trained model $A(S)$.

Algorithmic Stability and PL-Condition: The algorithmic stability ϵ_{stab} of gradient descent is depend on the speed of the parameter convergence and the flatness of loss landscape around the global minima as follows.

The algorithmic stability is related to the geometry of the empirical risk. Let g: $\mathcal{B} \to \mathbb{R}$ be a function on a set $\mathcal{B}$.

Definition 2. *The function g is α-Lipschitz if the following holds*

$$\forall b_1, b_2 \in \mathcal{B}, \quad |g(b_1) - g(b_2)| \leq \alpha \|b_1 - b_2\|. \tag{9}$$

Definition 3. *The function g is β-smooth if the following holds*

$$\forall b_1, b_2 \in \mathcal{B}, \quad \|\nabla g(b_1) - \nabla g(b_2)\| \leq \beta \|b_1 - b_2\|. \tag{10}$$

Definition 4. [4] *The function g satisfies the PL-condition with μ if the following holds*

$$\forall b \in \mathcal{B}, \quad \mu(g(b) - g_*) \leq \frac{1}{2} \|\nabla g(b)\|^2 \tag{11}$$

where g_* is the minimum value of g on $\mathcal{B}$.

Roughly speaking, the μ for the PL-condition measures the flatness of loss landscape around the global minima. If the empirical risk R_S satisfies the PL-condition with μ and is β-smooth for all $w \in \mathcal{W}$, we can derive

$$R_S(w) - R_{S*} \leq \frac{\beta^2}{2\mu} \|w - \prod_{\mathcal{W}_*}(w)\|^2 \tag{12}$$

where R_{S*} is the minimum value of R_S on $\mathcal{W}$ and $\prod_{\mathcal{W}_*}(w)$ is the projection of w onto the set of the global minima $\mathcal{W}_*$. This shows that an excess risk is smaller than a quadratic function of parameters w and the μ controls the flatness of loss landscape around the global minima (Fig. 3). Therefore, an empirical risk with a large μ has flat global minima.

If we apply a training algorithm which quickly converges parameters into the global minima to an empirical risk with flat global minima, its output achieves a small generalization gap.

Theorem 2. [4] *Assume that for all training set S and parameters $w \in \mathcal{W}$, the empirical risk R_S satisfies the PL-condition with μ and the loss function ℓ is α-lipschitz. If the training algorithm A converges to the global minima w_* with*

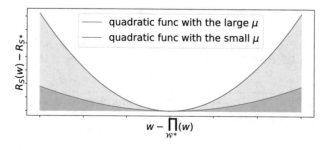

Fig. 3. An excess risk is smaller than a quadratic function of parameters w and the μ for the PL-condition controls the flatness of loss landscape around the global minima.

the speed of the parameter convergence $\|w_t - w_*\| \le \epsilon_t$, *the algorithm* A *is pointwise hypothesis stable:*

$$\epsilon_{\text{stab}} \le 2\alpha\epsilon_t + \frac{2\alpha^2}{\mu(N-1)}. \tag{13}$$

3   Main Results

This section shows that skipping two layers in ResNet makes the generalization gap smaller than skipping one or no layer after gradient descent updates parameters enough times by analyzing the two components related to the generalization gap: the speed of the parameter convergence ϵ_t and the flatness of loss landscape around the global minima measured by the μ for the PL-condition.

In order to confirm our analysis is valid, we ran numerical simulation. We trained linear neural networks with the modified skip-connections (explained in Sect. 2.2) on the MNIST dataset preprocessed by PCA whitening, which satisfies Assumption 1 and the weak version of Assumption 2.

3.1   The Speed of the Parameter Convergence by Gradient Descent

A speed of the parameter convergence of the MLP by gradient descent was analyzed [1]. So, we applied their proof strategy into the case of the ResNet1 and the ResNet2 as follows.

We can obtain the following special learning dynamics by initializing weights of MLP, ResNet1, and ResNet2: $W^l = U\bar{W}U^T$ where U are the eigenvectors of the output-input correlation matrix Σ^{yx} and $\bar{W}$ is a diagonal matrix.

Lemma 1. *Gradient descent updates only the diagonal elements of* $\bar{W}$ *and keeps the same diagonal matrix* $\bar{W}$ *among different layers.*

We calculated how fast each of the diagonal elements $a := \bar{W}_{ii}$ (mode strength) converges to the global minimum a_* (Table 1). Let a_t be the mode strength at t-th epoch.

Theorem 3. *For small enough learning rate* η, *a speed of the parameter convergence* $|a_t - a_*| \le \epsilon_t$ *of the ResNet2 is slower than the others. That of the ResNet1 is the same as the MLP (Table 1).*

Numerical experiments confirmed the analysis is valid (Fig. 4).

Table 1. Speeds of the parameter convergence. γ is the minimum mode strength during training. ϵ_0 is the difference between the initial mode strength and the global minimum.

Model	Speeds of the parameter convergence ϵ_t
MLP and ResNet1	$(1 - \eta L\gamma^{2(L-1)})^t \epsilon_0$
ResNet2	$(1 - \eta L \underbrace{\frac{\gamma^2 - 1}{2\gamma^2}}_{\le 1} \gamma^{2(L-1)})^t \epsilon_0$

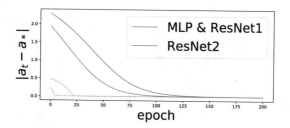

Fig. 4. Speeds of the parameter convergence of two layers (bold lines) and ten layers linear neural networks (dash lines).

3.2 The PL-Condition

The μ for the PL-condition of the MLP and the ResNet1 were analyzed [4,5]. We applied their proof strategy into the case of the ResNet2 and calculated its μ for the PL-condition.

Let a_{min}, a_{max} be the minimum and maximum absolute values of the mode strength during training. The special learning dynamics (Lemma 1) guarantees that the mode strength lies between initial mode strength a_0 and the global minima a_* during training for small enough learning rate [1]. Therefore, $0 \leq a_{min} \leq a_{max} \leq a_*$. Moreover, the global minima of ResNet1 and ResNet2 are proportional to inverse of its depth $a_* \propto \frac{1}{L}$ [5]. Therefore, enough deep neural networks guarantees $a_{max} \leq a_* \leq 1$.

Theorem 4. *Regarding the PL-condition of the MLP, a large a_{min} induces a large μ. On the other hand, regarding the PL-condition of the ResNet1, a small a_{max} induces a large μ. The ResNet2 has the above two advantages. In other words, the small condition number a_{max}/a_{min} induces a large μ (Table 2).*

In order to compare the μ for the PL-condition of the MLP, ResNet1, and ResNet2, we plotted the μ of ten layers neural networks (Fig. 5). The ResNet2 has a larger μ (flatter minima) than the others. Actually, the eigenvalues of the Hessian matrix of the ResNet2 after training are more concentrated on zero

Table 2. The μ for the PL condition of L layers linear neural networks. C is a constant.

Model	The μ for the PL-condition	
MLP	$\frac{L a_{min}^{2L-2}}{C}$	[4]
ResNet1	$\frac{L(1-a_{max})^{2L-2}}{C}$	[5]
ResNet2	$\frac{L(1-a_{max}^2)^{L-2} a_{min}^2}{C}$	

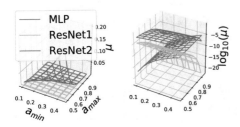

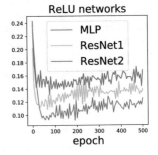

Fig. 5. The μ for the PL-condition of ten layers linear neural networks.

Fig. 6. The eigenvalues of the Hessian of ten layers linear networks after training.

than the others (Fig. 6). This implies loss landscape of the ResNet2 around global minima is flatter than the others.

3.3 The Stability

In the combination of the speed of the parameter convergence ϵ_t (Theorem 3) and the μ for the PL-condition (Theorem 4), we can find that stabilities of the MLP and the ResNet1 are better at early epoch due to faster speeds of the parameter convergence ϵ_t. But, at last, the ResNet2 achieves a better stability due to a larger μ for the PL-condition.

This implies skipping two layers in the ResNet makes a generalization gap smaller than skipping one or no layer after gradient descent updates parameters enough times. Our numerical experiments confirm the analysis is valid and is applicable to the case with the ReLU activation function (Figs. 7 and 8).

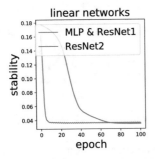

Fig. 7. Approximations of the algorithmic stabilities of ten layers neural networks: $\frac{1}{N}\sum_{n=1}^{N}|\ell(A(S), z^n) - \ell(A(S^n), z^n)|$.

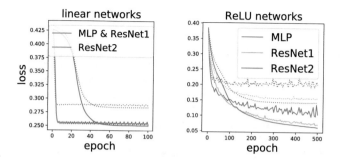

Fig. 8. Train losses (bold lines) and test losses (dash lines) of ten layers neural networks.

4   Discussion

Our results show that skipping two layers in the ResNet makes the generalization gap smaller than skipping one or no layer. This result was derived from the result that the ResNet2 has flatter global minima than the others. These theoretical results also support the following interesting experimental results. Loss landscape of deep neural networks around minima were visualized and shows that the ResNet2 has flat loss landscape around minima unlike the MLP which has spiky landscape around minima [12]. Another study experimentally shows that stochastic gradient descent with small mini-batch converges parameters into flat minima, which achieves small generalization gap [9]. These experimental results can be explained by our theoretical results. On top of that, derived algorithmic stability ϵ_{stab} can be used as a reasonable criteria for model selection. This value can be calculated before training due to the special learning dynamics (Lemma 1). Thus, there is no need to take a heavy time for training deep neural networks many times to find good structure of network.

5   Conclusion

In this paper, we analyzed effects of the skip-connections on the generalization gap in case of linear neural networks to clear reasons for the recent success of the ResNet. The stability based analysis shows that a generalization gap is related to the speed of the parameter convergence and the flatness of loss landscape around the global minima. We show that skipping two layers makes a speed of the parameter convergence slower but loss landscapes around the global minima flatter than skipping one or no layer. In the combination of these results, skipping two layers in the ResNet makes a generalization gap smaller than skipping one or no layer after gradient descent updates parameters enough times. Numerical experiments confirmed the analysis is valid and is applicable to the case with the ReLU activation function.

Acknowledgements. This work was supported by JSPS KAKENHI Grant Number JP18J15055, JP18K19821, and NAIST Big Data Project.

References

1. Bartlett, P.L., Helmbold, D.P., Long, P.M.: Gradient descent with identity initialization efficiently learns positive definite linear transformations by deep residual networks. In: International Conference on Machine Learning (2018)
2. Bengio, Y., et al.: Learning deep architectures for AI. Found. Trends® Mach. Learn. **2**(1), 1–127 (2009)
3. Bousquet, O., Elisseeff, A.: Stability and generalization. J. Mach. Learn. Res. **2**(Mar), 499–526 (2002)
4. Charles, Z., Papailiopoulos, D.: Stability and generalization of learning algorithms that converge to global optima. In: International Conference on Machine Learning (2018)
5. Hardt, M., Ma, T.: Identity matters in deep learning. In: International Conference on Learning Representations (2017)
6. He, K., Zhang, X., Ren, S., Sun, J.: Deep residual learning for image recognition. In: Proceedings of the IEEE Conference on Computer Vision and Pattern Recognition, pp. 770–778 (2016)
7. He, K., Zhang, X., Ren, S., Sun, J.: Identity mappings in deep residual networks. In: European Conference on Computer Vision, pp. 630–645. Springer (2016)
8. Kawaguchi, K.: Deep learning without poor local minima. In: Advances in Neural Information Processing Systems, pp. 586–594 (2016)
9. Keskar, N.S., Mudigere, D., Nocedal, J., Smelyanskiy, M., Tang, P.T.P.: On large-batch training for deep learning: generalization gap and sharp minima. In: International Conference on Learning Representations (2017)
10. LeCun, Y., Bengio, Y., Hinton, G.: Deep learning. Nature **521**(7553), 436 (2015)
11. LeCun, Y.A., Bottou, L., Orr, G.B., Müller, K.R.: Efficient backprop. In: Neural Networks: Tricks of the Trade, pp. 9–48. Springer (2012)
12. Li, H., Xu, Z., Taylor, G., Goldstein, T.: Visualizing the loss landscape of neural nets. In: International Conference on Learning Representations (2018)
13. Li, S., Jiao, J., Han, Y., Weissman, T.: Demystifying ResNet. arXiv preprint arXiv:1611.01186 (2016)
14. Montufar, G.F., Pascanu, R., Cho, K., Bengio, Y.: On the number of linear regions of deep neural networks. In: Advances in Neural Information Processing Systems, pp. 2924–2932 (2014)
15. Raghu, M., Poole, B., Kleinberg, J., Ganguli, S., Sohl-Dickstein, J.: On the expressive power of deep neural networks. In: International Conference on Machine Learning, pp. 2847–2854 (2017)
16. Russakovsky, O., Deng, J., Su, H., Krause, J., Satheesh, S., Ma, S., Huang, Z., Karpathy, A., Khosla, A., Bernstein, M., et al.: ImageNet large scale visual recognition challenge. Int. J. Comput. Vis. **115**(3), 211–252 (2015)
17. Saxe, A.M., McClelland, J.L., Ganguli, S.: Exact solutions to the nonlinear dynamics of learning in deep linear neural networks. In: International Conference on Learning Representations (2014)
18. Telgarsky, M.: Benefits of depth in neural networks. In: Conference on Learning Theory, pp. 1517–1539 (2016)

A Preference-Learning Framework for Modeling Relational Data

Ivano Lauriola[1,2(✉)], Mirko Polato[1], Guglielmo Faggioli[1], and Fabio Aiolli[1]

[1] Department of Mathematics, University of Padova,
Via Trieste, 63, 35121 Padua, Italy
ivano.lauriola@phd.unipd.it, {mpolato,gfaggiol,aiolli}@math.unipd.it
[2] Bruno Kessler Foundation, Via Sommarive, 18, 38123 Trento, Italy

Abstract. Nowadays large scale Knowledge Bases (KBs) represent very important resources when it comes to develop expert systems. However, despite their huge sizes, KBs often suffer from incompleteness. Recently, much effort has been devoted in developing learning models to reduce the aforementioned issue.

In this work, we show how relational learning tasks, such as link prediction, can be cast into a preference learning tasks. In particular, we propose a preference learning method, called REC-PLM, for learning low-dimensional representations of entities and relations in a KB. Being highly parallelizable, REC-PLM is a powerful resource to deal with high-dimensional modern KBs. Experiments against state-of-the-art methods on a large scale KB show the potential of the proposed approach.

Keywords: Preference learning · Embeddings · Knowledge base · Relational data · Relational learning

1 Introduction

As the era of the information continues, the size of Knowledge bases (KBs) keep growing. Broadly speaking a KB (a.k.a. Knowledge graph) encodes structured information of entities and their relations (also called *facts*), usually in the form of RDF triplets[1]. Some examples of KB are Freebase [1], DBpedia [2], and WordNet [3]. Despite their large sizes, for instance Freebase contains millions of entities and billions of facts, they are far from being complete. The task of inferring new facts in a KB is called *knowledge base completion* or more generally *relational learning*. Specifically, KB completion aims at predicting new relations between entities under supervision of the existing relations. KB completion is closely related to link prediction and it is very challenging mainly for two reasons: (*i*) nodes are heterogeneous, and (*ii*) edges represent relations of different types.

---

[1] https://www.w3.org/TR/rdf11-concepts/.

© Springer Nature Switzerland AG 2020
L. Oneto et al. (Eds.): INNSBDDL 2019, INNS 1, pp. 359–369, 2020.
https://doi.org/10.1007/978-3-030-16841-4_37

In the last years, many efforts have been invested in developing learning models that can scale to large knowledge bases. Tensor Factorization, Neural-embedding-based and translation-based models are the most popular kinds of approaches that have been extensively studied for relational learning purposes. In all these families of models, the learning process aims at encoding relational information using low-dimensional representations of entities and relations, usually called *knowledge graph embeddings*.

The wave of translation-based models finds inspiration from word2vec [4]. The core assumption is that a relation in the embedding space represents a fixed linear translation between the involved entities. One of the first proposed translation-based method has been Trans-E [5] which showed state-of-the-art performance. One of the strength of Trans-E is its simplicity, and thus its efficiency. However, it was soon demonstrated that Trans-E struggles in modeling one-to-many/many-to-one/many-to-many relations. For this reason new translation-based methods have been proposed, such as, Trans-H [6], Trans-M [7], Trans-R [8] and S-Trans-E [9], to name a few. All these approaches try to better model 1-n, n-1, and n-n relations by adding relation-dependent parameters to the model. In general, translation-based techniques have shown better performance w.r.t. tensor factorization methods (e.g., RESCAL [10]). State-of-the-art performance have also been shown by neural-embedding-based models, in which representations are learned using neural networks with energy-based objectives. For example, SME [11] and NTN [12] represent entities as low-dimensional vectors while relations are represented as operators that combine the representations of two entities. The core difference of these two methods lies in the way relation operators are parametrized. Generally, neural-embedding models have shown good scalability and strong generalization capability on large-scale KBs.

In this work we show how low-dimensional representations can be learned using a preference learning framework [13]. In particular, we adapt a label ranking approach to relational learning. This adaptation is made possible by converting the KB at hand into a set of preferences where existing relations are preferred to non existing ones. Our proposed method, dubbed REC-PLM, starting from a set of preferences over a single relation type, learns a ranking over the triplets. Differently from the aforementioned relational learning approaches, REC-PLM is based on similarities (defined in terms of dot-products) rather than distances, and it can handle a single relation at a time. This latter characteristic is a limitation, however, we show that REC-PLM is able to achieve comparable results with the compared methods even though it uses much less information during the training. Moreover, for each relation type a REC-PLM model can be learned in parallel which makes our approach easily scalable with the number of relations in the KB.

2   Related Work

In this work we consider KBs as graphs: each node of the graph corresponds to an element of the database, called *entity*, and each edge (a.k.a. link) defines a *relation* between the involved entities. Relations are considered directed, and a single graph generally contains different kinds of relations. Formally, a relation is denoted by a triplet (h, l, t), where h is the head (or left entity) of the relation, l is the type (label) of relation, and t is the tail (or right entity) of the relation (see Fig. 1). We assume that $h, t \in \mathcal{E}$ where $\mathcal{E}$ is the set of all possible entities and $l \in \mathcal{R}$, where $\mathcal{R}$ is the set of all possible relations. Knowledge base completion can be seen as a particular case of the more general link prediction task. In this work we focus on a link prediction related task, called *tail prediction*. Specifically, given a previously unseen golden triplet (h, l, t), the learning model fixes both relation and head, and it ranks all possible entities from the most likely of being tail to the least one. The goal is to rank t as higher as possible.

Fig. 1. Graphical depiction of a relation l between the entities h and t, which is represented by the triplet (h, l, t).

In the remainder we will assume that the plausible (a.k.a. positive/golden) training triplets are taken from the set $\mathcal{S}$, while the set of implausible (negative) triplets are taken from the set $\mathcal{S}' \equiv (\mathcal{E} \times \mathcal{R} \times \mathcal{E}) \setminus \mathcal{S}$.

There is a large body of research concerning the problem of modeling KBs. SE (Structured Embeddings) [14] has been one of the first methods to model entities of a KB as vectors. Specifically, the method assumes that entities can be modeled in a k-dimensional embedding space, i.e., $h \rightarrow \mathbf{h} \in \mathbb{R}^k$ and $t \rightarrow \mathbf{t} \in \mathbb{R}^k$, and in such space a relation type l defines a specific similarity measure. SE associates to each relation type l a pair of matrices $(\mathbf{R}_l^h, \mathbf{R}_l^t)$ defined in the space $\mathbb{R}^{k \times k}$. Given a relation type l, the scoring function is defined as $f_l(h, t) = \|\mathbf{R}_l^h \mathbf{h} - \mathbf{R}_l^t \mathbf{t}\|_p$, for a fixed p-norm. The idea is that a plausible relation (h, l, t) should have low value of $f_l(\mathbf{h}, \mathbf{t})$ (close to 0), while implausible ones should have higher values.

More recently, a family of translation-based methods has been proposed. In the context of multi-relation data, Trans-E [5] has been the first proposed translation-based method. The intuition behind Trans-E (and related approaches), inspired by [4], is that a relation corresponds to a linear translation in the embedding space. Likewise SE, Trans-E leverages on a k-dimensional embedding space, but differently from SE, it also models the relations in such a way that $\mathbf{h} + \mathbf{l} \approx \mathbf{t}$ when (h, l, t) exists, otherwise $\mathbf{t}$ should be far away from $\mathbf{h} + \mathbf{l}$. Let d be a dissimilarity function between vectors in the embedding space (usually an L_1 or L_2 norm), then the scoring function is defined as $f_l(h, t) = d(\mathbf{h} + \mathbf{l}, \mathbf{t})$. In order to learn these embeddings, Trans-E minimizes a margin-based ranking criterion

$$\mathcal{L} = \sum_{(h,l,t) \in \mathcal{S}} \sum_{(h',l,t') \in \mathcal{S}'_{(h,l,t)}} [f_l(h, t) - f_l(h', t') + \gamma]_+,$$

where $[x]_+$ denotes the positive part of x, and γ is a margin hyper-parameter. $\mathcal{S}'_{(h,l,t)}$ is the set of corrupted triplets composed by training triplets with either the head or tail replaced by a random entity (but not simultaneously).

The optimization problem is solved with the constraint that the L_2 norm of the embeddings (for both entities and relations) is unitary. Despite its efficiency and effectiveness in modeling one-to-one relations, Trans-E suffers in modeling reflexive/one-to-many/many-to-one/many-to-many relations. To overcome these limitations Trans-H [6] has been proposed which enables an entity to have a so-called *distributed representation* when involved in different relations. Trans-H considers linear translations not directly in the embedding space of the entities but in relation-specific spaces determined by relation-specific hyperplanes. Given a golden triplet (h, l, t), the embeddings $\mathbf{h}$ and $\mathbf{t}$ are first projected onto the relation-specific hyperplanes $\mathbf{w}_l$, i.e., $\mathbf{h}_\perp = \mathbf{h} - \mathbf{w}_l^\mathsf{T}\mathbf{h}\mathbf{w}_l$, $\mathbf{t}_\perp = \mathbf{t} - \mathbf{w}_l^\mathsf{T}\mathbf{t}\mathbf{w}_l$, with the constraint $\|\mathbf{w}_l\|_2 = 1$. Then, the scoring function is defined as $f_l(\mathbf{h}, \mathbf{t}) = \|\mathbf{h}_\perp + \mathbf{l} - \mathbf{t}_\perp\|_2^2$. In order to learn the parameters, Trans-H minimizes the same margin-based ranking loss $\mathcal{L}$ as Trans-E. Figure 2 provides a simple 2D illustration of the difference between Trans-E and Trans-H.

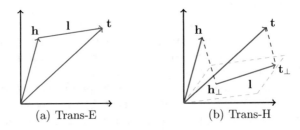

(a) Trans-E   (b) Trans-H

Fig. 2. Simple 2D illustration of the difference between (a) Trans-E, and (b) Trans-H.

Afterwards, other translation-based methods have been proposed in order to solve the limitations of Trans-E, such as, Trans-M [7], Trans-R [8] and S-Trans-E [9]. Despite the translation related methods, the task of modeling KBs has also been extensively present in the neural-embedding literature. Methods that deserve to be mentioned are SME [11], NTN [12], DISTMULT [15] and HolE [16].

In this preliminary work, we compare our approach against both Trans-E and Trans-H. In the future we aim to extend our analysis to at least all the above mentioned methods.

3   Preference Learning in a Nutshell

Preference Learning is a general framework to model a ranking problem based on the concept of *preferences*. Formally, a preference can be defined as a bipartite graph $g = (\mathcal{N}, \mathcal{A})$, where $\mathcal{N}$ is the set of nodes and $\mathcal{A} \subseteq \mathcal{N} \times \mathcal{N}$ is the set of arcs. In the Relational Learning setting, a node $n = (h, l, t) \in \mathcal{N} \subseteq (\mathcal{E} \times \mathcal{R} \times \mathcal{E})$ is a triplet representing a label l which connects a head entity h to the tail entity t. A node $n = (h, l, t)$ is said to be positive iff the triplet n exists in the KB, i.e., $n \in \mathcal{S}$, otherwise $n \in \mathcal{S}'$ is a negative node. An arc $a = (n_s, n_e) \in \mathcal{A}$ connects a starting (positive) node n_s to its ending (negative) node n_e. The direction of the arc indicates that the starting node must be *preferred* to the ending node.

In this work, preferences are used to model a ranking hypothesis f_Θ, which relates triplets to a probability score of belonging to the set $\mathcal{S}$. The margin of an arc $a = (n_s, n_e)$ is the difference between the application of the ranking function f_Θ to the starting and ending nodes, i.e.,

$$\rho_A(a, \Theta) = f_\Theta(n_s) - f_\Theta(n_e).$$

We say that an arc $a = (n_s, n_e)$ is consistent with the hypothesis f_Θ iff the score assigned to the node n_s is greater than the score assigned to the node n_e, i.e., $f_\Theta(n_s) > f_\Theta(n_e)$, thus the margin $\rho_A(a, \Theta) > 0$. The margin of a preference graph $g = (\mathcal{N}, \mathcal{A})$ is defined as the minimum margin of its arcs, i.e.,

$$\rho_G(g, \Theta) = \min_{a \in \mathcal{A}} \rho_A(a, \Theta).$$

The hypothesis f_Θ satisfies the preference graph iff all of its arcs are consistent, hence the preference has a positive margin $\rho_G(g, \Theta) > 0$.

Several architectures for preference graphs could be used. Figure 3, shows three examples of preference graphs, where: (left) there is only one fully connected bipartite graph which connects two positive nodes to a set of three negatives; (center) the previous graph is split into two simpler graphs, each of them with a single positive node; (right) there is a graph for each pair of positive and negative nodes. The last example of preferences has been considered in this work. We adopt the notation $(h, l, t) \to (h', l, t')$ to indicate a preference graph where the triplet (h, l, t) is preferred to the triplet (h', l, t').

Note that for each graph structure, the number of total arcs is the same, as well as the number of satisfied arcs, but the number of satisfied preferences is different.

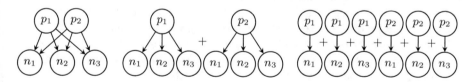

Fig. 3. Examples of preference graphs. p_i and n_j are the positive and negative nodes.

The last ingredient of a preference-based algorithm is the loss function $\mathcal{L}$ which penalizes the non-consistent preferences. The algorithm tries to find the hypothesis $\hat{f}$ from the hypothesis space $\mathcal{F}$ which minimizes $\mathcal{L}$

$$\hat{f} = \arg \min_{f_\Theta \in \mathcal{F}} \sum_{g \in \mathcal{V}} \mathcal{L}(\rho_G(g, \Theta)),$$

where $\mathcal{V}$ is the set of preference graphs. Usually, the loss functions are based on the margin of preferences. In this work the loss is defined as the number of non satisfied graphs, that is $\mathcal{L}(\rho) = [\![\rho \leq 0]\!]$, where $[\![\cdot]\!]$ is the indicator function.

4   The REC Preference Learning Machine

The Relational Embedding-Coding Preference Learning Machine (REC-PLM) is a preference-based algorithm. It learns a ranking function f_Θ which assigns a score to a input triplet (h, l, t), that is $f_\Theta : \mathcal{N} \to \mathbb{R}$.

The algorithm relies on two inner functions, the Embedding $W : \mathcal{E} \to \mathbb{R}^s$ and the Coding $M : \mathcal{E} \to \mathbb{R}^s$. The former aims at learning an s-dimensional embedding for the head entities of the input triplets, whereas the latter learns a different embedding for the tails of triplets in the same vector space. This means that an entity $e \in \mathcal{E}$ has two different embeddings depending on its role in the triplets. In fact, the role could emphasize different aspects of an entity [9].

In the REC-PLM setting, the scoring function ranks a relation between the entities h and t depending on the similarities of their associated embedding and coding, i.e., $W(h)$ and $M(t)$. We consider the dot-product as similarity score, that is

$$f_\Theta(h, l, t) = \langle W(h), M(t) \rangle.$$

The Embedding and Coding functions perform a linear transformation of their input, that is, $W(h) = h^\mathsf{T} W$, and $M(t) = t^\mathsf{T} M$, where h and t are the input representations of the entities h and t. W and M are $|\mathcal{E}| \times s$ weights matrices. Input representations consist of one-hot vectors of entities, hence, the matrices W and M could be seen as look-up tables.

The algorithm learns the two matrices M and W with an alternate optimization procedure. On each optimization step the algorithm fixes the matrix M and optimizes W, then it fixes W and optimizes M. This procedure is repeated until convergence.

When the algorithm optimizes W, a preference graph $(h, l, t) \to (h', l, t')$ is coded as $(t^\mathsf{T} M \otimes h) - (t'^\mathsf{T} M \otimes h')$, where $\otimes$ is the Kronecker product between two vectors. A batch of randomly selected preferences are used to optimize W with the Voted Perceptron algorithm [17]. To optimize the Coding matrix M instead, we use a different representation of the preferences, that is $(t \otimes h^\mathsf{T} W) - (t' \otimes h'^\mathsf{T} W)$. As in the previous case, a batch of preferences are built from the training KB, and M is optimized by means of a Voted Perceptron. The matrices M and Q are normalized after each training epoch.

We consider the number of perceptron updates as the training loss. In our experiments, the batch size has been set to 1000. Algorithm 1 contains the psudo-code of the proposed method.

4.1   REC-PLM as Neural Architecture

The REC-PLM intrinsically represents the same architecture of a Multiple Layer Perceptron (MLP) with a single hidden layer and linear activation functions.

Let us consider a fully connected MLP with a $|\mathcal{E}|$-dimensional input layer, which maps the input into a hidden s-dimensional layer by means of a dense $|\mathcal{E}| \times s$ linear connection. Then, the hidden layer maps information on a $|\mathcal{E}|$-dimensional output layer by using a dense $s \times |\mathcal{E}|$ linear connection.

Algorithm 1: The REC Preference Learning Machine

Input:
    KB: the set of triplets from the Knowledge Base
    s: the dimensionality of codes
    n_epochs: the number of training epochs
Output:
    $\boldsymbol{W}$: the embedding matrix
    $\boldsymbol{M}$: the coding matrix

1   $\boldsymbol{W}^{(0)} \leftarrow$ random $|\mathcal{E}| \times s$ matrix
2   $\boldsymbol{M}^{(0)} \leftarrow$ random $|\mathcal{E}| \times s$ matrix
3   **for** $t \in 1 \ldots n\_epochs$ **do**
4      $P_{\text{SET}} \leftarrow$ batch of preferences
5      $\boldsymbol{W}^{(i)} \leftarrow Voted\_Perceptron(P_{\text{SET}}, \boldsymbol{M}^{(i-1)})$
6      $P_{\text{SET}} \leftarrow$ batch of preferences
7      $\boldsymbol{M}^{(i)} \leftarrow Voted\_Perceptron(P_{\text{SET}} \boldsymbol{W}^{(i)})$
8   **end**
9   **return** $\boldsymbol{W}^{(t)}, \boldsymbol{M}^{(t)}$

The two mappings between layers correspond to the Embedding $\boldsymbol{W}$ and the Coding $\boldsymbol{M}$ learned in the REC-PLM setting.

Although REC-PLM can be mapped into a MLP and vice versa, the learning mechanisms are quite different. The MLP uses a back-propagation procedure whereas REC-PLM tries to optimize each input preference.

The MLP has one-hot heads of triplets as input instances, and one-hot tails as output. As is the case of the REC-PLM, this Neural Network is able to learn a single relation.

5   Experimental Assessment

In this work, the WordNet18 [3] database has been used to evaluate the proposed method. WordNet is a lexical database of English, containing nouns, verbs, adjectives and adverbs. The database consists of a network, which groups terms and concepts with proper relations, such as synonyms, hyponymy and so on. The KB includes 40943 entities, 18 relations and 151442 triplets.

The KB has been split into training ($\mathcal{S}_{\text{TR}}$) and test ($\mathcal{S}_{\text{TE}}$) sets, containing respectively the 70% and 30% of the available triplets. The validation set has been built with 5000 randomly extracted triplets from the training set. The test set has been augmented with $|\mathcal{S}_{\text{TE}}|$ additional implausible triplets from the set $\mathcal{S}'$, by swapping the tail entities.

REC-PLM has been compared against Trans-E and Trans-H on 8 relations from the WN18 database. These relations are described in Table 1.

We decided to compare one relation at a time for two main reasons. Firstly, relations are very different each other, and we want to observe the behavior of the algorithms on these different scenarios. The second reason is more practical,

Table 1. Relations description, including the number of triplets and the number of entities covered for each relation.

Relation	# triplets	# entity	type
part_of	5148	5445	n-1
has_part	5142	5444	1-n
similar_to	86	82	1-1
member_meronym	7928	8173	1-n
member_of_domain_topic	3341	3453	1-n
derivationally_related_form	31867	16737	n-n
also_see	1396	1061	n-n
verb_group	1220	1038	1-1

and it depends on the fact that REC-PLM learns a single model for each relation. However, Trans-E and Trans-H are able to exploit multi-relational information, and they are not suitable for a single relation task. Thus, in order to empirically analyze the advantages of multi-relational embeddings, two different models have been trained for Trans-E and Trans-H. The first considers all of the training triplets and relations, and the second uses only the 8 selected relations.

To better understand the convergence of REC-PLM, we analyze its behavior while increasing the number of training epochs. Figure 4 shows the training loss and the AUC score on the test set on two relations, namely PART_OF and MEMBER_OF_DOMAIN_TOPIC. On one hand, the training loss decreases with the increasing of the number of training epochs. On the other hand, the AUC score computed on the test set is less stable, and thus it requires a check-point procedure based on a validation loss. To this end, we plan to add the check-point procedure in the future.

We compared the proposed method against three baselines, which are Trans-E, Trans-H, and the Neural Network described in Sect. 4.1. The results of the comparison are available on the Table 2. The metric used for the evaluation is the AUC score. Results clearly show that Trans-E and Trans-H work better when using the complete set of available relations. This means that the information from multiple relations is important in these tasks. On one hand, the Trans-H algorithm is the one that achieves better results on average. On the other hand, REC-PLM reaches comparable results on several relations by using much less information. The MLP achieves on average the lowest results. We argue that the bad performance of the MLP could depend on both the sparsity of the input representations, and the number of parameters of the network w.r.t. the number of training examples.

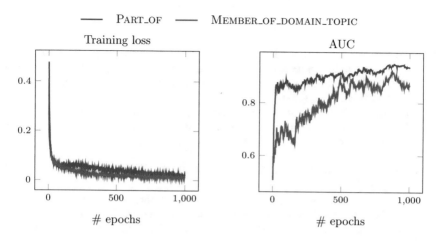

Fig. 4. Empirical convergence of the proposed algorithm.

Table 2. AUC score of the baselines and the REC-PLM on the tail prediction task.

Relation	Trans-E complete	Trans-H complete	Trans-E subset	Trans-H subset	MLP linear	MLP non linear	PLM
part_of	85.29	92.41	77.44	91.66	54.96	79.66	**93.71**
has_part	76.05	89.19	68.08	86.31	44.41	63.91	**90.00**
similar_to	73.10	81.16	67.60	**86.42**	65.93	65.93	77.78
member_meronym	75.29	**91.88**	65.54	51.81	52.52	90.76	88.10
member_of_domain_t	76.99	**87.67**	53.17	52.15	49.54	56.31	86.84
derivationaly_related.	**96.59**	95.30	96.03	94.41	61.13	59.93	76.61
also_see	64.45	87.80	56.83	86.63	52.33	79.66	**93.72**
verb_group	72.88	**98.32**	68.99	95.09	52.95	58.14	86.62

6   Conclusion and Future Work

In this paper we have proposed REC-PLM, a simple algorithm based on the preference learning framework to solve relational learning tasks. Unlike classical approaches based on distances, REC-PLM defines a relation as the similarity between two entities in a vector space. The proposed method has been compared with three baselines, showing promising results on the WordNet database.

However, in the near future we will focus on some weaknesses of the algorithm, such as the extension to multi-relations.

References

1. Bollacker, K., Evans, C., Paritosh, P., Sturge, T., Taylor, J.: Freebase: a collaboratively created graph database for structuring human knowledge. In: Proceedings of the 2008 ACM SIGMOD International Conference on Management of Data, SIGMOD 2008, pp. 1247–1250 (2008)
2. Lehmann, J., Isele, R., Jakob, M., Jentzsch, A., Kontokostas, D., Mendes, P.N., Hellmann, S., Morsey, M., van Kleef, P., Auer, S., Bizer, C.: DBpedia - a large-scale, multilingual knowledge base extracted from Wikipedia. Semant. Web J. **6**(2), 167–195 (2015)
3. Miller, G.A.: WordNet: a lexical database for English. Commun. ACM **38**(11), 39–41 (1995)
4. Mikolov, T., Sutskever, I., Chen, K., Corrado, G., Dean, J.: Distributed representations of words and phrases and their compositionality. In: Proceedings of the 26th International Conference on Neural Information Processing Systems - Volume 2, NIPS 2013, pp. 3111–3119 (2013)
5. Bordes, A., Usunier, N., Garcia-Durán, A., Weston, J., Yakhnenko, O.: Translating embeddings for modeling multi-relational data. In: Proceedings of the 26th International Conference on Neural Information Processing Systems - Volume 2, NIPS 2013, pp. 2787–2795 (2013)
6. Wang, Z., Zhang, J., Feng, J., Chen, Z.: Knowledge graph embedding by translating on hyperplanes. In: Proceedings of the Twenty-Eighth AAAI Conference on Artificial Intelligence, AAAI 2014, pp. 1112–1119 (2014)
7. Fan, M., Zhou, Q., Chang, E., Zheng, T.F.: Transition-based knowledge graph embedding with relational mapping properties. In: PACLIC (2014)
8. Lin, Y., Liu, Z., Sun, M., Liu, Y., Zhu, X.: Learning entity and relation embeddings for knowledge graph completion. In: Proceedings of the Twenty-Ninth AAAI Conference on Artificial Intelligence, AAAI 2015, pp. 2181–2187 (2015)
9. Nguyen, D.Q., Sirts, K., Qu, L., Johnson, M.: STransE: a novel embedding model of entities and relationships in knowledge bases. In: HLT-NAACL (2016)
10. Nickel, M., Tresp, V., Kriegel, H.-P.: A three-way model for collective learning on multi-relational data. In: Proceedings of the 28th International Conference on International Conference on Machine Learning, ICML 2011, pp. 809–816 (2011)
11. Bordes, A., Glorot, X., Weston, J., Bengio, Y.: A semantic matching energy function for learning with multi-relational data. Mach. Learn. **94**(2), 233–259 (2014)
12. Socher, R., Chen, D., Manning, C.D., Ng, A.Y.: Reasoning with neural tensor networks for knowledge base completion. In: Proceedings of the 26th International Conference on Neural Information Processing Systems - Volume 1, NIPS 2013, pp. 926–934 (2013)
13. Lauriola, I., Polato, M., Lavelli, A., Rinaldi, F., Aiolli, F.: Learning preferences for large scale multi-label problems. In: International Conference on Artificial Neural Networks, pp. 546–555. Springer (2018)
14. Bordes, A., Weston, J., Collobert, R., Bengio, Y.: Learning structured embeddings of knowledge bases. In: Proceedings of the Twenty-Fifth AAAI Conference on Artificial Intelligence, AAAI 2011, pp. 301–306 (2011)

15. Yang, B., Yih, W., He, X., Gao, J., Deng, L.: Embedding entities and relations for learning and inference in knowledge bases. In: Proceedings of the International Conference on Learning Representations (2015)
16. Nickel, M., Rosasco, L., Poggio, T.: Holographic embeddings of knowledge graphs. In: Proceedings of the Thirtieth AAAI Conference on Artificial Intelligence, AAAI 2016, pp. 1955–1961 (2016)
17. Freund, Y., Schapire, R.E.: Large margin classification using the perceptron algorithm. Mach. Learn. **37**(3), 277–296 (1999)

Convolutional Neural Networks
for Twitter Text Toxicity Analysis

Spiros V. Georgakopoulos, Sotiris K. Tasoulis, Aristidis G. Vrahatis,
and Vassilis P. Plagianakos$^{(\boxtimes)}$

Department of Computer Science and Biomedical Informatics,
University of Thessaly, Lamia, Greece
{spirosgeorg,stasoulis,arisvrahatis,vpp}@uth.gr

Abstract. Toxic comment classification is an emerging research field
with several studies that have address several tasks in the detection of
unwanted messages on communication platforms. Although sentiment
analysis is an accurate approach for observing the crowd behavior, it is
incapable of discovering other types of information in text, such as toxicity, which can usually reveal hidden information. Towards this direction,
a model for temporal tracking of comments toxicity is proposed using
tweets related to the hashtag under study. More specifically, a classifier
is trained for toxic comments prediction using a Convolutional Neural
Network model. Next, given a hashtag all relevant tweets are parsed and
used as input in the classifier, hence, the knowledge about toxic texts
is transferred to a new dataset for categorization. In the meantime, an
adapted change detection approach is applied for monitoring the toxicity trend changes over time within the hashtag tweets. Our experimental
results showed that toxic comment classification on twitter conversations can reveal significant knowledge and changes in the toxicity are
accurately identified over time.

Keywords: Convolutional neural networks · Toxic comments ·
Twitter conversations · Change detection

1 Introduction

Twitter sentiment analysis has covered a plethora of needs related to product opinion [1], stock market movement [22,24] and political influence on
crowed [11,18,25]. Twitter has already shown its impact on politics [4,30], especially on the electoral body constituting it as an important tool for the popularity
of a politician. As a sequence, twitter has been launched as the common communication platform for the politicians in the last decade [6]. Thus, an imperative
need is created for more accurate text mining and Machine Learning methods
towards twitter analysis.

The core of text mining and Machine Learning methods for text analysis is
based on the detection of the message substance through the words (or words
combination) that provide this information. The tendency of that field is the

© Springer Nature Switzerland AG 2020
L. Oneto et al. (Eds.): INNSBDDL 2019, INNS 1, pp. 370–379, 2020.
https://doi.org/10.1007/978-3-030-16841-4_38

sentiment analysis of text in order to analyze the message and to identify voters and crowd polarity. To achieve that, sentiment analysis tools based either on machine learning or lexical approaches are used. These tools estimate the sentiment content of the text [29]. The estimation is related to word count (binary - positive/negative, trinary - positive/neutral/negative, etc.) approaches using vocabularies with polarity influence.

Although the literature has been overwhelmed by such tools, there is a growing interest for online conversation analysis under different perspectives that may seem more appropriate for individual tasks. In this work, we focus on unveiling twitter comment toxicity based on a recent dataset provided within a Kaggle competition[1]. A toxic message is not only the regular scurrility, but also an aggressive message [33] or a message that occurs a personal attacks (such as treat and insult message) [32,34]. These types of behavior when they are appearing sequentially in time and not as outline message in the period of time (assuming that a regular voter do not present a such behaviour) may be malicious (bot, spam, etc.). The malicious messages may be part of politician rivals and have to be removed from the time series analysis of the text in order to extract accurate results of the voters behavior.

An indicative relative example is the project created by Google and Jigsaw, called Perspective, which uses machine learning to automatically detect toxic language [13]. More recently, toxic comment classification tools have been proposed using convolutional neural networks [9], recurrent neural networks [19] and deep learning approaches [26], but none of these have been applied on twitter data so far. Given the fact that twitter analysis has a major social impact and toxic comment classification is an emerging field for online conversations, identifying toxicity trend in twitter threads appears to be quite interesting. However, this task imposes two challenges that should be addressed. Characterization of twitter posts can be a hard task due to the short length of the documents, while the continuous stream of data necessitates the use of online methods for change detection. In this work, we propose a complete methodology that provides answers to both of these challenges.

2 Methodology

The proposed methodology is based on the construction of a classifier specifically trained for toxic comment classification, employing a recent and quite widespread labeled toxic comment dataset. The classifier is designed through a Convolutional Neural Network (CNN) model using the word2vec method [20] for mapping a physical text to a low-dimension representation. The gained knowledge of the classifier is exploited to classify a selection of twitter posts originated from a particular hashtag under study. The tweets are parsed online and a real time method for detecting trend changes is applied.

---

[1] https://www.kaggle.com/c/jigsaw-toxic-comment-classification-challenge.

Our methodology consists of three main steps, (i) design and implementation of CNN model for toxic comment classification, (ii) online parsing and characterization of tweets relevant to a pre-defined hashtag and (iii) real-time change detection analysis in the toxicity trend using an adapted Cumulative Sum (CUSUM) algorithm. A flowchart diagram of the methodology is presented in Fig. 1.

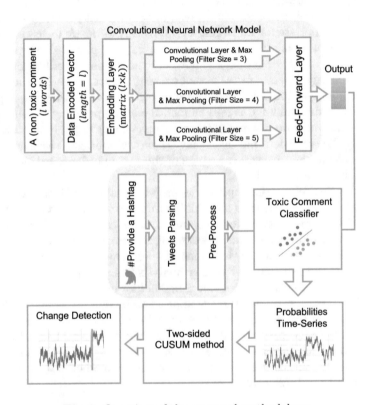

Fig. 1. Overview of the proposed methodology

2.1   CNN Model for Toxic Comments Classification

CNN models are appropriate for image classification, where the pixels of the images are represented by integer values within a specific range of values. On the other hand, the components of a sentence (the words) have to be encoded before being fed to the CNN [5]. To overcome this limitation, a vocabulary is applied as an index containing the words that appear in the set of document texts, mapping each word to an integer in $[0, 1]$. However, we have to deal also the variability in documents length (number of words in a document), since CNNs require a constant input dimensionality. To address it, we adopted the padding technique filling with zeros the document matrix to reach the maximum length amongst all documents in dimensionality.

In the next step, the encoded documents are transformed into matrices and each row corresponds to one word. The generated matrices pass through the embedding layer, where each word (row) is transformed into a low-dimension representation by a dense vector [7]. The procedure continues following the standard CNN methodology. In this work, we employ the fixed dense vectors for words, word2vec [20]. The word2vec embedding method has been trained on 100 billion words from Google News, producing a vocabulary of 3 million words. The embedding layer matches the input words with the fixed dense vector of the pre-trained embedding methods that have been selected. The values of these vectors do not change during the training process, unless there are words not already included in the vocabulary of the embedding method in which case they are randomly initialized.

The CNN model at hand is based on a selection of documents retrieved from Wikipedia by the Kaggle competition. The provided class labels were originally defined across different levels or types of toxicity for this study we consider the binary classification problem (document samples are characterized as toxic or non-toxic).

2.2   Tweets Streaming and Toxicity-Based Classification

Following the aforementioned training procedure, the toxic comment classifier is able to classify any set of tweets as toxic or non toxic. In our methodology, given a user-defined hashtag, all relevant tweets are streamed in real time using the Twitter stream Application Program Interfaces (API) proposed by Kalucki [14]. The streaming documents are consecutively cleaned removing noise, such as hashtags, spaces, numbers, punctuations, URLs etc. Following the procedure described in [28], before passed through the trained classifier, from which they receive a toxicity probability value. This is the value returned by the neuron of the output layer, that corresponds to the trained neuron on toxic comments indication. Subsequently, a time series of probability scores is generated, bounded by the softmax function performed on the output layer.

2.3   Toxicity Change Detection

To identify significant changes upon toxicity level, we utilize an online version of the Cumulative Sum (CUSUM) algorithm, focusing on a technique connected to a simple integration of signals with adaptive threshold [2,8,27]. To describe the change detection algorithm, we consider a sequence of independent random variables y_k, where y_k is a signal at the current time instant k (discrete time), with a probability density $p_\theta(y)$ depending only upon one scalar parameter θ. Before the unknown change time t, the parameter θ is equal to θ_0, and after the change it is equal to $\theta_1 \neq \theta_0$. Thus, the problem is to detect and estimate this parameter change. In this work, our goal is to detect the change assuming that the parameters θ_0 and θ_1 are known, which is a quite unrealistic assumption for many practical applications.

Usually parameters θ_0 and θ_1 can be experimentally estimated using test data, which we also do not consider available for the application at hand. However, for similar strongly bounded problems (signal values correspond to probabilities) parameter values can be empirically assumed. For example, we may consider θ_0 the state of toxicity level at the beginning of monitoring, which we may get by averaging over the first few samples. To this end, we are not interested in characterizing the current state (i.e. toxic, non toxic or neutral), but only in discovering negative or positive toxicity changes. Thus we only need to define the *the change magnitude*, which we arbitrary set to 0.3, taking into account that signal values lie between 0 and 1.

Since it is necessary to detect changes in each direction, discovering both increasing and decreasing toxicity of twitter posts, two one-sided algorithms were used for which $\theta_1^{pos} = \theta_0 + 0.3$ and $\theta_1^{neg} = \theta_0 - 0.3$, respectively. When a change is triggered for any of the two-sided functions at time t, the CUSUM algorithm is reset to zero and a re-initialization takes place. The algorithm restarts with a new value for θ_0 equal to the average of the few last observations ($[t - n \rightarrow t]$) and the new control state at time t is defined, while θ_1 is recalculated based on a fixed *change magnitude*.

To detect a change, as the samples (tweets) are arriving at each time instant, a decision rule is computed and compared against an adaptive threshold. This is the detection threshold h, a user-defined tuning parameter based on the average run length function that is defined as the expected number of samples before an action is taken [21]. More precisely, one has to set the mean time between false alarms ARL_0 and the mean detection delay ARL_1. These two specific values of the ARL function depend on the detection threshold h and can thus be used to set the performance of the CUSUM algorithm to the desired level for each particular application [10].

Under this perspective, we detect behavior changes under a dynamic estimation. This is crucial since a hashtag's toxicity could change several times across its lifetime and changes should be estimated based on its current state rather than its initial state. Towards this direction, our methodological framework can imprint on-line the change detections of a hashtag by updating its initial state.

3   Data Retrieval and Cleaning

The described methodological framework was applied on a Twitter data stream collected from 15-03-2018 to 24-03-2018 using the hashtag 'theresamay'. Obviously, this term refers to Theresa Mary May, Prime Minister of the United Kingdom and Leader of the Conservative Party since 2016. Our choice lies in the fact that politic-related Twitter hashtags offer a satisfying opportunity for testing toxicity changes. In addition, this hashtag was selected considering that during this period the 'Brexit' news topic attracted attention due to further discussions amongst high-level politicians regarding the relationships between the United Kingdom and European Union.

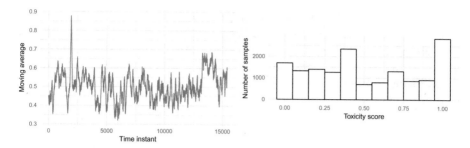

Fig. 2. The calculated moving average of window size 200 (left). Histogram of the toxicity score for the whole dataset (right).

Tweets are collected by the Twitter stream API, which is free to use and only requires a valid Twitter account for authentication. Withing the R-project, we used the "rtweet" package [15] to connect to the API and stream tweets filtered by keywords, such as Twitter hashtags (in our case 'theresamay'). Data are retrieved in JSON format and are parsed using the "rtweet" package. Then, the data cleaning needs to take place removing hashtags, spaces, numbers, punctuations, URLs etc. To achieve this, we employed functions from the "stringr" and "glue" packages respectively [12,31]. 15491 tweets have been streamed in total after discarding non-English language posts. Amongst them there are 11872 retweets (76%), which are not specially treated given that they still express an opinion. Then, generating a text corpus we may investigate most common words based on the Term Frequency-Inverse Document Frequency (TF-IDF) methodology [23], where the frequency of words is rescaled by how often they appear, penalizing most frequent words. The resulting words in descending order are: britain, look, russia, threat, strong, power.

4   Experimental Results

The first part of the proposed methodology is the training of a CNN model with the toxic comment dataset of Kaggle competition. The architecture that is used consist of three different convolutional layers simultaneously, with filter size width 128 and dense vector dimension 300. The width of the filters is equal to the vector dimension, while their height was 3, 4 and 5, for each convolutional layer, respectively. After each convolutional layer, a max-over-time pooling operation is applied. The output of the pooling layers is concatenated to a fully-connected layer, while the softmax function is applied on the final layer. The model is trained for 100 epochs using the Stochastic Gradient Descent algorithm [3] with mini-batches of 64 inputs and learning rate 0.005. The model achieved a classification accuracy 91.2%, when applied on unlabeled data from the same source [9].

In an attempt to visually inspect the toxicity trend changes, the corresponding moving average of the toxicity prediction probability scores retrieved by the CNN classifier is employed. Figure 2 (left) illustrates the calculated moving average for a window of total size 200. Subsequently, we investigate the histogram of tweet's distribution according to their toxicity level (see Fig. 2 (right)).

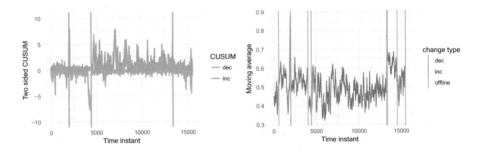

Fig. 3. The two sided CUSUM function along with the reported change points annotated by vertical lines corresponding to changes (positive and negative, respectively) (left). The reported change points with respect to the calculated moving average, where the additional green lines correspond to the change points reported by the off-line algorithm (right).

In what follows, we apply the change detection algorithm to the retrieved time series of probability scores and report the results. For the initialization of parameters values, we set $\theta_0 = 0.5$, while the *change magnitude* is set to 0.3; thus, we consider $\theta_1^{inc} = 0.8$ and $\theta_1^{dec} = 0.2$ for the two-sided CUSUM, respectively. In Fig. 3 (left) we show the CUSUM functions and the corresponding changes retrieved by the algorithm with vertical lines. Notice that each time a change is detected the algorithm restarts with an updated initialization. Selecting the value of the h parameter is usually subject to question and depends on the user requirements. For this particular topic of study in our analysis, we assume that a small number of reported changes per day are sufficient and thus we define $h = 10$. To visually investigate the reported change points we employed the calculated moving average presented in Fig. 2 (left). Vertical lines are depicted in the plot (see Fig. 3 (right)) with the corresponding colors from Fig. 3 (left). At this point, we may conclude that reported change points agree with a simple visual investigation of possible changes. To further verify this outcome, we employ a well established off-line methodology for change detection, which is capable to discover multiple change points [16,17] and estimate their number in the time series automatically if required. The penalty parameter of this algorithm is set to $1/2 * \log(n)$, where n is the total number of samples. The retrieved change points are also reported in Fig. 3 (right) with green vertical lines. We observe that this result accurately matches the change points retrieved by the adaptive CUSUM scheme, considering that the difference between reported changes mainly concerns the delay enforced by the online algorithm.

Fig. 4. Word-clouds of tweets belonging to different categories.

In the last part of our experimental analysis, we investigate the content of tweets according to the characterization by the classifier. For this purpose, we report word-clouds plotting the frequency of words appearing in the corresponding collection of Twitter posts. Again, the text is being preprocessed using the Term Frequency-Inverse Document Frequency (TF-IDF) (see Fig. 4). It appears that there exists a clear discrimination between the word-clouds. We observe that words like *coward* and *pathetic* appear in the toxic class validating the initial hypothesis. At this point we may also investigate example tweets from the two different categories that has been classified with strong confidence (see Fig. 5). The tweet on the right part of the Figure has been characterized as toxic with strong confidence and apparently is an ironic comment using also scurrility words that are not amongst the most common ones.

Remember when the *Brexiteers* made up the jokey lie about Cameron saying EURef result could lead to WWIII How we laughed Putin Trump Russia Brexit TheresaMay SalisburyAttack

You can thank you leaders and especially TheresaMay for the sad sorry state of your dingy *pathetic* little country or commonwealth or EuropeanUnion vassal state or whatever you are at this point *Coward* TheresaMay TheresaTheAppeaser

Fig. 5. A regular non-toxic tweet (left) and a toxic one (right) that contain some of the most frequent words from the two categories (annotated with red color).

5   Conclusion

Toxic comment classification is an emerging research field and although there are several approaches under this perspective, it remains at its infancy. Twitter's extensive data availability provides great potential for the development of Machine Learning research in social conversations. In this work, we study the comments toxicity under the perspective of a time series analysis to discover significant changes in real time. We may argue that this process increases the value of social media data by widening the understanding of community reaction in certain circumstances supporting immediate decision making. The experimental analysis provided pieces of evidence that toxicity identification may unravel significant knowledge, which is hidden from other sentiment-based approaches.

Acknowledgment. We gratefully acknowledge the support of NVIDIA Corporation with the donation of the Titan X Pascal GPU used for this research. This project has received funding from the Hellenic Foundation for Research and Innovation (HFRI) and the General Secretariat for Research and Technology (GSRT), under grant agreement No 1901.

References

1. Anastasia, S., Budi, I.: Twitter sentiment analysis of online transportation service providers. In: 2016 International Conference on Advanced Computer Science and Information Systems (ICACSIS), pp. 359–365, October 2016
2. Basseville, M., Nikiforov, I.V.: Detection of abrupt changes: theory and application (1993)
3. Bottou, L.: On-line learning and stochastic approximations. In: On-Line Learning in Neural Networks, pp. 9–42. Cambridge University Press, New York (1998). http://dl.acm.org/citation.cfm?id=304710.304720
4. Burgess, J., Bruns, A.: (Not) the Twitter election: the dynamics of the# ausvotes conversation in relation to the Australian media ecology. Journal. Pract. **6**(3), 384–402 (2012)
5. Collobert, R., Weston, J., Bottou, L., Karlen, M., Kavukcuoglu, K., Kuksa, P.: Natural language processing (almost) from scratch. J. Mach. Learn. Res. **12**, 2493–2537 (2011)
6. Enli, G.S., Skogerbø, E.: Personalized campaigns in party-centred politics: Twitter and Facebook as arenas for political communication. Inf. Commun. Soc. **16**(5), 757–774 (2013)
7. Gal, Y., Ghahramani, Z.: A theoretically grounded application of dropout in recurrent neural networks. In: Advances in Neural Information Processing Systems 29: Annual Conference on Neural Information Processing Systems 2016, Barcelona, Spain, 5–10 December 2016, pp. 1019–1027 (2016)
8. Georgakopoulos, S.V., Tasoulis, S.K., Plagianakos, V.P.: Efficient change detection for high dimensional data streams. In: 2015 IEEE International Conference on Big Data (Big Data), pp. 2219–2222, October 2015
9. Georgakopoulos, S.V., Tasoulis, S.K., Vrahatis, A.G., Plagianakos, V.P.: Convolutional neural networks for toxic comment classification. CoRR abs/1802.09957 (2018). http://arxiv.org/abs/1802.09957
10. Granjon, P.: The CUSUM algorithm a small review (2014)
11. Haselmayer, M., Jenny, M.: Sentiment analysis of political communication: combining a dictionary approach with crowdcoding. Qual. Quant. **51**(6), 2623–2646 (2017)
12. Hester, J.: glue: Interpreted String Literals (2017). https://CRAN.R-project.org/package=glue, r package version 1.2.0
13. Hosseini, H., Kannan, S., Zhang, B., Poovendran, R.: Deceiving Google's perspective API built for detecting toxic comments. arXiv preprint arXiv:1702.08138 (2017)

14. Kalucki, J.: Twitter streaming API (2010). http://apiwiki.twitter.com/Streaming-API-Documentation
15. Kearney, M.W.: rtweet: Collecting Twitter Data (2017). R package version 0.6.0
16. Killick, R., Fearnhead, P., Eckley, I.: Optimal detection of changepoints with a linear computational cost **107**, 1590–1598 (2012)
17. Killick, R., Haynes, K., Eckley, I.A.: changepoint: an R package for changepoint analysis (2016). https://CRAN.R-project.org/package=changepoint. R package version 2.2.2
18. Kušen, E., Strembeck, M.: Politics, sentiments, and misinformation: an analysis of the Twitter discussion on the 2016 Austrian presidential elections. Online Soc. Netw. Media **5**, 37–50 (2018)
19. Li, S.: Application of recurrent neural networks in toxic comment classification. Ph.D. thesis, UCLA (2018)
20. Mikolov, T., Sutskever, I., Chen, K., Corrado, G., Dean, J.: Distributed representations of words and phrases and their compositionality. In: Neural and Information Processing System (NIPS) (2013)
21. Page, E.S.: Continuous inspection schemes. Biometrika **41**(1/2), 100–115 (1954)
22. Pagolu, V.S., Reddy, K.N., Panda, G., Majhi, B.: Sentiment analysis of Twitter data for predicting stock market movements. In: 2016 International Conference on Signal Processing, Communication, Power and Embedded System (SCOPES), pp. 1345–1350, October 2016
23. Rajaraman, A., Ullman, J.D.: Mining of Massive Datasets. Cambridge University Press, Cambridge (2011)
24. Ranco, G., Aleksovski, D., Caldarelli, G., Grčar, M., Mozetič, I.: The effects of Twitter sentiment on stock price returns. PLoS One **10**(9), e0138441 (2015)
25. Ringsquandl, M., Petkovic, D.: Analyzing political sentiment on Twitter. In: AAAI Spring Symposium: Analyzing Microtext. AAAI Technical report, vol. SS-13-01. AAAI (2013)
26. Risch, J., Krestel, R.: Aggression identification using deep learning and data augmentation. In: Proceedings of the First Workshop on Trolling, Aggression and Cyberbullying (TRAC 2018), pp. 150–158 (2018)
27. Tasoulis, S., Doukas, C., Plagianakos, V., Maglogiannis, I.: Statistical data mining of streaming motion data for activity and fall recognition in assistive environments. Neurocomputing **107**, 87–96 (2013). Timely Neural Networks Applications in Engineering
28. Tasoulis, S.K., Vrahatis, A.G., Georgakopoulos, S.V., Plagianakos, V.P.: Real time sentiment change detection of Twitter data streams. CoRR abs/1804.00482 (2018)
29. Thelwall, M.: The heart and soul of the web? Sentiment strength detection in the social web with sentistrength, pp. 119–134. Springer, Cham (2017)
30. Wang, H., Can, D., Kazemzadeh, A., Bar, F., Narayanan, S.: A system for real-time Twitter sentiment analysis of 2012 US presidential election cycle. In: Proceedings of the ACL 2012 System Demonstrations, pp. 115–120. Association for Computational Linguistics (2012)
31. Wickham, H.: stringr: Simple, Consistent Wrappers for Common String Operations (2017). https://CRAN.R-project.org/package=stringr. R package version 1.2.0
32. Wulczyn, E., Thain, N., Dixon, L.: Ex machina: personal attacks seen at scale. In: Proceedings of the 26th International Conference on World Wide Web, WWW 2017, pp. 1391–1399. International World Wide Web Conferences Steering Committee, Republic and Canton of Geneva (2017)
33. Wulczyn, E., Thain, N., Dixon, L.: Wikipedia talk labels: aggression (2017)
34. Wulczyn, E., Thain, N., Dixon, L.: Wikipedia talk labels: personal attacks (2017)

Fast Spectral Radius Initialization for Recurrent Neural Networks

Claudio Gallicchio$^{(\boxtimes)}$, Alessio Micheli, and Luca Pedrelli

Department of Computer Science, University of Pisa,
Largo B. Pontecorvo 3, Pisa, Italy
{gallicch,micheli,luca.pedrelli}@di.unipi.it

Abstract. In this paper we address the problem of grounded weights initialization for Recurrent Neural Networks. Specifically, we propose a method, rooted in the field of Random Matrix theory, to perform a fast initialization of recurrent weight matrices that meet specific constraints on their spectral radius. Focusing on the Reservoir Computing (RC) framework, the proposed approach allows us to overcome the typical computational bottleneck related to the eigendecomposition of large matrices, enabling to efficiently design large reservoir networks and hence to address time-series tasks characterized by medium/big datasets. Experimental results show that the proposed method enables an accurate control of the spectral radius of randomly initialized recurrent matrices, providing an initialization approach that is extremely more efficient compared to common RC practice.

Keywords: Recurrent Neural Networks · Echo state networks · Spectral radius of random matrices · Time-series prediction

1 Introduction

Recurrent Neural Networks (RNNs) represent a powerful class of learning models suitable for time-series prediction and sequence classification. In particular, the Reservoir Computing (RC) framework [9] provides an efficient approach for the implementation of RNN architectures in which the recurrent part of the network is left untrained after initialization. Thereby, studies in the field of RC allow focusing the research investigations towards the analysis of the intrinsic characterizations of RNN models, independently from learning aspects. A major issue in this context is related to stability of recurrent network dynamics in presence of driving input signals, which is typically addressed by requiring the network to fulfill a stability property known as the Echo State Property (ESP) [9]. Literature conditions for the ESP link the stability behavior of the recurrent system to spectral properties, and in particular to the spectral radius, of the recurrent weight matrix in the RNN architecture, resulting in a network initialization procedure that requires to perform an eigendecomposition algorithm. Hence, despite their simplicity, commonly used RC initialization procedures can become impractical

© Springer Nature Switzerland AG 2020
L. Oneto et al. (Eds.): INNSBDDL 2019, INNS 1, pp. 380–390, 2020.
https://doi.org/10.1007/978-3-030-16841-4_39

and lead to a computational bottleneck in the case of real-world tasks for which a large number of recurrent units (in the order of thousands [6]) is needed. Besides, from a broader perspective, studies on RC-like initialization are of interest also for fully-trained RNNs [11], where a proper control of spectral properties of the recurrent weight matrices can be important to alleviate vanishing and exploding gradients [10].

In this paper, we propose an efficient approach to initialize the recurrent weights of an RNN in order to obtain an approximation of the desired spectral radius, avoiding to perform expensive algorithms. In particular, we propose two theoretically grounded algorithms stemming from Random Matrix theory and circular law [1,2,12] for the initialization of fully-connected and sparse matrices from a uniform distribution. The proposed approach is experimentally assessed in terms of both accuracy of spectral radius estimation and computational advantage with respect to common RC initialization practice. Moreover, we show how to exploit the proposed approach for implementing large RC networks (up to 10000 units) in challenging time-series tasks, considering both RC benchmarks and real-world datasets in the area of music processing.

2 Echo State Networks

The Echo State Network (ESN) [8] model is an efficient implementation of the RNN approach within the RC framework. The architecture is composed of two parts, a non-linear recurrent layer called reservoir and a linear output layer (the readout). The reservoir is randomly initialized and left untrained, while the readout computes the output of the network exploiting the temporal state representation developed by the reservoir part. Figure 1 shows the ESN architecture.

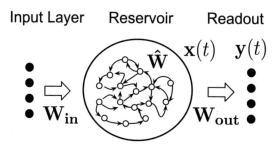

Fig. 1. The ESN architecture.

Omitting the bias terms in the following formula for the ease of notation, the state transition function of ESN is defined as follows:

$$\mathbf{x}(t) = \tanh(\mathbf{W}_{in}\mathbf{u}(t) + \hat{\mathbf{W}}\mathbf{x}(t-1)), \tag{1}$$

where $\mathbf{u}(t) \in \mathbb{R}^{N_U}$ and $\mathbf{x}(t) \in \mathbb{R}^{N_R}$ are respectively the *input* and the reservoir *state* at time t, $\mathbf{W}_{in} \in \mathbb{R}^{N_R \times N_U}$ is the matrix of the input weights, $\hat{\mathbf{W}} \in \mathbb{R}^{N_R \times N_R}$

is the matrix of the recurrent weights, and **tanh** represents the element-wise application of the hyperbolic tangent activation function.

Reservoir parameters are initialized according to the Echo State Property (ESP) [4,8]. Typically, following the necessary condition for the ESP, elements in $\hat{\mathbf{W}}$ are randomly initialized from a uniform distribution and then re-scaled such that $\rho(\hat{\mathbf{W}})$, which denotes the *spectral radius* of $\hat{\mathbf{W}}$ (i.e. its maximum absolute eigenvalue), is smaller than 1:

$$\rho(\hat{\mathbf{W}}) < 1. \tag{2}$$

The typical ESN initialization (ESNinit) is shown in Algorithm 1. The cost of the ESNinit procedure is dominated by the eigendecomposition algorithm computed in line 3 with an asymptotic cost of $\mathcal{O}(N_R{}^3)$. Concerning the input weights, the values in matrix $\mathbf{W}_{in}$ are randomly selected from a uniform distribution over $[-scale_{in}, scale_{in}]$, where $scale_{in}$ is the *input scaling* parameter.

Algorithm 1. ESNinit

1: **procedure** ESNINIT(N_R, $\hat{\rho}$)
2:     $\hat{\mathbf{W}} = \texttt{uniform}((N_R, N_R), -1, +1)$ ▷ generate matrix $\in \mathbb{R}^{N_R \times N_R}$ from u.d.
3:     $\rho = \texttt{eig}(\hat{\mathbf{W}})$ ▷ eigendecomposition of $\hat{\mathbf{W}}$
4:     $\hat{\mathbf{W}} = (\hat{\mathbf{W}}/\rho)*\hat{\rho}$ ▷ re-scale spectral radius of $\hat{\mathbf{W}}$ to $\hat{\rho}$
    return $\hat{\mathbf{W}}$

The output of the network at time t is computed by the readout as a linear combination of the reservoir units activations, i.e. $\mathbf{y}(t) = \mathbf{W}_{out}\mathbf{x}(t)$, where $\mathbf{y}(t) \in \mathbb{R}^{N_Y}$ is the *output* at time t, and $\mathbf{W}_{out} \in \mathbb{R}^{N_Y \times N_R}$ is the matrix of output weights. The readout layer is the only part of the network that is trained, typically in closed form by means of pseudo-inversion or ridge regression [9].

3   Fast Spectral Initialization

In this section, we introduce two efficient methods that we call Fast Spectral Initialization (FSI) and Sparse FSI (S-FSI). Assuming that the aim is to build a reservoir weight matrix with a desired spectral radius $\hat{\rho}$, FSI and S-FSI algorithms are defined in order to obtain a matrix $\hat{\mathbf{W}}$ such that $\rho(\hat{\mathbf{W}})$ is an approximation of the desired spectral radius $\hat{\rho}$. In the following, we propose a theoretical support rooted in the field of Random Matrix theory.

Theorem 1 *(FSI). Let $\hat{\boldsymbol{W}} \in \mathbb{R}^{N_R \times N_R}$ be the matrix of recurrent weights whose entries are i.i.d. copies of a random variable w with uniform distribution in $\left[-\frac{\hat{\rho}}{\sqrt{N_R}}\frac{6}{\sqrt{12}}, +\frac{\hat{\rho}}{\sqrt{N_R}}\frac{6}{\sqrt{12}}\right]$, then*

$$\rho(\hat{\boldsymbol{W}}) \to \hat{\rho} \text{ for } N_R \to \infty. \tag{3}$$

Proof. Let $\overline{\mathbf{W}} \in \mathbb{R}^{N_R \times N_R}$ be the matrix whose entries are i.i.d. copies of a random variable $\overline{w}$ with uniform distribution in $\left[-\frac{6}{\sqrt{12}}, +\frac{6}{\sqrt{12}}\right]$, hence $\overline{w}$ has mean 0 and variance 1. Then, from results in [1] and [12] (see [2] for details) we obtain that

$$\frac{\rho(\overline{\mathbf{W}})}{\sqrt{N_R}} \to 1 \text{ for } N_R \to \infty. \tag{4}$$

Let $\hat{\mathbf{W}} = \frac{\hat{\rho}}{\sqrt{N_R}} \overline{\mathbf{W}}$, hence random variables in $\hat{\mathbf{W}}$ are in $\left[-\frac{\hat{\rho}}{\sqrt{N_R}} \frac{6}{\sqrt{12}}, +\frac{\hat{\rho}}{\sqrt{N_R}} \frac{6}{\sqrt{12}}\right]$. Moreover, $\overline{\mathbf{W}} = \frac{\sqrt{N_R}}{\hat{\rho}} \hat{\mathbf{W}}$; by Property in Eq. (4) and linearity of spectral radius:

$$\frac{\rho(\frac{\sqrt{N_R}}{\hat{\rho}} \hat{\mathbf{W}})}{\sqrt{N_R}} = \frac{\sqrt{N_R}}{\hat{\rho}} \frac{\rho(\hat{\mathbf{W}})}{\sqrt{N_R}} \xrightarrow{N_R \to \infty} 1,$$

then by $\rho(\hat{\mathbf{W}}) \to \hat{\rho}$ for $N_R \to \infty$. $\qquad\square$

From Theorem 1 we can note that w has mean 0 and variance $\frac{\hat{\rho}^2}{N_R}$. Overall, Theorem 1 means that initializing the values of $\hat{\mathbf{W}}$ from a uniform distribution defined in $\left[-\frac{\hat{\rho}}{\sqrt{N_R}} \frac{6}{\sqrt{12}}, +\frac{\hat{\rho}}{\sqrt{N_R}} \frac{6}{\sqrt{12}}\right]$ (with mean 0 and variance $\frac{\hat{\rho}^2}{N_R}$) we obtain $\rho(\hat{\mathbf{W}}) \approx \hat{\rho}$ with an approximation accuracy that increases with the increasing of N_R. The FSI initialization is defined in Algorithm 2.

Algorithm 2. FSI initialization

1: **procedure** FSI(N_R, $\hat{\rho}$)
2:     $a = -\frac{\hat{\rho}}{\sqrt{N_R}} \frac{6}{\sqrt{12}}, b = +\frac{\hat{\rho}}{\sqrt{N_R}} \frac{6}{\sqrt{12}}$
3:     $\hat{\mathbf{W}} = \texttt{uniform}((N_R, N_R), a, b)$ ▷ generate matrix $\in \mathbb{R}^{N_R \times N_R}$ from u.d. [a,b]
      return $\hat{\mathbf{W}}$

Since we are interested to propose efficient approaches for medium/big data, we also take a step forward considering typical approaches for ESN based on sparse reservoirs. Accordingly, we introduce an extension of Theorem 1 for sparse matrices $\hat{\mathbf{W}}_C$, where C (called also connectivity) is the percentage of the incoming connections of each neuron in the recurrent layer.

Conjecture 1 (S-FSI). Let $\hat{\mathbf{W}}_C \in \mathbb{R}^{N_R \times N_R}$ be the matrix of recurrent weights with connectivity C whose non-null entries are i.i.d. copies of a random variable w with uniform distribution in $\left[-\frac{\hat{\rho}}{\sqrt{CN_R}} \frac{6}{\sqrt{12}}, +\frac{\hat{\rho}}{\sqrt{CN_R}} \frac{6}{\sqrt{12}}\right]$, then

$$\rho(\hat{\mathbf{W}}_C) \to \hat{\rho} \text{ for } N_R \to \infty. \tag{5}$$

The Conjecture 1 states that if we initialize a sparse recurrent matrix $\hat{\mathbf{W}}_C$ from a uniform distribution in $\left[-\frac{\hat{\rho}}{\sqrt{CN_R}} \frac{6}{\sqrt{12}}, +\frac{\hat{\rho}}{\sqrt{CN_R}} \frac{6}{\sqrt{12}}\right]$ (with mean 0 and variance $\frac{\hat{\rho}^2}{CN_R}$) we obtain $\rho(\hat{\mathbf{W}}) \approx \hat{\rho}$ with an accuracy of the approximation that increases with the increasing of N_R.

The insight that allowed us to formulate this conjecture is relatively simple. We considered a sparse matrix with connectivity C initialized from a uniform distribution as a matrix with non-null values of dimension $\lfloor CN_R \rfloor \times \lfloor CN_R \rfloor$. However, since the theoretical aspects of this case are not trivial, we decided to propose this enunciate as a conjecture delegating the theoretical proof to further works. The S-FSI initialization is defined in Algorithm 3. It is worth to note that in Algorithms 2 and 3 we avoid the cubic cost spent for eigendecomposition computation in Algorithm 1.

Algorithm 3. S-FSI initialization

1: **procedure** S-FSI(N_R, $\hat{\rho}$, C)
2:     $\hat{\mathbf{W}} = zeros((N_R, N_R))$ ▷ generate null matrix $\in \mathbb{R}^{N_R \times N_R}$
3:     $a = -\frac{\hat{\rho}}{\sqrt{CN_R}}\frac{6}{\sqrt{12}}$, $b = +\frac{\hat{\rho}}{\sqrt{CN_R}}\frac{6}{\sqrt{12}}$
4:     $N = \lfloor C{*}N_R \rfloor$ ▷ integer part of $C{*}N_R$
5:     **for** $irow$ **in** 1, ..., N_R **do**
6:         $P = \mathtt{randperm}(N_R, N)$ ▷ select only P indices to initialize (sparsity)
7:         $\hat{\mathbf{W}}(irow, P) = \mathtt{uniform}((1, N_R), a, b)$ ▷ generate vec. $\in \mathbb{R}^{N_R}$ from u.d. [a,b]
    return $\hat{\mathbf{W}}$

---

4   Experimental Assessments

In this section, we experimentally assess the FSI and S-FSI approaches introduced in Sect. 3 to efficiently exploit large RNNs and to improve models accuracy on medium/big time-series tasks. First, we empirically evaluate Theorem 1 and Conjecture 1. Then, we compare the computational times between our approaches and ESNinit. Finally, we show the benefits of the proposed method on RC benchmarks and real-world tasks characterized by medium/big dataset for time-series prediction.

4.1   Empirical Assessment

In this section, we perform an empirical evaluation of Theorem 1 and Conjecture 1 measuring the difference between the spectral radius of the recurrent matrices initialized by our proposed methods, i.e. $\rho(\mathtt{FSI}(N_R, \hat{\rho}))$ and $\rho(\mathtt{S\text{-}FSI}(N_R, \hat{\rho}, C))$, and the desired value of the spectral radius $\hat{\rho}$. Accordingly, we measure the error by the following equation:

$$ERR(\hat{\rho}) = \rho(\mathtt{FSI}(N_R, \hat{\rho})) - \hat{\rho}. \tag{6}$$

Analogously, we calculate the error for the S-FSI case as $\rho(\mathtt{S\text{-}FSI}(N_R, \hat{\rho}, C)) - \hat{\rho}$.

In the following experiments, for each reservoir hyper-parameterization, we independently generated 10 reservoir guesses, and the results has been averaged over such guesses. Figure 2(a) shows ERRs obtained at the increasing of N_R by FSI and S-FSI approaches. As expected from Theorem 1 we can note that the ERR obtained by FSI algorithm decreases with the increase of N_R, remaining below 0.03. Moreover, S-FSI obtained less that 0.02 ERR for all considered number of units. Note that these ERR results are very good for practical cases. Interestingly, the results achieved by FSI and S-FSI are very similar despite they work on different matrix structures. Overall, although Conjecture 1 (at the best of our knowledge) is not theoretically proved we think that these results represent a good empirical support in favor of the validity of the Conjecture 1.

Here, we assess the benefit in terms of computational time (on a CPU Intel Xeon E5, 1.80 GHz, 16 cores) offered by proposed methods. Accordingly, we measured the difference between the times (in seconds) required by ESNinit and our approaches. Figure 2(b) shows the time saved by FSI and S-FSI with respect to ESNinit at the increase of N_R.

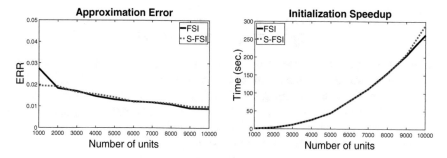

Fig. 2. Numerical results obtained for increasing number of units. Left: ERRs (with $\hat{\rho} = 1$). Right: Initialization speedup (in sec.) w.r.t. ESNinit.

In both fully and sparse reservoirs cases, the trend of curves shown in Fig. 2(b) confirms the cubic advantage in terms of computation time achieved by our algorithms as described in Sect. 3. In particular, in the case of fully-connected reservoirs composed by 10000 units, ESNinit spent 265 s while FSI required only 2.15 s. The best result is achieved by S-FSI that spent only 0.43 s to initialize a sparse matrix of size 10000×10000 with respect to ESNinit that required 290 s.

4.2 Experimental Assessment on Time-Series Prediction Tasks

In this section, we experimentally show how the proposed approaches are able to improve the model accuracy on medium/large datasets characterized by time-series prediction exploiting large reservoirs in an efficient way.

Experimental Setup. In order to show that very large reservoirs can be effectively exploited to obtain better performance, we performed a model selection on the validation set on a number of units N_R in $\{1000, 2000, 3000, 4000, 5000\}$ for RC benchmarks and N_R in $\{4000, 6000, 8000, 10000\}$ with a reservoir connectivity $C = 0.01$ for real-world tasks. Moreover, in the model selection we considered spectral radius ρ values in $\{0.5, 0.6, 0.7, 0.8, 0.9\}$, and $scale_{in} = 1$ for RC benchmarks and $scale_{in}$ values in $\{0.01, 0.1\}$ for real-world tasks. The training was performed through ridge regression [8,9] with regularization coefficient λ_r in $\{10^{-16}, 10^{-15}, \dots, 10^{-1}\}$ for RC benchmarks and λ_r in $\{10^{-4}, 10^{-3}, 10^{-2}, 10^{-1}\}$ for real-world tasks.

RC Benchmarks. First, we assessed the FSI approach on 3 RC benchmarks: Mackey-Glass time-series, 10th order NARMA system and Santa Fe Laser time-series, that we call in the following MG, NARMA and LASER tasks, respectively. For the RC benchmarks we measured the mean squared error (MSE).

The MG task is a standard benchmark [8] for chaotic time-series next-step prediction. The input sequence is obtained by a discretization of the following equation:

$$\frac{\partial u(t)}{\partial t} = \frac{0.2u(t-17)}{1 + u(t-17)} - 0.1u(t). \tag{7}$$

The input sequence was shifted by -1 and fed to **tanh** function as performed in [8]. We generated a time-series with 10000 time-steps, first $N_{train} = 5000$ time-steps were used for training, last $N_{validation} = 1000$ of the training set were used for validation and final $N_{test} = 5000$ for test. The first $N_{transient} = 1000$ time-steps was discarded before the readout training.

The LASER task [4] consists in a next-step prediction of a sequence obtained by sampling the intensity of a far-infrared laser in a chaotic regime. The input sequence consists in 10093 time-steps, first $N_{train} = 5000$ time-steps were used for training, last $N_{validation} = 1000$ of the training set were used for validation and final $N_{test} = 5093$ for test. The first $N_{transient} = 1000$ time-steps were used as initial transient.

The NARMA task [4] consists in the output prediction of a 10-th order non-linear autoregressive moving average system. The input sequence $u(t)$ is generated on a uniform distribution in $[0, 5]$. The target sequence is computed by the following equation:

$$\hat{y}(t) = 0.3\hat{y}(t-1) + 0.05\hat{y}(t-1)(\sum_{i=1}^{10} \hat{y}(t-i)) + 1.5u(t)(t-10)u(t-1) + 0.1. \tag{8}$$

Given $u(t)$ the aim of the task is to predict $\hat{y}(t)$. The input sequence consists in 4200 time-steps, first $N_{train} = 2200$ time-steps were used for training, last $N_{validation} = 1000$ of training set were used for validation and final $N_{test} = 2000$ for test. The initial transient was set at $N_{transient} = 200$.

The results obtained by FSI and ESN initialization through eigendecomposition (ESNinit) are shown in Table 1. The models obtained the best results with

$N_R = 5000$ recurrent units in all tasks. Therefore, in this cases, a fast approach to approximate the spectral radius of large recurrent matrices is very important to improve the performance in an efficient way. Moreover, we can see that FSI approach obtained very similar test MSEs to those obtained by ESNinit highlighting the effectiveness of the spectral radius approximation. Interestingly, FSI required only 0.51 s to initialize the recurrent layer with respect to 45 s required by ESNinit for each hyper-parameters configuration. In this concern, note that the computational advantage of the proposed approach is further amplified when considering a grid search for the hyper-parametrization. In the following, we take a step forward considering larger recurrent matrices and more challenging and bigger datasets.

Table 1. Test MSE and initialization time (Init. time column). $ERR(\hat{\rho})$ column: ERR obtained by FSI where $\hat{\rho}$ is the selected spectral radius by model selection. Training time column: time required for readout training, averaged over the FSI and ESNinit reservoir initialization methods (it does not depend on the initialization approach). Units column: number of recurrent units selected by model selection.

Approach	Test MSE	$ERR(\hat{\rho})$	Init. time	Training time	Units
MG					
FSI	8.32e−15 (1.06e−14)	0.012 (0.002)	0.51 sec	222 sec	5000
ESNinit	8.97e−15 (1.44e−14)	–	45 sec		
LASER					
FSI	3.90e−3 (1.83e−04)	0.007 (0.001)	0.51 sec	207 sec	5000
ESNinit	4.14e−3 (2.96e−04)	–	45 sec		
NARMA					
FSI	2.80e−4 (2.02e−05)	0.012 (0.002)	0.51 sec	83 sec	5000
ESNinit	2.78e−4 (2.30e−05)	–	45 sec		

Polyphonic Music Tasks. Here we consider S-FSI for sparse reservoirs on challenging real-world tasks characterized by medium/big datasets. The sparse connections implemented in large recurrent layers allow models to perform efficient computations despite the big quantity of units. The approach is assessed on polyphonic music tasks defined in [3]. In particular, we consider the following 4 datasets[1]: Piano-midi.de, MuseData, JSBchorales, and Nottingham. A polyphonic music task is defined as a next-step prediction on 88-, 82-, 52- and 58-dimensional sequences for Piano-midi.de, MuseData, JSBchorales and Nottingham datasets, respectively. The datasets comprise a large number of timesteps (in the order of hundreds of thousands, see further details in [3,6]). Training,

---

[1] Piano-midi.de (www.piano-midi.de); MuseData (www.musedata.org); JSBchorales (chorales by J. S. Bach); Nottingham (ifdo.ca/~seymour/nottingham/nottingham. html).

validation and test sets of piano-rolls used in [3] are publicly available[2]. Further details concerning the polyphonic music datasets (omitted here for the sake of conciseness) are reported in [6].

Table 2. Test MSE and initialization time (Init. time column). $ERR(\hat{\rho})$ column: ERR obtained by S-FSI where $\hat{\rho}$ is the selected spectral radius by model selection. Training time column: time required for readout training, averaged over the S-FSI and ESNinit reservoir initialization methods (it does not depend on the initialization approach). Units column: number of recurrent units selected by model selection.

Approach	Test ACC	$ERR(\hat{\rho})$	Init. time	Training time	Units
Piano-midi.de					
S-FSI	28.80 (0.01)%	0.009 (0.003)	0.43 sec	124 sec	10000
ESNinit	28.80 (0.03)%	–	290 sec		
MuseData					
S-FSI	33.79 (0.01)%	0.005 (0.002)	0.43 sec	101 sec	10000
ESNinit	33.80 (0.03)%	–	290 sec		
JSBchorales					
S-FSI	32.18 (0.26)%	0.005 (0.002)	0.43 sec	29 sec	10000
ESNinit	32.14 (0.44)%	–	290 sec		
Nottingham					
S-FSI	70.28 (0.02)%	0.009 (0.003)	0.43 sec	280 sec	10000
ESNinit	70.25 (0.04)%	–	290 sec		

The prediction accuracy of the models is measured in expected frame-level accuracy (ACC) adopted in polyphonic music tasks [3], and computed as follows:

$$\text{ACC} = \frac{\sum_{t=1}^{T} TP(t)}{\sum_{t=1}^{T} TP(t) + \sum_{t=1}^{T} FP(t) + \sum_{t=1}^{T} FN(t)}, \tag{9}$$

where T is the total number of time-steps, while $TP(t)$, $FP(t)$ and $FN(t)$ respectively denote the numbers of true positive, false positive and false negative notes predicted at time-step t.

Table 2 shows the results obtained by S-FSI and ESNinit. Similarly to results obtained in RC benchmarks, here the models obtained the best results with $N_R = 10000$ units. Moreover, the test ACCs obtained by S-FSI and ESNinit are very similar. These results show that there is no need to compute the exact spectral radius to obtain good performance. Moreover, the use of sparsity operation enables to further improve both initialization and training times for the models. Indeed, S-FSI resulted significantly more efficient than FSI. Note that the order of magnitudes of times required by ESNinit is comparable with the

---

[2] www-etud.iro.umontreal.ca/~boulanni/icml2012.

ones required by readout training. Overall, S-FSI is extremely more efficient than ESNinit requiring only 0.43 s to initialize a recurrent matrix of size 10000×10000 with respect to 290 s required by ESNinit for each hyper-parameters configuration. In conclusion, the proposed methods allowed us to design extremely efficient ESN models able to reach a performance that is close to state-of-the-art results present in literature on polyphonic music tasks [6,7].

5 Conclusions

In this paper, we introduced a fast initialization approach for RNNs able to accurately control the spectral radius of recurrent matrices. First, we provided theoretical support and then we defined FSI and S-FSI algorithms for fully-connected and sparse matrices. Moreover, we empirically verified the validity of the theoretical tools as the number of recurrent units increases. Finally, we evaluated the proposed approaches in large RC networks on RC benchmarks and real-world tasks. The results show that the proposed approaches are extremely more efficient than typical spectral radius initialization methods allowing reservoir models to exploit many units and to improve accuracy. Overall, the proposed procedures overcome the computational bottleneck typical of ESNs initialization providing practical tools for the rapid design of such models.

The methodologies proposed in this paper open the way to further works based on the exploitation of RC in Big Data tasks especially for what regards the synergy with the deep learning paradigm such as in Deep Echo State Network [5] models. More in general, future studies can regard the assessment of our initialization approaches for fully-trained deep RNNs.

References

1. Bai, Z., Yin, Y.: Limiting behavior of the norm of products of random matrices and two problems of Geman-Hwang. Probab. Theory Relat. Fields **73**(4), 555–569 (1986)
2. Bordenave, C., Caputo, P., Chafaï, D., Tikhomirov, K.: On the spectral radius of a random matrix. arXiv preprint arXiv:1607.05484 (2016)
3. Boulanger-Lewandowski, N., Bengio, Y., Vincent, P.: Modeling temporal dependencies in high-dimensional sequences: application to polyphonic music generation and transcription. In: Proceedings of the 29th International Conference on Machine Learning (2012)
4. Gallicchio, C., Micheli, A.: Architectural and Markovian factors of echo state networks. Neural Netw. **24**(5), 440–456 (2011)
5. Gallicchio, C., Micheli, A., Pedrelli, L.: Deep reservoir computing: a critical experimental analysis. Neurocomputing **268**, 87–99 (2017)
6. Gallicchio, C., Micheli, A., Pedrelli, L.: Design of deep echo state networks. Neural Netw. **108**, 33–47 (2018)
7. Gallicchio, C., Micheli, A., Pedrelli, L.: Comparison between DeepESNs and gated RNNs on multivariate time-series prediction. In: Proceedings of the 27th European Symposium on Artificial Neural Networks (ESANN) (in press)

8. Jaeger, H., Haas, H.: Harnessing nonlinearity: predicting chaotic systems and saving energy in wireless communication. Science **304**(5667), 78–80 (2004)
9. Lukoševičius, M., Jaeger, H.: Reservoir computing approaches to recurrent neural network training. Comput. Sci. Rev. **3**(3), 127–149 (2009)
10. Pascanu, R., Mikolov, T., Bengio, Y.: On the difficulty of training recurrent neural networks. In: Proceedings of International Conference on Machine Learning (ICML), vol. 28, pp. 1310–1318 (2013)
11. Sutskever, I., Martens, J., Dahl, G., Hinton, G.: On the importance of initialization and momentum in deep learning. In: International Conference on Machine Learning, pp. 1139–1147 (2013)
12. Tao, T., Vu, V., Krishnapur, M., et al.: Random matrices: universality of ESDs and the circular law. Ann. Probab. **38**(5), 2023–2065 (2010)

Author Index

A
Aiolli, Fabio, 48, 359
Alamudun, Folami, 89
Anastasovski, Dimitar, 120
Angelov, Plamen P., 257
Anguita, Davide, 120, 136, 142, 152, 206
Arandjelović, Ognjen, 330

B
Bacciu, Davide, 236
Becker, Stephen, 58
Bharath, Anil A., 246
Bontempi, Gianluca, 78
Bouchachia, Abdelhamid, 296
Bruno, Antonio, 236
Buchmueller, Juri, 206
Buselli, Irene, 136, 142
Byrne, Jonathan, 17

C
Cakmak, Eren, 206
Cambria, Erik, 180
Campese, Stefano, 48
Canepa, Renzo, 120, 136, 142, 206
Cantwell, Chris C., 246
Castiglia, Giuliano, 206
Cerquitelli, Tania, 216
Cetto, Tomaso, 17
Chadha, Gavneet Singh, 196
Chadha, Gavneet, 226
Chen, Ziqi, 267
Cheung, Siu, 267
Christian, J. Blair, 89
Cipollini, Francesca, 152
Colotte, Vincent, 186

Cooper, Jessica, 330
Crescimanna, Vincenzo, 99

D
da Silva, Eduardo Costa, 109
du Plessis, Mathys C., 38

F
Fabrizio, Nadia, 120
Faggioli, Guglielmo, 359
Feng, Shoubo, 158
Firmani, Donatella, 320
Franchini, Giorgia, 286
Furusho, Yasutaka, 341, 349

G
Galinier, Mathilde, 286
Gallicchio, Claudio, 380
Garasto, Stef, 246
Gastaldo, Paolo, 27
Georgakopoulos, Spiros V., 370
Gnecco, Giorgio, 1
Gorban, Alexander, 276
Graham, Bruce, 99
Green, Stephen, 276
Gu, Xiaowei, 257
Guo, Ping, 158

H
Hajewski, Jeff, 7
Han, Min, 158
He-Guelton, Liyun, 78
Hinkle, Jacob, 89
Huang, Zhishen, 58

© Springer Nature Switzerland AG 2020
L. Oneto et al. (Eds.): INNSBDDL 2019, INNS 1, pp. 391–392, 2020.
https://doi.org/10.1007/978-3-030-16841-4

I
Ikeda, Kazushi, 341, 349

J
Jentner, Wolfgang, 206
Jouvet, Denis, 186

K
Keim, Daniel A., 206
Kůrková, Věra, 309

L
Lauriola, Ivano, 48, 359
Le Borgne, Yann-Aël, 78
Lebichot, Bertrand, 78
Li, Yanli, 267
Liu, Tongliang, 349
Lulli, Alessandro, 142

M
Makarenko, Alexander, 126
Manzini, Raffaella, 169
Merello, Simone, 180
Merialdo, Paolo, 320
Meydani, Elnaz, 196
Micheli, Alessio, 380
Miglianti, Fabiana, 152
Mnasri, Zied, 186
Mohamad, Saad, 296
Moloney, David, 17

N
Nieddu, Elena, 320
Nutarelli, Federico, 1

O
Oblé, Frédéric, 78
Oliveira, Suely, 7
Oneto, Luca, 120, 136, 142, 152, 169, 180, 206

P
Palatnik de Sousa, Iam, 109
Pedrelli, Luca, 380
Perotti, Alan, 216
Petralli, Simone, 136, 142, 206
Pinto, Viviana, 216
Plagianakos, Vassilis P., 370
Polato, Mirko, 359
Ponta, Linda, 169
Popov, Anton, 126
Puliga, Gloria, 169

Q
Qiu, John X., 89

R
Ragusa, Edoardo, 27
Ratto, Andrea Picasso, 180
Rehlaender, Philipp, 226

S
Sanetti, Paolo, 136
Sanguineti, Marcello, 309
Sartori, Giuseppe, 48
Sayed-Mouchaweh, Moamar, 296
Scardapane, Simone, 320
Scarpazza, Cristina, 48
Schlegel, Udo, 206
Schroeer, Maik, 226
Schwung, Andreas, 196, 226
Sorteberg, Wilhelm E., 246
Spigolon, Roberto, 120
Swiatek, Marie, 120

T
Tani, Giorgio, 152
Tasoulis, Sotiris K., 370
Tourassi, Georgia, 89
Tyukin, Ivan, 276

V
Vellasco, Marley Maria Bernardes Rebuzzi, 109
Verucchi, Micaela, 286
Viviani, Michele, 152
Vrahatis, Aristidis G., 370

W
Watt, Nathan, 38

X
Xu, Xiaofan, 17

Y
Yoon, Hong-Jun, 89

Z
Zangar, Imene, 186
Zhao, Dongbin, 158
Zunino, Rodolfo, 27

Printed in the United States
By Bookmasters